QP
399
D43
1997

110555

Deacon, Terrence William
The Symbolic species

DATE DUE MAY 7 1999			
MAR 28 2001			

The
Symbolic
Species

The Symbolic Species

THE CO-EVOLUTION OF LANGUAGE AND THE BRAIN

Terrence W. Deacon

W. W. NORTON & COMPANY ▶ NEW YORK   LONDON

For information about permission to reproduce selections from this book, write to
Permissions, W. W. Norton & Company, Inc., 500 Fifth Avenue, New York, NY 10110.

The text of this book is composed in New Caledonia with the display set in Trade Gothic
Condensed Bold. Compostion and manufacturing by the Haddon Craftsmen, Inc.

BOOK DESIGN BY BTD / ROBIN BENTZ.

LIBRARY OF CONGRESS CATALOGING-IN-PUBLICATION DATA
Deacon, Terrence William.
      The symbolic species: the co-evolution of language and the brain /
by Terrence W. Deacon.
          p. cm.
      Includes bibliographical references and index.
      ISBN 0-393-03838-6
      1. Neurolinguistics. 2. Brain—Evolution. I. Title.
✓ QP399.D43 1997
      153.6—DC20 96-31115
CIP

W. W. Norton & Company, Inc.
500 Fifth Avenue, New York, N.Y. 10110
http://www.wwnorton.com

W. W. Norton & Company Ltd.
10 Coptic Street, London WC1A 1PU

1 2 3 4 5 6 7 8 9 0

In memory of Harriet Deacon, my grandmother and first mentor, who taught me to recognize the miraculous in everyday things.

Very special thanks are due to my family, Cris, Anneka, and John, who supported me even as I stole precious time away from them to complete this work, and to my parents, whose encouragement I could always count on. Thanks are also due to many who have directly inspired, assisted, and endured this project. These include: Joseph Marcus, who read, edited, and commented extensively on earlier drafts and who has consistently held me to the high standards he expects from a mentor; my own mentors, who have knowingly or unknowingly contributed their insights and valuable criticisms, but who are far too numerous to list; my former student Alan Sokoloff, whose Ph.D. research on oral tract innervation underlies my thoughts on vocal evolution; David Rudner, Sandra Kleinman, Alan Aronie, and many other friends whose feedback has helped clear away some of the fog; my colleagues in the lab, who understood and picked up more of the load as my attentions were diverted from the lab bench; Robyn Swierk and Julie Criniere, who helped enter the endless corrections, and copyeditor Ann Adelman; Hoover the seal, who opened my ears to the mystery of speech; and my editor, Hilary Hinzmann, who was patient through my long spells of writer's block and overcommitment, and who helped nurse a rough collection of ideas and notes into a narrative. I can never adequately repay these many gifts.

Contents

Part Three: Co-Evolution

Preface

*The conviction persists, though history shows it to be a
hallucination, that all the questions that the human mind has asked
are questions that can be answered in terms of the alternatives that
the questions themselves present. But in fact, intellectual progress
usually occurs through sheer abandonment of questions together
with both of the alternatives they assume, an abandonment that
results from their decreasing vitalism and a change of urgent
interest. We do not solve them, we get over them.*

—John Dewey

A number of years ago I was giving a brief talk about the evolution of
the brain when someone asked a question I couldn't answer. It was not
a complicated question. It didn't come from a colleague who had
found a weakness in my theory or a graduate student who had read about
a new experiment that contradicted my data. It came from a child in my
son's elementary school class. I had given school talks on being a scientist
before, and I thought I knew what to expect. I never expected an eight-year
old to stump me.

I was talking about brains and how they work, and how human brains
are different, and how this difference is reflected in our unique and com-

plex mode of communication: language. But when I explained that only humans communicate with language, I struck a dissonant chord.

"But don't other animals have their own languages?" one child asked.

This gave me the opportunity to outline some of the ways that language is special: how speech is far more rapid and precise than any other communication behavior, how the underlying rules for constructing sentences are so complicated and curious that it's hard to explain how they could ever be learned, and how no other form of animal communication has the logical structure and open-ended possibilities that all languages have. But this wasn't enough to satisfy a mind raised on Walt Disney animal stories.

"Do animals just have SIMPLE languages?" my questioner continued.

"No, apparently not," I explained. "Although other animals communicate with one another, at least within the same species, this communication resembles language only in a very superficial way—for example, using sounds—but none that I know of has the equivalents of such things as words, much less nouns, verbs, and sentences. Not even simple ones."

"Why not?" asked another child.

At this I hesitated. And the more I thought about it, the more I recognized that I didn't really have an answer. As far as I could tell no other scientists had seriously considered the question in this form. Why are there no simple languages, with simple forms of nouns, verbs, and sentences? It is indeed a counterintuitive fact. Myths, fables, fairy tales, animated cartoons, and Disney movies portray what common sense suggests ought to be the case: that other animals with their simpler minds communicate and reason using simpler languages than ours. Why isn't it so?

I'm not sure why I hadn't noticed this paradox before, or why other scientists hadn't. Most mammals aren't stupid. Many are capable of quite remarkable learning. Yet they don't communicate with simple languages, nor do they show much of a capacity to learn them—if our pets are any indication. Perhaps we have been too preoccupied with trying to explain our big brains, or too complacent with the metaphoric use of the term *animal language,* to notice this contradictory little fact. But the question may also have been unconsciously avoided because of the intellectual costs of considering it seriously. Indeed, the more deeply I have pursued this question, the more it seems like a Pandora's box that unleashes troubling doubts about many other questions that once seemed all but settled. This isn't the question we had been asking, but maybe it should have been. As Dewey suggests, the alternatives we pose in our scientific questions may not even address the most crucial issues.

This book starts with this curious question, because it supersedes many of the questions we thought were more important, and because it stubbornly refuses to resolve itself as a side effect of the superiority of human intelligence or the savantlike language ability of young children. But in my efforts to answer it, I am forced to reopen many questions long thought to have been resolved, or at least reduced to a few alternatives which now appear less informative than we once thought.

In the chapters that follow, I investigate how language differs from other forms of communication, why other species encounter virtually intractable difficulties when it comes to learning even simple language, how human brain structure has evolved to overcome these difficulties, and what forces and conditions initiated and steered us along this unprecedented evolutionary course. What results is a detailed reappraisal of human brain and language evolution that emphasizes the unbroken continuity between human and nonhuman brains, and yet, at the same time, describes a singular discontinuity between human and nonhuman minds, or to be more precise, between brains that use this form of communication and brains that do not. My somewhat unprecedented approach to these questions unfolds as a step by step argument, in which each chapter builds on the questions, analyses, and evidence provided in prior chapters. At almost every step of the argument, I arrive at different interpretations from what might loosely be called the accepted theories in the field. So I suggest approaching this narrative as one might approach a mystery novel, where the order and presentation of clues are critical to the plot, because encountering the clues and unexpected conclusions out of context might require some tricky mental gymnastics to discover how they fit back together.

The presentation is broken up into three major sections. The first part of the book—Language—focuses on the nature of language and the reasons that it is virtually confined to the human species. The second part of the book—Brain—tackles the problem of identifying what is unusual about human brain structure that corresponds with the unique problems posed by language. The third part of the book—Co-Evolution—examines the peculiar extension of natural selection logic that is behind human brain and language evolution, and tries to identify what sort of communication "problem" precipitated the evolution of our unprecedented mode of communication. The book ends with some speculations on the significance of these new findings for the understanding of human consciousness.

A major intent of the book is to engage the reader in a reexamination of many tacit assumptions that lie behind current views. In service of these aims I have tried to make my presentation accessible to the broadest possible scientific audience and, I hope, to a scientifically interested lay audience. Whenever possible I have tried to explain technical points in non-technical terms, and although I have not avoided introducing biological and neuroanatomical terminology when it is relevant, I have tried to illustrate some of the more technical points in graphic form. Some readers may find the middle section of the book—Brain—a bit rough going, but I believe that a struggle with this material will be rewarded by seeing how it leads to the novel reassessment of human origins and human consciousness that I offer in the more accessible and imaginative last section of the book.

Almost everyone who has written on the origins of language recounts with a sense of irony how the Société Linguistique de Paris passed a resolution in 1866 banning all papers on the origins of language. This resolution was meant to halt a growing flow of purely speculative papers, which the Société deemed to be contributing little of substance and occupying otherwise valuable time and resources. A young discipline eager to model itself after other natural sciences could ill afford to sponsor research in a topic that was almost entirely without empirical support. But the shadow of suspicion that looms over language origins theories is not just due to this historical reputation. Language origins scenarios no more empirically grounded than their banned predecessors still abound in the popular science literature, and they provide a perennial topic for cocktail party discussions. What's worse, assumptions about the nature of language and the differences between nonhuman and human minds are implicit in almost every philosophical and scientific theory concerned with cognition, knowledge, or human social behavior. It is truly a multidisciplinary problem that defies analysis from any one perspective alone, and where the breadth of technical topics that must be mastered exceeds even the most erudite scholars' capabilities. So it is hard to overestimate the immensity of the task or the risks of superficial analysis, and it is unlikely that any one account can hope to achieve anything close to a comprehensive treatment of the problem.

I take this as a serious caution to my own ambitions, and must admit from the outset that the depth of coverage and degree of expertise invested in the topics considered in this book clearly reflect my own intellectual biases, drawing primarily on my training in the neurosciences and in evolutionary anthropology and supplemented by a dilettante's training in other important areas. Consequently, the book focuses on the various implications of only this one human/nonhuman difference in mental abilities—particularly neurological implications—and ignores many other aspects of the brain-language relationship. I have not attempted in any systematic way to review or compare the many alternative explanations proposed for the phenomena I consider, and I have only discussed specific alternative theories as they serve as counterpoints to make my own approach clearer. An exhaustive review of competing explanations would require another book at least as long as this one. I apologize to my many scientific colleagues, who also have labored over these issues, for making this a rather personal exploration that does not do full justice to other theories, and does not explain why I do not even mention many of them. My own contributions are only possible because of the labors of untold dozens of previous researchers whose work has informed and influenced my own, and to whose contributions I have added only a handful of new findings. For those interested, I have tried to provide a reference to other approaches to these same puzzles in the end notes.

In what follows, it may appear as though I am a scientist with a naturally rebellious nature. I must admit that I have an attraction to heresies, and that my sympathies naturally tend to be with the cranks and doubters and against well-established doctrines. But this is not because I enjoy controversy. Rather, it is because, like Dewey, I believe that the search for knowledge is as often impeded by faulty assumptions and by a limited creative vision for alternatives as by a lack of necessary tools or critical evidence. So I will have achieved my intent if, in the process of recounting my thoughts on this mystery, I leave a few unquestioned assumptions more questionable, make some counter-intuitive alternatives more plausible, and provide a new vantage point from which to reflect upon human uniqueness.

The
Symbolic
Species

PART ONE

Language

1

The Human Paradox

. . . the paradox is the source of the thinker's passion, and the thinker without a paradox is like a lover without feeling: a paltry mediocrity.

—Søren Kierkegaard

An Evolutionary Anomaly

As our species designation—*sapiens*—suggests, the defining attribute of human beings is an unparalleled cognitive ability. We think differently from all other creatures on earth, and we can share those thoughts with one another in ways that no other species even approaches. In comparison, the rest of our biology is almost incidental. Hundreds of millions of years of evolution have produced hundreds of thousands of species with brains, and tens of thousands with complex behavioral, perceptual, and learning abilities. Only one of these has ever wondered about its place in the world, because only one evolved the ability to do so.

Though we share the same earth with millions of kinds of living creatures, we also live in a world that no other species has access to. We inhabit a world full of abstractions, impossibilities, and paradoxes. We alone brood about what didn't happen, and spend a large part of each day musing about the way things could have been if events had transpired differently. And we alone ponder what it will be like not to be. In what other species could individuals ever be troubled by the fact that they do not recall the way things were before they were born and will not know what will occur after they die? We tell stories about our real experiences and invent stories about imagined ones, and we even make use of these stories to organize our lives. In a real sense, we live our lives in this shared virtual world. And slowly, over the millennia, we have come to realize that no other species on earth seems able to follow us into this miraculous place.

We are all familiar with this facet of our lives, but how, you might ask, could I feel so confident that it is not part of the mental experience of other species—so sure that they do not share these kinds of thoughts and concerns—when they cannot be queried about them? That's just it! My answer, which will be argued in detail in the following chapters, has everything to do with language and the absence of it in other species. The doorway into this virtual world was opened to us alone by the evolution of language, because language is not merely a mode of communication, it is also the outward expression of an unusual mode of thought—symbolic representation. Without symbolization the entire virtual world that I have described is out of reach: inconceivable. My extravagant claim to know what other species cannot know rests on evidence that symbolic thought does not come innately built in, but develops by internalizing the symbolic process that underlies language. So species that have not acquired the ability to communicate symbolically cannot have acquired the ability to think this way either.

The way that language represents objects, events, and relationships provides a uniquely powerful economy of reference. It offers a means for generating an essentially infinite variety of novel representations, and an unprecedented inferential engine for predicting events, organizing memories, and planning behaviors. It entirely shapes our thinking and the ways we know the physical world. It is so pervasive and inseparable from human intelligence in general that it is difficult to distinguish what aspects of the human intellect have not been molded and streamlined by it. To explain this difference and describe the evolutionary circumstances that brought it about are the ultimate challenges in the study of human origins.

The question that ultimately motivates a perennial fascination with human origins is not who were our ancestors, or how they came to walk upright, or even how they discovered the use of stone tools. It is not really a question that has a paleontological answer. It is a question that might otherwise be asked of psychologists or neurologists or even philosophers. *Where do human minds come from?* The missing link that we hope to fill in by investigating human origins is not so much a gap in our family tree, but a gap that separates us from other species in general. Knowing how something originated often is the best clue to how it works. And we know that human consciousness had a beginning. Those features of our mental abilities that distinguish us from all other species arose within the handful of million years since we shared a common ancestor with the remaining African apes, and probably can mostly be traced to events that took place only within the last 2 million. It was a Rubicon that was crossed at a definite time and in a specific evolutionary context. If we could identify what was different on either side of this divide—differences in ecology, behavior, anatomy, especially neuroanatomy—perhaps we would find the critical change that catapulted us into this unprecedented world full of abstractions, stories, and impossibilities, that we call human.

It is not just the origins of our biological species that we seek to explain, but the origin of our novel form of mind. Biologically, we are just another ape. Mentally, we are a new phylum of organisms. In these two seemingly incommensurate facts lies a conundrum that must be resolved before we have an adequate explanation of what it means to be human.

Advances in the study of human evolution, the brain, and language processes have led many scientists confidently to claim to be closing in on the final clues to this mystery. How close are we? Many lines of evidence seem to be converging on an answer. With respect to our ancestry, the remaining gaps in the fossil evidence of our prehistory are being rapidly filled in. Within the last few decades a remarkably rich picture of the sizes and shapes of fossil hominid bodies and brains has emerged. It is probably fair to say that at least with respect to the critical changes that distinguish us in this way from other apes, there are few missing links yet to be found, just particulars to be filled in. That crucial phase in hominid evolution when our ancestors' brains began to diverge in relative size from other apes' brains is well bracketed by fossils that span the range. As for the inside story, the neurosciences are providing powerful new tools with which it has become possible to obtain detailed images from working human brains performing language tasks, or to investigate the processes that build our brains during development and distinguish the brains of different species, or even to model neural processes outside of brains. Finally, linguists' analyses of the logical structure of languages, their diversity and recent ancestry, and the patterns that characterize their development in children have provided a wealth of information about just what needs to be explained, and comparative studies of animals' communications in the wild and their languagelike capacities in the laboratory have helped to frame these questions with explicit examples.

Despite all these advances, some critical pieces of the puzzle still elude us. Even though neural science has pried ever deeper into the mysteries of brain function, we still lack a theory of global brain functions. We understand many of the cellular and molecular details, we have mapped a number of cognitive tasks to associated brain regions, and we even have constructed computer simulations of networks that operate in ways that are vaguely like parts of brains; but we still lack insight into the general logic that ties such details together. On the whole, most neuroscientists take the prudent perspective that only by continuing to unmask the details of simple neural processes in simple brains, and slowly, incrementally, putting these pieces together, will we ever be able to address such global theoretical questions as the neural basis for language. We must add to this many new problems arising out of the comparisons of animal communication to language. If anything, these problems have become more complex and more confusing the more we have learned about the sophistication of other species' abilities and the paradoxes implicit in our own abilities. But the most critical missing piece of the puzzle is access to the brains in question: ancestral hominid brains. Though we have considerable information about brain sizes in fossil species, and a little information about brain shapes, the relevant anatomical information, the internal microarchitecture of these brains, has left no fossil trail. With respect to fossil brains, we will never find the "smoking gun"—the first brain capable of language. We will only have access to circumstantial information.

So, what business do we have speculating about the beginnings of language? Given the complexity of the human brain, our current ignorance of many of its crucial operating principles, and the fact that neither languages nor the brains that produce them fossilize, there would appear to be many more immediate questions to be answered before even considering this one. There seem to be too many loose ends and gaps in the supportive evidence to provide solid leads in the search for clues to the nature of the human mind in the origins of language.

But this ignores the significance of the fact that language is a one-of-a-kind anomaly. Often the most salient and useful hints about the underlying logic of nature's designs are provided when unique or extreme features in two different domains are found to be correlated. Some notable examples include the correlation between superconductivity and extreme cold; between greater cosmic distances and the increasing redness of starlight; between the massive extinctions of fossil species and evidence of extraterrestrial impacts; between the peculiarity of haplo-diploid genetics and war, suicidal defense, and infertile castes in social insects; and so on. Each of these correlations cried out for an explanation and in so doing offered a critical clue to a more general principle. The more two related features diverge from what is typical in their respective domains, the more penetrating the insight that can be gleaned from their underlying relationship.

In this context, then, consider the case of human language. It is one of the most distinctive behavioral adaptations on the planet. Languages evolved in only one species, in only one way, without precedent, except in the most general sense. And the differences between languages and all other natural modes of communicating are vast. Such a major shift in behavioral adaptation could hardly fail to have left its impression on human anatomy. Even superficial appearances bear this out. We humans have an anomalously large brain and a uniquely modified vocal tract. Though these clues offer no more than a starting point, they suggest that the structural and functional relationships underlying these superficial correlations are likely to be robust and idiosyncratic to us.

Ironically, then, the problem of language origins may actually offer one of the most promising entry points in the search for the logic linking cognitive functions to brain organization. To the extent that the unique mental demands of language are reflected in unique neuroanatomical differences, we may find an unequivocal example of how nature maps cognitive differences to brain structure differences. Though the details of this mystery are challenging, no critical pieces of this puzzle lie buried in the deep evolutionary past or inaccessible to current technology. They are observable in the differences in cognitive abilities and brain structures to be found in living species.

I think that the difficulty of the language origins question is not to be blamed on what we don't know, but rather on what we think we already know. We think we know that what keeps language from being a widespread phenomenon is its byzantine complexity and the incredible demands it places on learning and memory. We think we know that language became possible for our ancestors when these impediments to language learning were overcome by some prior change in the brain. Depending on which aspects of language are deemed to be most complex, different prior adaptations are invoked to explain how language became possible. Perhaps it required an increase in intelligence, a streamlining of oral and auditory abilities, a separation of functions to the two sides of the brain, or the evolution of a sort of built-in grammar. I think we can be sure of none of these things. In fact, I think that the problem is more basic and far more counterintuitive than is assumed by any of these accounts.

There are a few common assumptions shared by all of these explanations that I think are the root of a deeper problem. In general, these arguments parallel many others that continually resurface along that age old divide between nature and nurture. Is language imposed from the outside or does it reflect what is already inside? For decades, the superficiality of this stale dichotomy has been evident, exposed by research in the psychological and biological sciences that demonstrates how truly complex and interdependent the biological and environmental contributions to development can be; but we still find it difficult to conceive of these phenomena in other terms. We have reinvented the same old answers in new guises in each generation, stubbornly insisting that the answer to the question of language knowledge must be found in one of just a few major alternative paradigms (depicted in cartoon fashion in Figure 1.1).

At one end of this spectrum is the assumption that the architecture of language originates entirely outside (simple associationism); at the other end is the assumption that it originates entirely inside (mentalese). What other alternatives could there be, that are not captured between these extremes? And if there are no other alternatives, then shouldn't answering this question also point to the solution to the language origins question? Discovering which aspects of language knowledge are contributed by nature and which by nurture ought to tell us what difference in us was necessary to bridge the original language acquisition gap. If the answer lies more toward the associationist end of the spectrum, then evolution must have given us language by endowing us with exceptionally powerful learning and memory. If the answer lies more toward the mentalese end, then evolution must have endowed us with remarkably sophisticated instinctual knowledge of language that made learning completely unnecessary.

In light of these intuitively compelling alternatives, the approach I am about to take may seem misguided. Not only do I think that these alternatives confuse the nature/nurture problem more than they illuminate it, I think that the whole question of where language knowledge originates during development is secondary. Though a young child's almost miraculous development of language abilities is indeed a remarkable mystery—one that will be considered in some detail later (in Chapters 4 and 11)—I think that the cause of language origins must be sought elsewhere, and by pursuing some very different kinds of research questions. While we have been worrying about where knowledge of language comes from, we have been avoiding a more basic question: What sort of thing is knowledge of language

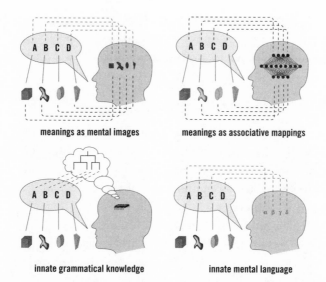

Figure 1.1 *Four cartoon depictions of some of the major theoretical paradigms proposed to explain the basis of human language. Top left: The notion that word meaning is created when the perception of the sound of a spoken word is associated with an object both as perceived and as stored in the mind in the form of a mental image. In this simple common sense view, stringing together words in a sentence leads the listener to bring together images in the mind. Top right: The notion that both word meaning and knowledge of language structure are learned by internalizing patterns of the associative probabilities linking words to one another, and linking words and objects. B. F. Skinner was the most prominent defender of this view, but recently more sophisticated versions of this basic idea have been reformulated with the aid of insights gained by studying parallel distributed learning processes. Knowledge of language is depicted as analogous to the distributed connection patterns in a neural net. Bottom left: One of the most influential views of grammatical knowledge conceives of it as built in prior to language experience, like firmware in a desktop computer (depicted as a computer chip inserted in the brain). The structure of language is imposed on strings of words (that presumably would still be meaningful, just less useful, without this structure). This view was first explicitly formulated by the linguist Noam Chomsky. Bottom right: The extreme innatist view of knowledge of language conceives of it as an external reflection of an internal lingua franca of the brain called "mentalese." In Steven Pinker's words* (The Language Instinct, *p. 82), "Knowing a language, then, is knowing how to translate mentalese into strings of words and vice versa. People without language would still have mentalese, and babies and many nonhuman animals presumably have simpler dialects. Indeed, if babies did not have a mentalese to translate to and from English, it is not clear how learning English could take place, or even what learning English would mean." None of these views provides a satisfactory explanation of the paradox explored in this chapter.*

anyway? Before turning to this question, however, it is worth while reflecting on some of the equally misleading evolutionary assumptions that reinforce the traditional theoretical alternatives.

Technical Difficulties and Hopeful Monsters

One of the most common views about language evolution is that it is the inevitable outcome of evolution. Evolution was headed this way, our way. As the only species capable of conceiving of our place among all others, we see what looks like a continuous series of stages leading up to one species capable of such reflections. It goes without saying that a more articulate, more precise, more flexible means of communicating should always be an advantage, all other things being equal. In terms of cooperative behaviors, a better ability to pass on information about distant or hidden food resources, or to organize labor for a hunt, or to warn of impending danger, would be advantageous for kin and the social group as a whole. Better communication skills might also contribute to more successful social manipulation and deception. The ability to convince and mislead one's competitors or cooperate and connive with one's social and sexual partners could also have provided significant reproductive advantages, particularly in social systems where competition determines access to defendable resources or multiple mates. In fact, it's difficult to imagine any human endeavor that would *not* benefit from better communication. Looked at this way, it appears that humans have just developed further than other species along an inevitable progressive trend toward better thinking and better communicating.

Surely we must be part of a trend of better communication in some form? It seems to be an unstated assumption that if biological evolution continues long enough, some form of language will eventually evolve in many other species. Are chimpanzees the runners-up, lagging only a little behind on the road to language? As in *Planet of the Apes*, a science fiction movie in which our more hairy cousins catch up to a human level of mental and liguistic abilities, we imagine that if given sufficient time, something like language is prefigured in evolution. We even imagine that if there is life on other planets, and if it has been evolving as long as life on earth, or longer, there will be "intelligent" species with whom we may someday converse.[1] The Renaissance notion of a "Great Chain of Being" gave rise to nineteenth-century theories of phylogeny that ranked species from lower to higher, from mechanism to godly, with humans just below angels. Though later nineteenth- and twentieth-century evolutionists rejected the static ranking of phylogeny and replaced it with the theory of evolutionary descent, the anthropocentric perspective was simply rephrased in evolutionary terms. Humans were presumably the most "advanced" species. Carrying this notion to its extreme, some people now suspect that there may be spaceships visiting earth, carrying beings that are "more highly evolved" than we are.

On the surface, progress seems to be implicit in natural selection. Gradual improvement of adaptations seems to imply that the longer evolution continues the better the design will become. Indeed, many scientists talk as though a special kind of retrograde selection would be necessary to halt the progress of inevitably increasing intelligence. Small-brained species are often considered primitive or throwbacks to earlier forms, left out of the main trend. From an anthropocentric perspective, it seems unquestionable that more intelligent species will outcompete less intelligent ones. Intelligence is always an advantage, right? Brain over brawn. We rank genius and mental retardation on a single scale, and presumably rank chimpanzees, dogs, and rats on the low end of the same scale. Human evolution is often termed an "ascent" to imply a climb from lower to higher intelligence. And from this it seems to follow that humans are just the pinnacle example of an inevitable trend. The winner in a war of neurons.

The apparent reasonableness of this view reflects our familiarity with technological progress in Western societies. The interchangeability of terms like *consciousness expansion, social progress*, and *evolution* is now almost commonplace in the popular press, and these ideas are seldom entirely disentangled even in the most sophisticated accounts of human evolution. But the idea of progress in evolution is an unnoticed habit left over from a misinformed common sense, from seeing the world in terms of *design*. The problem is that our intuitive model for evolution is borrowed from the history of technological change, which has been a cumulative process, adding more and more tidbits of know-how to the growing mass of devices, practices, and records each day. In contrast, biological evolution is not additive, except in some very limited ways. The human repertoire of expressed genes is about the same as that in a mouse or frog, and the body plans of all vertebrates seem to be mostly modifications of the same shared plan—even for the brain. Though we are on the large end of the range of body and brain sizes, this is not the result of adding new organs but merely enlarging existing ones with slight modifications.

Evolution is an irreversible process, a process of increasing diversification and distribution. Only in this sense does evolution exhibit a consistent direction. Like entropy, it is a process of spreading out to whatever possibilities are unfilled and within reach of a little more variation. Evolution does not continue to churn out ever better mousetraps, even if it has produced some remarkable examples along the way. But this pattern of spreading into unfilled niches does place us in one of the more extreme niches.

Evolution is diversification in all directions, but there are more options available in some directions than others. Organisms started out small and short-lived and couldn't get much smaller, but they could always get larger and more long-lived. For the smallest organisms, the resources that can be devoted to internal representations of the world are limited, though even bacteria appear able to use their one information-storage system, their genes, to take in information from around them and modify their behaviors appropriately. But the upper end of the range of information-handling abilities was not similarly bounded, and so the difference between the low end and the upper end of this range has increased over the hundreds of million years of animal evolution as part of this diversification. Nevertheless, the number of small-brained creatures has not diminished because of competition with those with big brains, and the no-brainers—all the plants and single-celled organisms—vastly outnumber the rest of us. It just happens that one very, very minor evolutionary direction is toward niches where doing a lot of information processing during one's life is a good way to pass on genes. Inevitable? Well, it's about as inevitable a direction in evolution as the development of arctic fish with antifreeze in their blood or electric eels who use electricity to sense their way through muddy Amazonian waters. The niche was just there, and was eventually filled. Still, in some measure, we are near the extreme of this distribution.

The question, then, is whether the evolution of language was somehow prefigured in this trend. Is there a general trend toward better communication? It's easy in hindsight to arrange the history of long-distance communication from telegraphs to telephones to cellular phones to *Star Trek* communicators. It is not so easy to determine if animal communication has been steadily getting better and if human language is a part of such a trend. Certainly, there were advances in distance and signal clarity in evolution, but even if we narrow our comparison to vocal communication, there is no evidence in living species that some inevitable progressive trend leads to us. Apparently simple species can use highly complex methods of sound communication, and some highly complex species can be oblivious to their advantages. There are also many great sound tricks, such as echolocation, that are completely beyond human ability. Among our closest relatives, the great apes, there are both highly vocal (chimp) and nearly silent (orangutan) species. In fact, most birds easily outshine any mammal in vocal skills, and though dogs, cats, horses, and monkeys are remarkably capable learners in many domains, vocalization is not one of them. Our remarkable vocal abilities are not part of a trend, but an exception.

We also tend to underestimate the complexity and subtlety of much non-human social communication. In recent decades, field studies of social communication in nonhuman species have demonstrated that many birds, primates, and social carnivores use extensive vocal and gestural repertoires to structure their large social groups. These provide a medium for organizing foraging and group movement, for identifying individuals, for maintaining and restructuring multidimensional social hierarchies, for mediating aggressive encounters, for soliciting aid, and for warning of dangers.[2] Indeed, even in our own species, a rich and complex language is still no substitute for a shocked expression, a muffled laugh, or a flood of silent tears, when it comes to communicating some of the more important messages of human social relationships.

However, although they are complex, these elaborate repertoires of calls, displays, and gestures do not seem to map onto any of the elements that compose languages. Although various researchers have suggested that parallels to certain facets of language are to be found in the learned dialects of birdsong, the external reference evident in vervet monkey alarm calls or honeybee dances, and the socially transmitted sequences of sounds that make up humpback whale songs (each of which will be considered in some detail in Chapter 2), these and many other examples like them only exhibit a superficial resemblance to language learning, word reference, or syntax, respectively. Even if we were to grant these parallels, no nonhuman species appears to put these facets of language together into a coordinated, rule-governed system.

Could we have missed recognizing nonhuman languages because they are as alien to us as our speech is to them? People have long entertained this possibility at least in mythology and children's stories. They offer the fantasy that we might someday overcome the communication boundaries that separate humans and other animals and share memories, beliefs, hopes, and fears with them. In the popular children's book made into a movie, Dr. Doolittle enlists the aid of a "multilingual" parrot to translate between animal and human speech. But is even a very superficial "translation" possible? What do you tell a child who asks, "What is the kitty saying?" Do animals' vocalizations and gestures explain, describe, ask, or command? Do they argue, disagree, bargain, gossip, persuade, or entertain one another with their thoughts? Are there any corresponding elements in animal communication that map onto the elements of human language? Unfortunately, animal calls and displays have nothing that corresponds to noun parts or verb parts of sentences, no grammatical versus ungrammatical strings, no marking of singular or plural, no indications of tense, and not even any elements that easily map onto words, except in the most basic sense of the beginning and ending of a sound.

One quite reasonable caution against making strong claims about the absence of nonhuman languages is that our study of other species' communication systems is still in its infancy. Isn't it more prudent to remain agnostic until we have learned a great deal more about other species communications? It is always wise not to prejudge the evidence, especially with respect to a subject about which we have undeniable prejudices. And there are far more species about whose communicative behavior we know next to nothing than there are whose communication has been studied. Nevertheless, I think that we have sufficient information to make a reasonably confident claim even about species whose communicative behaviors have been only superficially studied. What makes this a fairly safe guess is not the sophistication of our behavioral analyses, but rather the striking characteristics that would be evident in a nonhuman language. Where the differences should be glaring, the sensitivity and sophistication of observations and tests can be minimal.

What would be the characteristics of a nonhuman language that would allow us instantly to recognize it as a languagelike form of communication, even if it were quite alien with respect to all human languages? This is the sort of question that scientists scanning the heavens with radio telescopes listening for signals from unearthly species must ask themselves, or that might be asked by those engaged in electronic surveillance interested in distinguishing the transmission of coded or encrypted signals from random noise. Interestingly, many of these characteristics are exhibited in the surface structure of the signals, and require no special insight into meaning or referential function, and no obvious correspondence with natural language grammars, to discern.[3] Here are a few general features that ought to stand out. A languagelike signal would exhibit a combinatorial form in which distinguishable elements are able to recur in different combinations. It would exhibit a creative productivity of diverse outputs and a rather limited amount of large-scale redundancy. And although there would be a high degree of variety in the possible combinations of elements, the majority of combinatorial possibilities would be systematically excluded. In terrestrial examples, where it would be possible to observe the correlations between signals and contextual events, there should be another rather striking and counterintuitive feature. The correlations between individual signals and the objects and events that form the context of their production should not exhibit a simple one-to-one mapping. The correlations between sign elements and their pragmatic context should differ radically yet systematically from occasion to occasion, depending on how the signals are arranged in combination with respect to one another. These, of course, are all general features associated with syntax, though not just language syntax. Human games, mathematics, and even cultural customs exhibit these features.

If a radio-telescope observer identified a signal emanating from distant space with these characteristics it would make world headlines, despite the fact that the meaning of the signal would remain completely undecodable. With far more to go on than this, in even superficially studied animal communications, we can be reasonably sure that for the vast majority of likely candidate species such a signal has not yet been observed. Instead, though highly complex and sophisticated, the communicative behaviors in other species tend to occur as isolated signals, in fixed sequences, or in relatively unorganized combinations better described by summation than by formal rules. And their correspondences with events and behavioral outcomes, in the cases where this can be investigated, inevitably turn out to be of a one-to-one correlational nature. Though an as yet undescribed example of an animal communication system that satisfies these criteria cannot be ruled out, it seems reasonable to conclude that the chances are poor that it would have gone unobserved in common animal species, any more than we would miss it in a cosmic radio signal.

My point is not that we humans are better or smarter than other species, or that language is impossible for them. It is simply that these differences are not a matter of incommensurate *kinds* of language, but rather that these nonhuman forms of communication are something quite different from language. For this reason I think that the comparison is misguided, and useful only at a very superficial level. This fact should not be too difficult to appreciate because we all have personal experience of just the sort of incommensurability I am talking about. There are numerous human counterparts to other animals' nonlinguistic communicative behaviors. We too have a wide range of innately produced and universally understood facial expressions, vocalizations, and gestures. As in other species, they are an irreplaceable component of human social communication. Yet this is not analogous to being bilingual. This other repertoire of communicative behaviors is not a language of gestures instead of words. It is something else. And although these human calls and gestures comprise an entirely familiar system, we find the same difficulty translating them into word equivalents as we do with animal calls and gestures with which we are far less familiar. The problem is not their unfamiliarity but rather that it simply makes no sense to ask what kind of word a laugh is, whether a sob is expressed in past or present tense, or if a sequence of facial gestures is correctly stated. The problem isn't a difficulty mapping human to nonhuman languages, but rather a difficulty mapping languages to any other form of naturally evolved communication, human or otherwise.

Of no other natural form of communication is it legitimate to say that "language is a more complicated version of that." It is just as misleading to call other species' communication systems *simple* languages as it is to call them languages. In addition to asserting that a Procrustean mapping of one to the other is possible, the analogy ignores the sophistication and power of animals' nonlinguistic communication, whose capabilities may also be without language parallels. Perhaps we are predisposed to see other species' communications through the filter of language metaphors because language is too much a natural part of our everyday cognitive apparatus to let us easily gain an outside perspective on it. Yet our experience of its naturalness, its matter-of-factness, belies its alien nature in the grander scheme of things. It is an evolutionary anomaly, not merely an evolutionary extreme.

This lack of precedent makes language a problem for biologists. Evolutionary explanations are about biological continuity, so a lack of continuity limits the use of the comparative method in several important ways. We can't ask, "What ecological variable correlates with increasing language use in a sample of species?" Nor can we investigate the "neurological correlates of increased language complexity." There is no range of species to include in our analysis. As a result, efforts to analyze the evolutionary forces responsible for language have often relied on crude substitutes to make up for the lack of homology between language and nonhuman forms of communication. It is tempting to try to conceive of language as an extrapolated extreme of something that other species produce, such as calls, grunts, gestures, or social grooming.[4] It is also tempting to turn to some other feature of human anatomy that can be more easily compared to other species as an index of language evolution. Humans can, for example, be ranked along with other species with respect to brain size, group size, social-sexual organization, foraging strategy, etc. But even if humans are at the extreme in many of these measures, the correlations among these attributes are not obvious, and their linkage to language is dubious, since trends in these attitutes in other species occur irrespective of language.

Interpreting the discontinuity between linguistic and nonlinguistic communication as an essential distinction between humans and nonhumans, however, has led to an equally exaggerated and untenable interpretation of

This leads to the apparently inescapable conclusion tha[...] mation must already be "in the brain" before the process [...] for it to be successful. Children must already "know" what c[...] allowable grammar, in order to be able to ignore the innumerab[...] potheses about grammar that their limited experience might other[...] gest.

This device, a "language organ" unique to the human brain, coula[...] account for the failure of other species to acquire language. The appea[...] this scenario is that it eliminates many troublesome questions in one fe[...] swoop: the discontinuity between human and nonhuman communication, the larger human brain (adding a new part enlarges it), the systemic inter-dependent nature of grammatical rules (they all derive from one neuro-logical source), the presumed universal features of language structure (ditto), the intertranslatability of languages (ditto), and the ease with which language is initially acquired despite an insufficient input and a lack of grammatical error correction by adults.

Another appeal of the hopeful monster story is that it promises a defi-nite and dramatic transition from one stage to another in the evolutionary sequence. It offers a single-step evolutionary account that is much easier to comprehend and organize in one's thinking than continuous changes in-volving multiple factors interacting and overlapping in time in complex ways. It tantalizes the imagination to hear that the story of human origins was written in the course of a single dramatic and decisive prehistoric event. In one step, some ancestor crossed the threshold into humanity. But such a crucial transition could hardly occur without leaving a trail of evidence at-testing to its discontinuity. If modern language abilities appeared all of a sud-den in human prehistory, then we ought to find numerous other correlates of a radical reorganization of human behavior and biology. Inspired by this possibility, researchers in many fields have combed through their own data for signs of sudden transitions that might be the result of such an incredi-ble language mutation. Not surprisingly, many have been "discovered" in [t]he record of human prehistory. They include: abrupt technological transi-[ti]ons (e.g., the first appearance of stone tools or of extensive cultural vari-[ati]ons in tool design); possible punctuated speciation events (e.g., the [ori]gination of anatomically modern humans from a "mitochondrial Eve"); [rapi]d population changes (e.g., the demise of the Neanderthals); and signs [of m]ajor innovations in cultural artefacts (the first appearance of durable [repr]esentative art, such as carvings and cave paintings in Europe). But be-[caus]e they offer evidence that is indirect, at best, and so sparse and frag-[ment]ary, paleontological finds can appear irregular for many other reasons, nguage origins: the claim that language is the product of a unique one-of-a-kind piece of neural circuitry that provides all the essential features that make language unique (e.g., grammar). But this does not just assume that there is a unique neurological feature that correlates with this unique behavior, it also assumes an essential biological discontinuity. In other words, that language is somehow separate from the rest of our biology and neurology. It is as though we are apes *plus* language—as though one handed a language computer to a chimpanzee.

This reminds me of a wonderful piece of modern mythology from a recent film entitled *Short Circuit*. A sophisticated robot is accidentally transformed from a mechanism that "just runs programs" into a conscious, self-aware being as a result of being struck by lightning. The power surge damaged its circuits in just the right way. The now conscious robot, of course, does not think of this as "damage." From his perspective, the lightning bolt corrected a design limitation. As a cinematic device, the bolt of lightning accomplishes two important things. The catastrophic and unpredictable nature of lightning provides a vehicle for invoking drastic and unprecedented change, and its intrinsically chaotic—and, by tradition miraculous—character obviates any possibility of describing exactly wh alterations changed a computer mechanism into a human-type mind. the sake of the story, we suspend critical analysis and allow this miracu accident to stand in place of an otherwise inexplicable transformatio an allegory of human mental evolution, it offers a paradigm example biologists call a "hopeful monster" theory: the evolutionary theoris terpart to divine intervention, in which a freak mutation just happe duce a radically different and serendipitously better-equipped o

The single most influential "hopeful monster" theory of hum evolution was offered by the linguist Noam Chomsky, and ha echoed by numerous linguists, philosophers, anthropologists ogists. Chomsky argued that the ability of children to acquir of their first language, and the ability of adults effortlessly t mar, can only be explained if we assume that all grammar a single generic "Universal Grammar," and that all hu with a built-in language organ that contains this languag offered as the only plausible answer to an apparently in ing problem. Grammars appear to have an unparallel tematic logical structure, the individual grammatical evident in the information available to the child, and first language children are still poor at learning m these limitations children acquire language know not least of which is our predisposition to organize the evidence in categorical terms.

An accidental language organ requires no *adaptive* explanation for the structure of language. If this hypothetical organ was plugged into the brain in a single accident of prehistory, rather than evolving bit by bit with respect to its functional consequences, then no functional explanations would be necessary. If it was just an accident, any utility would be entirely accidental as well, discovered after the fact. This too might account for the many little idiosyncrasies of language and its discontinuities when compared with other nonhuman forms of communication. But I think that this story is far too neat and tidy precisely because it suggests that so many questions do not need to be addressed. The accidental language organ theory politely begs us to ignore the messy details of language origins, abandon hope of finding precedents in the structure of ape brains or their cognitive abilities, and stop looking for any deep design logic to the structural and functional relationships of language grammars and syntax. This is a lot to ignore. What does this hypothesis provide instead?

One of the characters in Molière's play *The Imaginary Invalid*,[6] is asked by his physician-examiners to explain the means by which opium induces sleep. He replies that it induces sleep because it contains a "soporific factor." This answer is applauded by the doctors. The playwright is, of course, satirizing the false expertise of these apparently learned men by showing their knowledge to be no more than sophistry. The answer is a nonexplanation. It merely takes what is in need of explanation and gives it a name, as though it were some physical object. Like phlogiston, the substance once hypothesized by pre-atomic chemistry to be the essence that determined flammability, the "soporific factor" fails to reduce the phenomenon in need of explanation to any more basic causal mechanisms.

For many linguists, grammatical knowledge *is* what needs to be explained, and what is lacking is an adequate account of the source of children's grammatical and syntactic abilities in terms of antecedents in the child's experience of language. We are thus like the characters in Molière's play, who know *what* is produced but don't know how it is produced. Failing to discover a satisfactory explanation for how grammatical knowledge could be impressed upon children's minds from the outside, we naturally turn to the possibility that it does not come from the outside at all. But simply assuming that this knowledge is already present, and so doesn't need to pass from outside to inside, only restates this negative finding in positive terms. A grammar instinct or a universal grammar serve as place holders for whatever could not be learned. The nature of this presumed innate knowledge of language is described only in terms of its consequences. Linguists have progressively redefined what supposedly cannot be learned in ever more formal and precise terms, and so we may have the feeling that these accounts are approaching closer and closer to an explanation. But although the description of what is missing has gotten more precise, ultimately it is only a more and more precise version of what is missing. These "explanations" of the nature of a language instinct are inevitably presented in the guise of elaborate definitions of grammatical principles or else as something akin to computer programs, and in this way they are only more formal restatements of the problem of the missing information. Saying that the human brain alone produces grammar because it alone possesses a grammar factor ultimately passes the explanatory buck out of the hands of linguists and into the hands of neurobiologists.

To be fair, the intent of language organ theories is not to address the question of initial language origins, but rather to explain the source of language competence in development. For this reason, it is not wedded to the hopeful monster assumption. Steven Pinker, a proponent of the Universal Grammar view of language abilities and an articulate champion of many of Chomsky's original insights about the uniqueness of language, argues in a recent book (*The Language Instinct*) that innate grammatical knowledge is not at all incompatible with an adaptationist interpretation of its origins. He argues that a language instinct could have gradually evolved through the action of natural selection. On the one hand, this is a far more biologically plausible alternative to miraculous accidents and it challenges us to face some of the difficult problems ignored by theories relying on miraculous accidents to fill in the gaps. On the other hand, an adequate formal account of language competence does not provide an adequate account of how it arose through natural selection, and the search for some new structures in the human brain to fulfil this theoretical vacuum, like the search for phlogiston, has no obvious end point. Failure to locate it in such a complex hierarchy of mechanisms can always be dismissed with the injunction: look harder.

A full evolutionary account cannot stop with a formal description of what is missing or a scenario of how selection might have favored the evolution of innate grammatical knowledge. It must also provide a functional account of why its particular organization was favored, how incremental and partial versions were also functional, and how structures present in nonhuman brains were modified to provide this ability. The language instinct theory provides an end point, an assessment of what a language evolution theory ultimately needs to explain. It rephrases the problem by giving it a new name. But this offers little more than the miraculous accident theory provided: a formal redescription of what remains unexplained. Unfortunately, I think it also misses the forest for the trees, even in this endeavor. I don't think that children's grammatical abilities are the crucial mystery of language.

The Missing Simple Languages

The two dominant paradigms for framing the language origins question—the evolution of greater intelligence versus the evolution of a specialized language organ—have one thing in common: both are stated in terms of the problem of learning a very large and complex set of rules and signs. They assume that other species are poor language learners because language is just too complicated for them to learn, and too demanding for them to perform. It requires rapid and efficient learning, demands immense memory storage, takes advantage of almost supernatural rates of articulation and auditory analysis, and poses an analytic problem that is worthy of a linguistic Einstein. Both approaches assume that the difficulty for other species is the complexity of language, but they disagree on the source of this difficulty and on what is required to overcome it. Do human children merely need to be very much more intelligent than other species in order to learn language, or is language *so* complicated that it is impossible to learn without some built-in language information to "jump start" the process? Accepting one or the other assumption leads to opposite claims about the nature of the evolution of language and the human mind. If language is just difficult to learn, then the neural adaptation that supports it could be quite general in its effect on cognitive abilities. If language is, for all practical purposes, impossible to learn, then the neural adaptation necessary to support it would need to be quite specific. However one looks at these problems, it appears that overcoming the limitations imposed by the obvious complexity of language is a prerequisite to language evolution. I say it "appears" this way, because I think that something has been missed in both views of the problem, something fundamental. These alternatives, and many plausible intermediates, only address *one* of the main problems in need of explanation, and it is not the critical one.

A task that is physically too difficult to perform may exceed our strength, our endurance, our rate of performance, our precision of action, our capacity to do many things at the same time, etc. In cognitive terms, these correspond to our ability to focus attention, the persistence of our memories, our rate of learning, and our short-term memory span, etc. When we say that a skill is difficult to learn, we mean that the desired movement sequence severely taxes our ability precisely to time, control, or coordinate the component movements. When we say that a perceptual task is difficult to learn, we mean that it requires utilizing cues too subtle or fleeting to detect, too irregular to discover their commonalities, or embedded in too many distracting cues to sort out. And when we say that a cognitive task is difficult to learn, we mean that there are too many associations to be held in working memory at one time or too many to be considered in too little time or simply too many to be remembered. Each demands too much, from too little, in insufficient time. Both the complexity of the task and the resources one has available will determine its relative difficulty.

Clearly, language is complicated in all these ways. Linguistic communication requires us to learn and perform some remarkably complicated skills, both in the production of speech and in the analysis of speech sounds. In addition, there is a great deal to learn: thousands of vocabulary items and an intricate system of grammatical rules and syntactic operations. And it's not enough that language is complicated. According to many linguists, we aren't even offered sufficient outside support to deal with it. We are forced to figure out the underlying implicit rules of grammar and syntax without good teaching and with vastly inadequate examples and counterexamples. This apparent lack of adequate instruction adds insult to injury, so to speak, by making a too complicated task even harder. The degree to which the support for language learning undershoots this need is (exponentially) proportional to how complicated the task is to begin with, and so the complexity of language is doubly limiting.

How could anyone doubt that language complexity is the problem? Languages are indeed complicated things. They are probably orders of magnitude more complicated than the next-most-complicated communication system outside of the human sphere. And they are indeed almost impossibly difficult for other species to acquire. The question is whether this complexity is the source of the difficulty that essentially limits the use of language to our species alone. Although this would seem to be the obvious conclusion, it is not quite so obvious as it might first appear. The most crucial distinguishing features of language cannot be accounted for merely in terms of language complexity.

The challenge to the complexity argument for human language origins rests on a simple thought experiment. Imagine a greatly simplified language, not a child's language that is a fragment of a more complicated adult language, but a language that is logically complete in itself, but with a very limited vocabulary and syntax, perhaps sufficient for only a very narrow range of activities. I do not mean "language" in a metaphoric sense, the way that all communication systems are sometimes glossed as languages. But I also do not restrict my meaning to speech, or to a system whose organizational principles are limited to the sorts of grammatical rules found in modern languages. I mean language in the following very generic sense: a mode of communication based upon symbolic reference (the way words refer to things) and involving combinatorial rules that comprise a system for representing synthetic logical relationships among these symbols. Under this definition, manual signing, mathematics, computer "languages," musical compositions, religious ceremonies, systems of etiquette, and many rule-governed games might qualify as having the core attributes of language. More important, no more than a tiny "vocabulary" of meaningful units and only two or three types of combinatorial rules would be necessary to fulfill these criteria. A five- or ten-word vocabulary and a syntax as simple as toddlers' two- and three-word combinations would suffice. Reducing the definition of language to such minimal conditions allows us to conceive of languagelike systems that are far simpler even than the communicative repertoires found to occur in the social interactions of many other species.

So this is the real mystery. Even under these loosened criteria, there are no simple languages used among other species, though there are many other equally or more complicated modes of communication. Why not? And the problem is even more counterintuitive when we consider the almost insurmountable difficulties of teaching language to other species. This is surprising, because there are many clever species. Though researchers report that languagelike communication has been taught to nonhuman species, even the best results are not above legitimate challenges, and the fact that it is difficult to prove whether or not some of these efforts have succeeded attests to the rather limited scope of the resulting behaviors, as well as to deep disagreements about what exactly constitutes languagelike behavior. Both the successes and failures that have come of this research are nevertheless highly informative with regard to both what animals can and can't do and how we conceive of language itself, but the few arguable successes must be seen against the background of domesticated animals and family pets that never seem to catch on, despite being raised in a context where they are bombarded with a constant barrage of commands, one-sided conversations, and "rhetorical" questions. For the vast majority of species in the vast majority of contexts, even simple language just doesn't compute. This lack of simple languages in the wild and inability to learn simple languages under human tutelage don't make sense! Many of these species engage in natural communicative behaviors that are far more complex than a simple language, and they are capable of learning larger sets of associations than are necessary for constructing a simple language. So why is language such a problem? The difference cannot be simple versus complex.

The complexity of language is important. It demands an explanation, as does the ability of young children to make sense of it, seemingly without sufficient feedback or time at their disposal. These are remarkable aspects of the language mystery, but they are secondary to a more basic mystery that has a lot more to do with the human/nonhuman difference. Despite the intelligence of other species, and the fact that they engage in communicative behaviors that are as complex in other ways as a simple language might be, no other language systems exist. And it's not just a matter of their not being needed. For some reason even a simple language seems impossibly difficult for nonhumans. This poses a profound riddle. So why has it been ignored? Perhaps we have been too preoccupied by the details to recognize this simpler problem. Or maybe we have been too eager to cast the problem in terms of progress in communication, with humans in the lead. Whatever the reason, it's time we recognized that the questions we thought needed to be explained by a theory of language origins were secondary to a more fundamental mystery: Why aren't there any simple languages? And why are even simple languages so nearly impossible for other species to learn?

This changes everything. If complexity is not the problem, then theories that purport to explain language evolution in terms of overcoming complexity lose their justification. A small vocabulary should not require vast intelligence or memory capacity or articulatory skill to master. Lower intelligence of our primate and mammalian relatives cannot, then, be the reason they don't catch on. A simple grammar and syntax should also be a trivial matter to learn. No special built-in encoder-decoder for grammars should be necessary if the combinatorial analyses are simple and the possible alternatives are relatively few. Even minimal powers of inductive learning would suffice. The whole *raison d'être* of an innate Universal Grammar or language organ evaporates when it comes to simple languages. Finally, complex phonology, rapid articulation, and automated speech sound analysis are equally unnecessary. The learning problems addressed by all these theories do not explain the absence of nonhuman languages, they only explain why nonhuman languages should not be as complicated as human ones. They point to issues that are relevant to complex modern human languages, but they do not illuminate the phenomenon we originally thought they explained. They don't provide any clue to why language evolved in the human lineage and nowhere else. This is an apples-versus-oranges problem, not a complicated-versus-simple one. It's not just curious that other species haven't started on this evolutionary path; it defies common sense.

What is left that is difficult about learning language, if its complexity is not at issue? When we strip away the complexity, only one significant difference between language and nonlanguage communication remains: the common, everyday miracle of word meaning and reference.

Neither grammar, nor syntax, nor articulate sound production, nor a huge vocabulary have kept other species from evolving languages. Just the simple problem of figuring out how combinations of words refer to things. Why should this be so difficult? Why should the curiously different way that languages represent things have imposed such an almost impenetrable barrier in evolution? If we succeed in explaining this one paradoxical difficulty, we may catch a glimpse of the critical evolutionary threshold that only our own ancestors managed to surmount.

The first major task of this book, then, is to describe precisely the difference between this unique human mode of reference, which can be termed *symbolic reference*, and the forms of nonsymbolic reference that are found in all nonhuman communication (and in many other forms of human communication as well). The second task is to explain why it is so incredibly difficult for other species to comprehend this form of reference. And the third task is to provide an explanation for how we humans (and a few other animals in carefully structured language learning experiments) have managed to overcome this difficulty. Even though this aspect of the language origins mystery is only a part of the story of language evolution, and understanding this difference offers no immediate answers to why languages are as complex as they are today, or why they obey seemingly inexplicable design rules, or how it is possible for human children to make sense of these otherwise byzantine and atypical details, none of these other questions can be answered without taking symbolic reference as a given. But it is *not* a given. Grammatical rules and categories are symbolic rules and categories. Syntactic structure is just physical regularity when considered irrespective of the symbolic operations it is intended to encode. Theories of language and mind that fail to address this issue head on, or suggest that it needs no explanation, ultimately assume what they set out to explain. We must explain the curious difficulty of symbolic reference first.

In hindsight, the centrality of this problem was recognized all along, at least implicitly. It would not be an exaggeration to suggest that more philosophical ink has been spilt over attempts to explain the basis for symbolic reference than over any other problem. Yet despite its intuitive familiarity (or because of it), and notwithstanding the efforts of some of the greatest minds of each century, it remains curiously unresolved. Linguists, too, have struggled with this problem in the form of semantic theories, with parallel difficulties. For this reason, we should not be surprised to find that it resurfaces as the central riddle in the problem of language origins. Linguists, psychologists, and biologists cannot be blamed for failing to solve this basic mystery of mind before turning their efforts to other aspects of the language problem. Grammar and syntax can be studied and compared from language to language and the correlations between language processes and brain functions can also be identified irrespective of solving the problem of symbolic reference. Even many facets of language learning can be studied without considering it in any depth. But theories that purport to explain the human/nonhuman difference in language abilities cannot ignore it, nor can accounts of what makes human minds different from nonhuman minds.

But if the way language represents things has been the primary barrier to language evolution in other species, then we will also need to rethink a great many other aspects of human mental evolution. If language complexity is a secondary development with respect to this more primary cognitive adaptation, then most theories have inverted the evolutionary cause and effect relationships that have driven human mental evolution. They have placed the cart (brain evolution) before the horse (language evolution). If neither greater intelligence, facile articulatory abilities, nor prescient grammatical predispositions of children were the keys to cracking this symbolic barrier, then the evolution of these supports for language complexity must have been consequences rather than causes or prerequisites of language evolution. More important, these adaptations could not have been the most critical determinants of brain evolution in our species. Approaching the language origins mystery from this perspective is like stepping out of a mirror to find everything to be the reverse of what we assumed.

From this perspective language must be viewed as its own prime mover. It is the author of a co-evolved complex of adaptations arrayed around a single core semiotic innovation that was initially extremely difficult to acquire. Subsequent brain evolution was a response to this selection pressure and progressively made this symbolic threshold ever easier to cross. This has in turn opened the door for the evolution of ever greater language complexity. Modern languages, with their complex grammars and syntax, their massive vocabularies, and their intense sensorimotor demands, evolved incrementally from simpler beginnings. Though simple languages exist in no society found today, they almost certainly existed at some point in our prehistory. These simple languages were superseded by modern complex languages, and the brains that originally struggled to support simple languages were replaced by brains better suited to this awkward adaptation.

Somehow, despite their cognitive limitations, our ancestors found a way to create and reproduce a simple system of symbols, and once available, these symbolic tools quickly became indispensable. This insinuated a novel mode of information transmission into the evolutionary process for the first time in the billions of years since living processes became encoded in DNA sequences. Because this novel form of information transmission was partially decoupled from genetic transmission, it sent our lineage of apes down a novel evolutionary path—a path that has continued to diverge from all other species ever since. This inversion of cause and effect has enormous consequences. If the human predisposition for language has been honed by evolution for a significant fraction of our prehistory, then our unique mentality must also be understood in these terms. The incessant demands of efficiently reconstituting a symbolic system in each generation would have created selection pressures to reshape our lineage's ape brains to fit this new function. The implications for brain evolution are also profound. The human brain should reflect language in its architecture the way birds reflect the aerodynamics of flight in the shape and movements of their wings. That which is most peculiar about language processing should correspond to that which is most peculiar about human brains. So if what is most unusual about language is its symbolic basis, what is most unusual about our brains?

Human brains are unusually large: three times larger than they should be for an ape with our size body. But large brain size is only the most superficial symptom of a substantial reorganization at deeper levels. Unpacking this complicated anatomical problem and mapping it onto the special computational demands posed by language is the purpose of the middle section of this book. Looking more closely, we will discover that a radical re-engineering of the whole brain has taken place, and on a scale that is unprecedented. Interpreting these differences as consequences of the functional demands imposed by eons of language processing may offer new insight into the relationship between differences in cognitive function and differences in large-scale brain organization. In the co-evolution of the brain and language two of the most formidable mysteries of science converge, and together they provide a substantial set of clues about their relationship to one another.

Though neuroanatomists have been searching for a "Rosetta Stone" of human brain function for centuries, it has been far from a trivial task to sort out the significant from the incidental differences in brain structure and it will take considerable effort just to identify exactly what has changed and how. As with the problem of determining what is fundamentally different about language, an analysis of how the brain has responded to these influences will require us to delve well below brain sizes and superficial differences in brain structures, to probe the processes that build brains in embryos. Brains are the most intricate and powerful computing devices on the planet. Linguistic communication is the most complex behavior known. Evolution is the epitome of inscrutability, indirectness, and opportunism—seldom following an obvious or elegant path. Now we must throw into this already daunting mix of problems the equally perplexing problem of explaining symbolic reference. A puzzle of such magnitude is unlikely to have an easy solution, nor do I imagine that what little evidence we have is sufficient to do any more than begin the process of hunting in the right places for more clues. But arranging the clues in the appropriate order is a first step, and considering an old problem from a novel perspective is often the best way to escape the maze of assumptions that prevents us from recognizing the obvious. Perhaps by juxtaposing these linked mysteries from different domains we will recognize the common thread of logic that runs through them all. Like the famous Rosetta Stone, on which the same text was written in three radically different scripts, these pieces of the cognitive and neural puzzle, aligned side by side, may enable us to discover how each translates the other.

A Loss for Words

If the only tool you have is a hammer, you tend to treat everything as if it were a nail.

—Abraham Maslow

Gymnastics of the Mind

We often find that an apparently simple task is difficult, not because it is complicated but because it is awkward. Sometimes, for instance, you just lack the right tools for the job. No matter how easy it is in principle to tighten a screw, if it has a slotted head (-) and all you have is a Phillips screwdriver (+), forget it. This is a familiar source of difficulty in physical activities. We humans are unprepared to perform a number of tasks that other species perform with ease. Other species can boast such adaptations as streamlining and fins for swimming, large flexible mobile surfaces for gliding and flying, claws for clinging to the trunks of trees or snagging prey, or sharp canines for tearing flesh. Attempt a task that you are poorly suited for and your performance is at best inelegant and clumsy, and often it just doesn't succeed.

Such awkwardness is, in essence, the opposite of preadaptation—that lucky chance of evolution when preexisting body parts coincidentally happen to be predisposed for a novel adaptational challenge. In comparison, this kind of built-in ineptness might be called "pre-maladaptation." Pre-maladaptation accounts for our difficulties trying to sleep standing up, cut meat without a knife, or distinguish friends from foes by their smell, though certain other species do each well. In the same way that we find certain actions or movements to be impossible unaided, or find ourselves maladroit at manipulating certain objects, certain kinds of mental tasks can also be unwieldy for a brain predisposed to different sorts of analyses. In addition, certain mental predispositions that serve well in some domains can get in the way of accomplishing otherwise trivial tasks that require a new perspective or logic. Cognitive pre-maladaptations might also include predispositions to behave inappropriately, in ways that are opposite to how events tend to occur, or preferences that lead us to pay attention to irrelevant or misleading details. This is, of course, the secret to many magicians' tricks and the basis for animals' protective coloration. These could make an otherwise simple task difficult by misdirecting attention and interfering with appropriate actions.

Learning is not any one general process. Learning always occurs in a particular context, involving particular senses and types of motor actions, and certain ways of organizing the information involved. The process of learning is also not just committing an association to memory under the influence of reinforcement. Learning involves figuring out what is relevant and then figuring out how the relevant variables are associated. It involves sorting and organizing, and sometimes recoding what we have already learned. When a pigeon learns that pecking at a red-lighted button will cause food to be delivered and that pecking at a blue-lighted button will shut off an aversive noise, it does not just commit these links to memory, it also learns to ignore a great deal that is irrelevant: the talking in the background, the time of day, the temperature of the room, the pecking of neighboring pigeons, odors that periodically waft by, and the other lights and buttons and structures of the cage. Success or failure at learning and problem solving depends on habits of attention, what we find salient and what we don't tend to notice, and how easily one can override and reprogram these tendencies.

There are also internal distractions, in particular those that arise from past learning experiences. Interference effects of this sort are a common experience. Old associations provide a sort of repository of hypotheses. In many circumstances in the real world, events with similar features will produce similar outcomes. Generalizing from resemblances of present to past associations can often provide useful shortcuts that avoid wasting long periods of trial-and-error learning. Sometimes, however, superficial similarities are just that, and the responses they elicit are off track. If the habits are strong, the opportunities to compare different habits are few, or the feedback about incorrect responses is weak, forgetting the old in order to learn the new can become a practical impossibility. As a result, certain problems can be difficult because they require us to think in ways that are not typical: to infer what's missing, to work back from an end, to assume the opposite of what we want to demonstrate, and so on. Trick questions and riddles stump us, not because the question is complicated but because the solution is *counterintuitive.* Jokes are endlessly fascinating because they provide an insidiously logical punch line that we are unable to anticipate or predict. A successful punch line cannot be just a non sequitur, it must follow inevitably and obviously, if only after the fact. The minor cognitive implosion summed up by "Now I get it" is in part an experience of admiration at being fooled so well, so simply, and yet so logically. As with a magic trick, the key to a good joke is misdirection; and the most effective misdirection is self-imposed. The ways we naturally think about the subject of the joke are precisely the causes of our capacity to be fooled.

Intelligence tests and "brain teasers" often challenge our capability to think outside of self-imposed and overly narrow·contexts, and we consider the facility to make leaps of logic a mark of intelligence. Indeed, the criterion for calling something a work of genius is seldom its complexity, but rather how innovative the approach and how many others were unable to see it. This is why some of the great discoveries in many fields have been made by people outside the inner circle of true experts, because the experts are often too acquainted with a problem to see it in a novel way. As the great biologist Thomas Huxley is said to have exclaimed after learning of Darwin's theory of natural selection, "How stupid of me not to have thought of that!" Sometimes what we do best works against us.

The difficulties that a problem poses are a function of both its intrinsic complexity and the learner's predispositions to attend to its most relevant aspects. Comparative psychologists have long struggled to untangle questions of general learning ability from specific learning abilities in different species. As they have become more sensitive to the problem of understanding a species' capabilities in the context of its natural environment, they are finding correlated differences in learning. Animal breeders have noticed these patterns for generations. Some breeds of dogs make better shepherds, some make better hunters, some make better Seeing Eye dogs. And there are almost always trade-offs: behavioral predispositions that are well suited to one sort of task often conflict with others. This sort of complementarity is also evident in studies of the effects of brain damage on animal learning. It is not uncommon to find that rather than producing impairment across the board, localized brain damage enhances learning of some tasks at the same time that it impairs learning of others. Such paradoxical enhancement shows that learning is not a monolithic brain process. Learning rates are species- and task-specific, and depend on the subject's particular balance between alternative predispositions. The notion that one can be better or worse at all forms of learning ignores the intrinsically competitive nature of different learning paradigms. The best attentional and mnemonic strategies for one type of task may be exactly wrong for another. Just as our past experiences and accumulated knowledge can in some cases pose impediments to solving a novel problem, so can the evolutionary heritage of a species set it up to handle some cognitive problems well and others poorly. To the extent that members of a species are innately biased to attend to irrelevant details and to ignore critical elements of a problem, they are pre-maladapted.

This suggests a very different way to approach the language mystery. Could there be something about the way even a simple language must be learned that is just awkward for other species? Could symbolic reference be naturally counterintuitive?

The fact that language is an unprecedented form of naturally evolved communication, compared to other forms, suggests that it likely requires a rather different problem-solving orientation in order to learn it. The point of the "missing languages paradox" is that the difficulty most species have in learning a language does not appear to be sufficiently diminished, even when the language being taught is vastly simplified. Somehow, they don't "get" what the language problem is about. This suggests that it is not only difficult but also goes against certain other predispositions that are quite strong. Language may require learning in ways that run counter to other more typical learning strategies. Just getting started, getting what it means for a word or sign to represent something—not simply point to it or bring it to mind by association but symbolize it—may require a kind of mental gymnastics for which most nonhuman brains are pre-maladapted. Even with considerable human social experience or specific training and support for their vocal limitations, only a select few nonhuman animals have ever come to grasp the symbolic relationships we have tried to teach them. Instead, most exhibit a remarkable capacity for anticipating our responses, mimicking our actions, or committing large sets of paired associations to memory—sophisticated abilities all, but not symbolic abilities.

In Other Words

How could something as simple as word meaning be counterintuitive and awkward for other species to grasp? This question has its own special awkwardness. The explanation of the nature of word meaning has challenged thinkers since before recorded philosophy, and it continues to plague every field where explanations of thought processes are important. Thousands of years and thousands of texts later, we still do not fully understand the basis of the relationship that invests words with their meanings and referential capacity. To be blunt, we do not really understand one of our most commonplace experiences. We know how to use a word to mean something and to refer to something. We know how to coin new words and to assign new meanings to them. We know how to create codes and artificial languages. Yet we do not know how we know how to do this, nor what we are doing when we do. Or rather, we know on the surface, but we do not know what mental processes underlie these activities, much less what neural processes are involved. The fact that a conceptual problem this basic has gone unresolved for so many generations suggests that something more than mere technical difficulty stands in the way of our understanding. It is not just a difficult puzzle; the concept seems to be as counterintuitive for us to understand as it is simple for us to use.

Few topics have generated as much debate and confusion. So one might think that this question is hardly the place to choose to begin an analysis of human language origins. Don't we risk getting stuck in the philosophical mire even before getting started? Yes, but there is something to be said for framing the most difficult and critical questions first, in order to avoid following pointless leads.

Dozens of theories purport to explain the many aspects of the relationship between a word and its meaning. Philosophers, psychologists, and linguists have been arguing for centuries over such matters, and the debates have approached a new intensity in recent years because of the rise of computer technology and the hint of a possibility of building "intelligent" machines. In this context, I find it curious that scholars and engineers fail to take into account the anomalous nature of this form of reference. Shouldn't the lack of counterparts to words and sentences in the rest of the biological world play a significant role in our thinking about the problem? Of all the enormously powerful computing devices that we find in the heads of birds and mammals, only one uses a symbolic mode of reference. What hints about its nature have we missed because we ignored its rarity?

The main reason that this seems to have been ignored is that it is generally assumed that other species *do* exhibit counterparts to words and sentences in their natural repertoires. Ambiguous definitions of what constitutes words, sentences, and languages on the one hand, and reference, meaning, and understanding on the other, allow researchers to stretch metaphorical uses of these terms to fit. Isn't any family dog able to learn to respond to many spoken commands? Doesn't my dog *understand* the word "come" if he obeys this command? There doesn't seem to be anything special about learning to associate a sound and an object, and isn't that the whole basis of word reference? Just multiply such associations and add syntax to string them together into different combinations and you have a simple language, right? Not exactly. We think we have a pretty good idea of what it means for a dog to "understand" a command like "Stay!" but are a dog's understanding and a person's understanding the same? Or is there some fundamental difference between the way my dog and I understand the same spoken sounds? Common sense psychology has provided terms for this difference. We say the dog learned the command "by rote," whereas we "understand" it. But this is a notoriously difficult difference to specify. Our uncertainty about this not only makes it difficult to figure out what animals are capable of, and what they are not, it also blurs the distinction between animal communication and language.

One reason we have such difficulty is that we don't know how to talk about communication apart from language. We look for the analogues of words and phrases in animal calls, we inquire about whether gestures have meanings, we consider the presence of combination and sequencing of calls and gestures as indicating primitive syntax. On the surface this might seem to be just an extension of the comparative method: looking for the evolutionary antecedents of these language features. But there is a serious problem with using language as the model for analyzing other species' communication in hindsight. It leads us to treat every other form of communication as exceptions to a rule based on the one most exceptional and divergent case. No analytic method could be more perverse. Social communication has been around for as long as animals have interacted and reproduced sexually. Vocal communication has been around at least as long as frogs have croaked out their mating calls in the night air. Linguistic communication was an afterthought, so to speak, a very recent and very idiosyncratic deviation from an ancient and well-established mode of communicating. It cannot possibly provide an appropriate model against which to assess other forms of communication. It is the rare exception, not the rule, and a quite anomalous exception at that. It is a bit like categorizing birds' wings with respect to the extent they possess or lack the characteristics of penguins' wings, or like analyzing the types of hair on different mammals with respect to their degree of resemblance to porcupine quills. It is an understandable anthropocentric bias—perhaps if we were penguins or porcupines we might see more typical wings and hair as primitive stages compared to our own more advanced adaptations—but it does more to obfuscate than clarify. Language is a derived characteristic and so should be analyzed as an exception to a more general rule, not vice versa.

This inversion of evolutionary logic leads to persistent attempts to model animal communication as some version of language *minus* something. I often hear animal behaviorists or linguists remarking that "animal languages just lack a grammar and syntax." In short, we analyze animal communication systems as though they were truncated or degenerate languages. Words without syntax, or names without semantics. Moreover, we often imagine the early stages of language evolution as though our ancestors spoke a sort of crippled language or child's language. Serious, well-received theories have suggested that we consider the speech of linguistically impaired brain-damaged adults (Broca's aphasics) or the speech of very young infants as models of the early stages of language evolution.[1] Notice, however, that just by treating other species' communication as partial languages, modern language becomes the implicit end point, as adult language is the end point of children's language development and recovered language is the ideal goal of rehabilitation after brain damage. Of course, this apparent "final causality" is an illusion created by a sort of Orwellian rewriting of the evolutionary past in terms of the present.

Treating animal calls and gestures as subsets of language not only reverses the sequence of evolutionary precedence, it also inverts their functional dependence as well. We know that the nonlanguage communication used by other animals is self-sufficient and needs no support from language to help acquire or interpret it. This is true even for human calls like sobbing or gestures like smiling. In contrast, however, language acquisition depends critically on nonlinguistic communication of all sorts, including much that is as innately prespecified as many nonhuman counterparts. Not only that, but extensive nonverbal communication is essential for providing the scaffolding on which most day-to-day language communication is supported. In conversations, demonstrations, and explanations using words we make extensive use of prosody, pointing, gesturing, and interactions with objects and other people to disambiguate our spoken messages. Only with the historical invention of writing has language enjoyed even partial independence from this nonlinguistic support. In the context of the rest of communication, then, language is a dependent stepchild with very odd features.

But if language evolved subsequent to hundreds of millions of years of success with these nonlinguistic modes of vocal communication, and even now functions only by virtue of their presence, this should warn against inverting the relationship and treating language as the measure of these other forms. Language did not supersede or replace other forms of communication. Language evolved in a parallel, alongside calls and gestures, and dependent on them—indeed, language and many human nonlinguistic forms of communication probably co-evolved (a dynamic that will be discussed in later chapters). This is demonstrated by the fact that innate call and gesture systems, comparable to what are available to other primates, still exist side by side with language in us. Their complementarity with and distinction from language are exemplified by the fact that they are invariably produced by very different brain regions than are involved in speech production and language comprehension (details of these differences will be reviewed in Chapters 8–10). In many ways, these human counterparts to innate calls and gestures in other species offer the best source of intuitions concerning this difference. Who, for example, would even bother to consider the possibility that somehow smiles, grimaces, laughs, sobs, hugs, kisses, and all the rest of our panhuman nonlinguistic communications were words without syntax?

The popular notion that the calls and gestures constituting the communications of other species are like words and sentences can mostly be traced to misconceptions about the concept of reference. The problem of reference has always been a major topic of debate in the study of animal communication (see Figure 2.1). On the one extreme, some animal behaviorists have argued that calls and gestures are merely external correlates of internal states, and so have no external reference; on the other extreme, some cognitive ethologists have argued that many animal calls, grunts, and gestures should be considered the equivalents of words that name specific objects in the world. One study, in particular, played a central role in rekindling debate on the role of reference in animal communication. In the mid-1980s, Robert Seyfarth, Dorothy Cheney, and their colleagues reported that vervet monkeys produced alarm calls that appeared to act something like names for distinct predators.[2] Their observations demonstrated that distinctly different calls were produced to warn other troop members of the presence of either eagles, leopards, or snakes (an even wider range of calls is now recognized). In response to hearing one of these calls, other troop

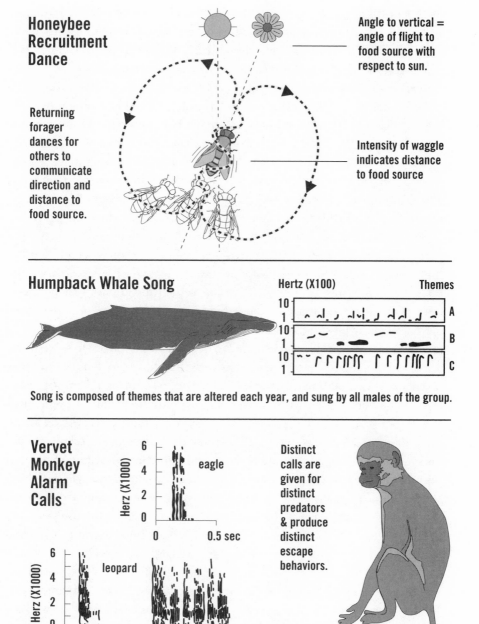

Honeybee Recruitment Dance

Angle to vertical = angle of flight to food source with respect to sun.

Returning forager dances for others to communicate direction and distance to food source.

Intensity of waggle indicates distance to food source

Humpback Whale Song

Hertz (X100)   Themes

A
B
C

Song is composed of themes that are altered each year, and sung by all males of the group.

Vervet Monkey Alarm Calls

Herz (X1000)

eagle

leopard

Distinct calls are given for distinct predators & produce distinct escape behaviors.

Figure 2.1 *Three often-cited examples of animal communication systems: honeybee recruitment dance, humpback whale songs, and vervet monkey alarm calls (discussed in the text). Each exhibits features that are similar to features thought to be special to language.*

members either raced out of the trees (eagle), climbed into the trees (leopard), or just rose up to peer into the bushes around them (snake). Thus, distinct calls referred to distinct classes of predators, not simply to some state of the caller (though they indicated a fearful state of mind as well).

It is not difficult to reconstruct the evolutionary processes that produced such distinctive calls and referential relationships. The key can be found in the behaviors that the calls induce in other troop members. These predators attack using very different approaches, and the appropriate defense behaviors for each turn out to be mutually exclusive. The worst place to be when a leopard is prowling nearby is on the ground. But since leopards can also climb trees, it is best to wait out on the thinner branches of trees. Unfortunately, this is the worst place to be when eagles are threatening. Best to be hiding under a tree on the ground. Imagine, then, the dilemma if this species had only a single type of alarm call. To ascend or descend, that is the question. Just to sit frozen with indecision or stand up and peer around is the worst response of all (unless a snake is the predator, for which there is another call), since it leaves you vulnerable to both. Consequently, predation selects against individuals that have trouble determining which call is which, and it will select against the kin of animals that do not in some way provide distinct information that helps with this choice, such as sound differences. This evolutionary logic is what is generally termed *disruptive selection:* selection against the intermediate (compromise) value of a trait and favoring the extremes. The referential specificity of these calls evolved over time, then, as the consequences of warning and escape provided selection pressures that changed the calling and responding predisposition of members of the species. Not surprisingly, a similar evolutionary logic has also shaped the alarm calls of many other species, and other design factors are also often involved, such as the localizability of the sounds themselves.[3]

Cheney and Seyfarth originally suggested that vervet alarm calls were analogous to "names" for these predators, and might be used in the way we shout the warning: "Fire!" This led many to argue that this system of calls was like a very simple language. Some even suggested it was comparable to the way that infants just beginning to learn to speak will use single words like "juice," to request a drink, or "doggy," to indicate that they want to pet a dog, and so on. Such human examples of single word sentences lacking overt syntax (though often with characteristic gestural support) have been called holophrastic utterances, though there is considerable debate about how much potential syntax can be read into these. The core of the alarm call argument is that these calls are different from a pain cry or a grimace, in the same way that words are different. In other words, they refer to something other than the animal's mental and physical state. This of course includes an unmentioned assumption that other sorts of calls or gestures like pain cries and grimaces can't be referential.

This interpretation implicitly invites us to imagine a species with many more of these distinctive types of calls, some for foods, some for important objects, maybe even sounds to identify specific individuals (for example, dolphins appear to use distinctive "signature whistles" to identify themselves). Would an elaborate repertoire of this kind constitute a kind of protolanguage? Are the calls essentially like a vocabulary? This even suggests a compelling language evolution scenario: Individual calls evolved first, they increased in number and in variety, they were combined in various ways, and eventually a grammar and syntax evolved to systematize the patterns of combinations.[4] Unfortunately, the entire house of cards is dependent on call reference being equivalent to word reference, and this resemblance is not complete in some important ways. Let's see if we can be more precise about how they are not the same.

Reference itself is not unique to language, in fact it is ubiquitous throughout animal communication. Even a symptom can refer to something other than itself and other than the state of the body that produces it. Take, for example, human laughter: a symptom of being in a highly amused state of mind. Laughter is an excellent example of a human innate call (I will return to analyze both the evolution and physiology of this call in later chapters). Like other calls it need not be intentionally produced; it often erupts spontaneously even when we would rather suppress it, even though it can also be faked (with variable success) if the social context demands. For the most part we tend to think of it as a way to work off feelings inspired by a joke or an awkward social situation. But laughter also refers to things as well. For example, when someone walks into a room laughing, it suggests that they probably heard or saw something funny just outside, before entering. The laughter points to this something else that caused it. And it specifies some of the characteristics that this cause probably exhibits; specifically, it was not a source of sorrow, not a disgusting or repulsive scene, not a real threat, and so on. It categorizes the event that induced it by virtue of what the laughter tells us about the state of the laugher. It points to a definite class of experiences that are deemed funny. But notice how different the reference to the same event is when the person stops laughing and says, "I just heard a great joke." Alarm calls refer to objects the way laughter does, not the way words do.

Another difference, subsequently noted by Cheney and Seyfarth,[5] has to do with the way we use words and sentences to transmit information from one to another. Specifically, they were interested in determining whether these calls were used intentionally to communicate information or just incidentally communicated information, and how this might relate to what individuals know about what others know and don't know.

Some critical features that distinguish automatic (unintended) forms of communication from intentional communications are characteristic of laughter. Laughter provides others with information about the laugher's state of mind and recent history, but it also exerts a more direct effect, a sort of compulsion to laugh along. We often acknowledge this by saying that laughter is contagious. Sitting in a room full of laughing people, one finds it difficult not to laugh as well, even though the reason for their laughter may not be entirely clear. Indeed, so strong is this odd compulsion, that artificial laughter produced by a mechanical device (e.g., a laugh box or the laugh track on a TV sitcom) can induce us to laugh, even though we are fully aware that it is not real laughter. To add insult to injury to our sense of self-control, the faked laughter actually induces us to experience things as funnier as a result. This involuntary power of laughter is shared by many other innate social signals as well, including sobbing, smiling, grimacing, etc., and contrasts sharply with the absence of such an echoic tendency in normal language communication. Not only do we seldom parrot what we have just heard from another, such a response is generally oddly annoying, as most children at some point discover to their glee and their siblings' and parents' distress. How odd and unnatural it would feel to enter a room where people were echoing each other's speech in the same way that they tend to echo each other's laughter! This may be why certain ritual practices that employ such patterns of language use are at the same time both disturbing and powerful, depending on whether one feels included or excluded.

In general, even casual conversation takes a certain degree of conscious effort and monitoring. Following another's speech takes at least a modicum of attention and intentionally controlled analysis—something that quickly becomes obvious when one conversant's mind begins to wander. Part of this derives from the fact that what one says is typically influenced to some degree by assumptions about what the other already knows. So a common factor in the use of language is an intention to convey something that the other person presumably doesn't know. The influential philosopher H. P. Grice has even argued that a sort of reflexive logic of the form "I believe that you believe that I believe x" is an essential component of communicating meaning in language.[6] But in this regard, both laughter and vervet monkey alarm calls ultimately fail the test in part because both are involuntary and contagious. When one vervet monkey produces an alarm call, others in the troop both escape and echo the call again and again, and all will join in until the general excitement is gone. This redundancy provides a minimum of new predator information for any troop members subsequent to the first presentation (though it may indicate that danger is still present). It almost certainly has more to do with communicating and recruiting arousal than predator information, as is analogously the case with laughter and humor. Knowing that another saw a predator does not appear to inhibit the vervet's tendency to call and knowing that another also got a joke does not apparently inhibit a person's tendency to laugh—indeed, both may be potentiated by this knowledge.

So, in this regard, the example of vervet alarm calls offers a false lead. Reference is not the difference between alarm calls and words. Both can refer to things in the world and both can refer to internal states, but there is a difference. This difference is the source of the most common misunderstanding about the nature of linguistic versus nonlinguistic communication. It is a difference in the *kind* of reference. We tend to confuse different forms of reference with one another or else dichotomize referential versus nonreferential communication, instead of recognizing that modes of reference may differ and may depend on one another in complicated ways. We cannot hope to make appropriate comparisons between different forms of human communication, much less comparisons between human language and other species' forms of communication, unless we can clear up this confusion and discover what exactly constitutes this difference.

The Reference Problem

So what is the difference between the way a word refers to things and the way a vervet monkey alarm call, a laugh, or a portrait can refer to something else? Word meaning has always fascinated people because it is at once so simple and yet so elusive in the way it works. On the surface it seems to be no more than a mapping or pairing between one thing and another—a sound or conventional set of markings (the "signifier") on the one hand, and an object, process, or state of things (the "signified") on the other (see Figure 2.2). How the thing signified is brought into correspondence with the signifier is thought to distinguish different forms of reference. The difference between words and other means of referring to things appears to be the arbitrariness and conventionality of the linguistic link. But a little further probing into these relationships demonstrates that there must be more to it.

Precisely identifying this difference has probably nowhere been more

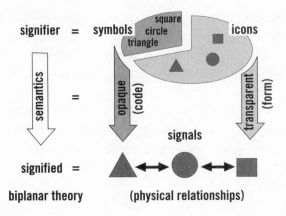

Figure 2.2 *Depiction of the classic distinctions between different forms of referential and meaningful relationships. In this view, two sets (or "planes") of elements—signifiers (e.g., signs, words, pictures) and signified objects—are associated by a conceptual relationship (semantics) that maps individual elements in one set to those in the other. Most theories recognize at least two sorts of semantic or meaning relationships, transparent and opaque: those that link signifiers and signifieds by virtue of their similarity are "transparent" to the extent that they require no additional knowledge to "see" the one through experience of the other; and those that link them by virtue of some arbitrary coding or mapping are "opaque" because they require knowledge of the code. The transparent kind of signifier is often referred to as an icon; the opaque variety is often referred to as a symbol (though these terms have a history of widely varying uses). Considered irrespective of semantics, in terms of merely physical or mechanical correlations with other objects, signifiers are often called signals. Thus, for example, a light on the dashboard of a car might be a signal that the engine is low on oil by virtue of its being electrically linked with an oil pressure sensor. Later, I will refer to this as an index. This biplanar account, based on an oppositional categorical scheme (transparent/opaque; correspondingly patterned/arbitrary; natural/conventional; semantic/nonsemantic), obscures the underlying interdependence of these forms of referential relationship.*

troublesome than in studies that have attempted to teach other species to use language. No one suspects that when a parrot says, "Pretty bird!" he really "knows" that these sounds are supposed to refer to his (or another bird's) appearance. He has learned to mimic the sound of the words, that's all. Sometime in the past he was rewarded for producing this phrase (even if only by the intrinsic reward of mimicking a sound, something parrots do in the wild) and he now produces it spontaneously. But what if he is taught to say, "Wanna cracker!" and rewarded with a cracker every time he says it?

Presumably, when he wants a cracker, he will say so. Is this different? Should we now say that he knows what the words mean?

Referring to the interpretation of a call or gesture as its "meaning" again betrays a Procrustean linguistic bias in our analysis of communication. Though words and sentences have meanings, there is something odd about saying that laughter has a meaning. A more accurate statement is that a person's laughter indicates something, both about the person and about the context. We would only say it means something (except in a sort of metaphorical looseness of phrase) under rather special circumstances, when it is used as a prearranged signal; e.g., for someone to enter the room. Here too, we would tend to say that it *indicates* the right moment to enter, not that it *means* "Enter!"

The late nineteenth-century mathematician-philosopher Gottlob Frege[7] provided a concise distinction between these two often-confused aspects of word meaning. He distinguished between the *sense* of a term and its *reference*. Its sense is the idea one has in mind that corresponds with considering a particular word or phrase. This is distinguished from the reference of the same word or phrase, which is something in the world which corresponds with this term and its sense. The logic of this distinction has influenced most subsequent theories and is only slightly redefined by such complementary terms as *intension* (as distinct from intention) and *extension*, respectively, in later philosphical discussions. One might summarize this as "sense is something in the head" and "reference is something in the world." Frege offers the example of "the morning star" and "the evening star," which refer to the same physical object, the planet Venus, but have different senses. In this case, these different senses trace back to a historical context when this common reference was not recognized.[8]

Numerous philosophers of language and mind have built on this original insight. One classic view argues that sense (intension) is used to determine reference (extension). This fits the common sense insight that what a word refers to results from the idea that it invokes. This invites the obvious questions: what do we mean by idea in this context, and how does this determine reference? Candidate interpretations of "idea" include mental images, stimulus-stimulus associations, something like a dictionary definition or encyclopedia entry, or checklists of "features" or qualities of objects. All may play some role. But how such mental objects pick out external physical objects, or whether they do, remains a thorny philosophical problem.

A number of critics of this classic conception have demonstrated that the referential power of words can be otherwise independent of their sense in many cases. They point out that there are cases (particularly in the example of names, which have somewhat minimalistic senses) where we discover that the sense of a term does not correspond to the reference we all along accepted. For example, if we were to discover that William Shakespeare was a pseudonymous proxy for Sir Francis Bacon, who actually wrote the famous plays and sonnets (a claim that has been argued by some), it would not change the reference of either name to a particular historical figure, just statements made about the men the names refer to. Other philosophers have employed more exotic examples to demonstrate this same point because they show how radical falsifications of meaning still do not change reference.[9] For example, imagine someday discovering that all mosquitos were not animals but really stealth devices designed by extraterrestrials for obtaining human DNA samples. This would not undermine the reference of the term *mosquito*. The proponents of this view explain it by arguing that the reference in such a case was previously determined in a more concrete and causal fashion by the correlation in space and time of the use of the word with the presence of the physical objects to which it refers. The word and the object must have co-occurred at least at some point in the past, and all modern uses derive their reference by virtue of an unbroken causal-historical link to some such reference-establishing event or events. The ideas we entertain about this link between a name and what it names may thus help perpetuate and refine its use, but they are not its crucial determinants.

It is interesting, however, that a more subtle version of this logic can probably account for the evolution of a link between alarm calls and predators and between laughter and humorous experiences, via biological evolutionary history. This demonstrates that reference in general does not require some conscious concept or meaning to determine it. But whereas a smile may exhibit only this aspect of reference, it seems that words exhibit this and more. In other words, reference may have a sort of hierarchic character. The way we so often need nonverbal gestures to get across the meaning of things or to explain what we have in mind is another reflection of this dependence of word reference on more basic forms of reference.

A more complicated terminology is necessary, then, to begin to differentiate between the way that words, as opposed to laughter and other non-language signs, refer to things. We need terms that cut beneath word reference and from which word reference can be derived as a *special* case, since that is the way it evolved and the way it develops in each of us. Words are not just sounds, configurations of ink on paper, or light on a computer screen. What endows these otherwise inanimate things with the capacity to refer to other things is an interpretive process, a critical part of which (though not all) occurs "in the head." If I am unable to produce an appropriate symbolic interpretive response to an unfamiliar foreign word, I can only use physical co-occurrence to guess at what objects or events might be relevant to what is being said. Clearly, what distinguishes a dog's ability to get one sort of reference from a phrase and a person's to get that and something more is the result of something additional that is produced in the head of the person. So maybe taking this approach to the problem can help define the distinction.

Ultimately, reference is not intrinsic to a word, sound, gesture, or hieroglyph; it is created by the nature of some response to it. Reference derives from the process of generating some cognitive action, an interpretive response; and differences in interpretive responses not only can determine different references for the same sign, but can determine reference in different ways. We can refer to such interpretive responses as *interpretants*, following the terminology of the late nineteenth-century American philosopher Charles Sanders Peirce. In cognitive terms, an interpretant is whatever enables one to infer the reference from some sign or signs and their context. Peirce recognized that interpretants can not only be of different degrees of complexity but they can also be of categorically different kinds as well; moreover, he did not confine his definition only to what goes on in the head. Whatever process determines reference qualifies as an interpretant. The problem is to explain how differences in interpretants produce different kinds of reference, and specifically what distinguishes the interpretants required for language.

So, what are some of the interpretants of words? Probably the most common view of word meaning is that a word is interpreted when one generates a mental image of something that it refers to: for example, an image of a familiar dog for "dog" or of someone throwing a baseball for "pitch." Though once treated as a sort of fairy tale of introspectionism, mental imagery has in recent years become recognized as an experience that has clear neural and behavioral correlates. The relative "locations" of features on an imagined object, or its size, shape, and movement in imagination, or other changes, can have a direct effect on such factors as the time and effort required to consider these features and what parts of the brain might be involved. But a mental image (or the neural process that constitutes it) is only one sort of interpretive response that a word might elicit, and it may not be the most important one.

A word also might bring to mind something like a dictionary definition, or another word that has a related meaning, or it might induce us to act out some behavior, or it even might produce a vague visceral feeling correlated with past experiences of what is referred to. All of these are interpretants, but the way they bring a particular word-reference relationship into being can be quite diverse, and of course many can be present simultaneously. The kind of interpretive response determines the nature of the reference relationship. The interpretant is the mediator that brings a sign and its referent together. Differences in the form of reference are due to differences in the form of this mediation process.

Though producing a mental image may be an inevitable product of comprehending certain words, it is not the mediator that distinguishes symbolic reference. A mental image may also be the primary interpretive action in numerous nonsymbolic processes of reference. For example, smelling the scent of a skunk while walking in the woods might bring to mind the mental image of a skunk, as it might also do for a dog who previously had a nasty skunk experience. A more experienced dog or dog owner might even find that the odor produces a revulsion response, which, like the mental image, could also contribute to the reference relationship. In this way the scent refers to the animal by virtue of a mental image. For both the human and the dog the scent would have acquired this reference from past skunk experiences. To a naive puppy or child, however, with no past experience to base an interpretation on, it is just a strong odor. Simply reading these words, of course, might also bring to mind a visual image or a skunk "odor image." This emphasizes that it is not the thing that is ultimately brought into reference, nor even the common images that are elicited, that determines the difference in the ways that words and odors can derive their reference; it is rather *how* these responses are produced. In the case of these words, there is something more, some additional interpretants that are crucial to their symbolic capacity, and a great deal of additional learning has intervened to make their production possible.

The symbolic basis of word meaning is mediated, additionally, by the elicitation of other words (at various levels of awareness). Even if we do not consciously experience the elicitation of other words, evidence that they are activated comes from priming and interference effects that show up in word association tests. Words referring to abstract qualities, such as "justice," "false," and "peculiarity," that don't easily lend themselves to imagery, may produce word association effects that are just as robust as more concrete words. But there are function words for which we seem unable to think up either kind of interpretant. Words such as "that," "which," and "what" function to point to other words and phrases, but not to specific categories of meanings, and don't evoke mental images. Nevertheless, they produce certain expectations about the grammatical structure of what is to follow that we recognize when they are violated. Though we hesitate to call these interpretants "meanings" in the same sense as for common nouns and verbs, they are functionally equivalent. Finally, consider the complicated mixtures of interpretants that are produced in response to whole phrases, sentences, and larger narratives or arguments. These can be too abstract to elicit clear imagery and yet have clear meanings. So what does it matter if the interpretive responses were learned and reproduced by rote, like the parrot saying, "Wanna cracker!" when it's hungry, or learned in some other way, or not learned at all? What matters that these learning differences will produce different patterns of mental action, so to speak, and although what is referred to may be the same, this difference in the interpretive process will dictate the nature of the referential link that results. So, to distinguish forms of reference, we need to understand the learning processes that produce the competence to interpret things differently.

It is probably possible to train almost any intelligent mammal to use a complicated arbitrary sign system, so long as the medium of expression is appropriate to its sensorimotor abilities. All that is necessary is to train individuals both to produce certain behaviors under specific stimulus circumstances and to respond to these same signals produced by others with an appropriate behavior, and so on. Depending on the mnemonic limitations of the animals, the repertoire could become arbitrarily large. This is, in skeletal form, the recipe for learning language that the famous American behaviorist B. F. Skinner imagined more than fifty years ago, and although it has been challenged as inadequate to produce grammar and syntax, many still implicitly conceive of word reference this way. Recently, in a study involving pigeons in separate training cages linked electronically so that the responses of the one pigeon could be registered as signals in the cage of the other, the Harvard psychologist Richard Herrnstein and his colleagues demonstrated that one could set up a pattern of linked associative learning tasks for multiple subjects so that the resultant behaviors resembled communication with learned arbitrary signs.[10] Using pigeons trained in adjacent cages, he set up the relationships between the stimuli and the responses in each so that only one bird got a signal for food availability but was not able to access it directly. Instead, its response would become a signal transmitted to the second bird in the other cage, who by responding to this could make the food accessible to both. In the end, the first bird transmitted the crucial information to the second bird via an arbitrary code. One could easily imagine complicating the system to include more subjects and signals. Hernnstein offered this experiment as a challenge to primate language researchers at the time who had shown similar interindividual languagelike communication in chimpanzees (though it is not clear to me whether he intended to show that this form of communication wasn't equivalent to language or rather that linguistic reference was just this simple).

To me, this experiment demonstrates the simplicity and mechanical nature of this form of reference. And how its key features—learned associations, arbitrarity, reference, and transmission of information from one individual to another—are not sufficient to define symbolic reference. Any bright undergraduate could write a short computer program or build a simple mechanical device to stand in for one of the pigeons. Nevertheless, a system of dozens of signals arranged in such interlocking relationships to one another, and with respect to events and objects of interest to all, could be a powerful communication system. Probably a significant fraction of the communications used by many highly social animals are either partially or wholly dependent upon use of signs in this way. It doesn't matter whether they are learned and arbitrary, as were those used by the pigeons, or innate and physically linked to some state of arousal. Many animal groups in the wild exhibit regionally variable social behaviors and displays for communication, much of which may be learned and passed from individual to individual by mimickry and association. But such a system is not just words without syntax.

There is something mechanical about innate calls as well as behaviors learned by rote. In our own experiences of learning, we have a sense for the difference between what we have learned by rote and what we say we "understand." At various stages of learning mathematics, we often find ourselves manipulating symbols and numbers according to certain specific instructions, and although we come up with the correct answer if we follow the instructions exactly, in the end we know what we did without knowing *what* we did. We were unclear on the concept. Actually, this experience is becoming more common for me as a result of using computers and calculators, which have now become my necessary prostheses. I type in a bunch of numbers and select a few computational operations and a string of values and graphs comes back. At one time I knew why certain operations produced the results that I now produce with a few keystrokes—I learned them the hard way, by rote, until I figured out the significance—but much of that support has faded in my memory. I am left with knowing that pushing certain buttons in certain orders does what I need, and have stopped worrying why.

This is the same intuition we have about words. Kids, trying to impress their friends (or scholars trying to impress their colleagues), may repeat a technical phrase they have heard in conversation, without really knowing what they mean. It often works if the context is right, and so long as no one asks too many questions, but the application is quite limited. One way we learn about new meanings is to figure out what the right contexts are; but knowing five or ten more contexts in which the same phrase works does not really change the superficial nature of the reference. Learning more and more appropriate contexts does not in itself constitute understanding the meaning or significance. Yet, when we know what the phrase means, the problem of remembering all the applicable contexts becomes irrelevant, and innumerable novel contexts can be immediately recognized as appropriate. In between these alternatives there is not just a quantitative increase, but a radical change in cognitive strategy.

Many have suggested that the key to this flexibility of word reference is arbitrarity. Innate calls and gestures have some features built in from birth, but learned vocalizations and movements can be freely associated with different external stimuli. For this reason, we might argue that what makes an alarm call or a laugh different and limited in its referential ability is the fact that there is a built-in association between the production of these calls and a specific emotional state. However, there is a sense in which even an alarm call isn't *necessarily* linked to its referent. As in a learned association, each vervet alarm call repeatedly co-occurred with a distinct class of predator and escape and fear responses during evolution. The apparent inflexibility of their relationship is just a momentary stage in evolution. The difference between this link and one based on an arbitrarily learned behavior is only a matter of degree. Both are, in one sense, internalizations of external correlations of events. One is built in before birth, one after birth.

But there is a sense in which a degree of necessary association is involved in all the nonlanguage examples we have discussed, and this is an important clue. If the parrot stopped getting fed when he squawked the words, "Wanna cracker!" or the dog stopped being let out when he nosed the doorknob, eventually both would probably stop producing these signs. If I started going out the door every time I said to my dog, "Do you want to eat?", I suspect it wouldn't take too long for him to reverse his old habits of interpretation. And if the vervet monkeys' predators disappeared from Africa, then future evolution would see vervet monkey alarm calls disappear from their repertoire (or perhaps become coopted for some other purpose). All rely on a relatively stable correlation with what they refer, in order to refer.

This is not true of words. Or not in the same way. If our use of words failed to correspond in *some* way with things in the rest of the world, they would be of little use, but there is something rather odd about this correspondence when we compare it to each of the examples above. If my use of the word "skunk" to refer to a certain animal was sustained by this critter being present, even a small percentage of the times that I used the word (in other words if there had to be a physical correlation), then the association would have been extinguished long ago. A learned association will tend to get weaker and weaker if some significant degree of co-occurrence of stimuli is not maintained. I very seldom find myself in the company of members of this species, if I can help it, and yet I read and talk about them often. Despite this, I don't have the impression that the strength of the referential link between the animals and the name is any less strong than that between the word "finger" and my flesh-and-blood finger, which is always present. There is some kind of word-object correspondence, but it isn't based on a physical correlational relationship.

To understand this difference, then, we need to be able to describe the difference between the interpretive responses that are capable of sustaining associations between a word and its reference, irrespective of their being correlated in experience, and those rote associations that are established and dissolved as experience dictates. When we interpret the meaning and reference of a word or sentence, we produce something more than what a parrot produces when it requests a cracker or what a dog produces when it interprets a command. This "something more" is what constitutes our symbolic competence.

Symbols Aren't Simple

> Alice laughed. "There's no use trying," she said: "one can't believe impossible things."
>
> "I daresay you haven't had much practice," said the Queen. "When I was your age I always did it for half-an-hour a day. Why, sometimes I've believed as many as six impossible things before breakfast."
>
> —Lewis Carroll, *Alice Through the Looking-Glass*

The Hierarchical Nature of Reference

The assumption that a one-to-one mapping of words onto objects and vice versa is the basis for meaning and reference was made explicit in the work of the turn-of-the-century French linguist Ferdinand de Saussure. In his widely influential work on semiology (his term for the study of language),[1] he argued that word meaning can be modeled by an element-by-element mapping between two "planes" of objects: from elements constituting the plane of the signifiers (e.g., words) to elements on the plane of the signified (the ideas, objects, events, etc., that words refer to). On this view, the mapping of vervet monkey alarm calls onto predators could be considered a signifier-signified relationship. But how accurately does this model word reference? Although it is natural to imagine words as labels for objects, or mental images, or concepts, we can now see that such correspondences only capture superficial aspects of word meaning. Focusing on correspondence alone collapses a multileveled relationship into a simple mapping relationship. It fails to distinguish between the rote understanding of words that my dog possesses and the semantic understanding of them that a normal human speaker exhibits. We also saw that the correspondence of words to referents is not enough to explain word meaning because the actual frequency of correlations between items on the two planes is extremely low. Instead, what I hope to show is that the relationship is the reverse of what we commonly imagine. The correspondence between words and objects is a secondary relationship, subordinate to a web of associative relationships of a quite different sort, which even allows us reference to impossible things.

In order to be more specific about differences in referential form, philosophers and semioticians have often distinguished between different forms of referential relationships. Probably the most successful classification of representational relationships was, again, provided by the American philosopher Charles Sanders Peirce. As part of a larger scheme of semiotic relationships, he distinguished three categories of referential associations: *icon, index,* and *symbol.*[2] These terms were, of course, around before Peirce, and have been used in different ways by others since. Peirce confined the use of these terms to describing the nature of the formal relationship between the characteristics of the sign token and those of the physical object represented. As a first approximation these are as follows: icons are mediated by a similarity between sign and object, indices are mediated by some physical or temporal connection between sign and object, and symbols are mediated by some formal or merely agreed-upon link irrespective of any physical characteristics of either sign or object. These three forms of reference reflect a classic philosophical trichotomy of possible modes of associative relationship: (a) similarity, (b) contiguity or correlation, and (c) law, causality, or convention. The great philosophers of mind, such as John Locke, David Hume, Immanuel Kant, Georg Wilhelm Friedrich Hegel, and many others, had each in one way or another argued that these three modes of relationship describe the fundamental forms by which ideas can come to be associated. Peirce took these insights and rephrased the problem of mind in terms of communication, essentially arguing that all forms of thought (ideas) are essentially communication (transmission of signs), organized by an underlying logic (or *semiotic,* as he called it) that is not fundamentally different for communication processes inside or outside of brains. If so, it might be possible to investigate the logic of thought processes by studying the sign production and interpretation processes in more overt communication.

To get a sense of this logic of signs, let's begin by considering a few examples. When we say something is "iconic" of something else we usually mean that there is a resemblance that we notice. Landscapes, portraits, and pictures of all kinds are iconic of what they depict. When we say something is an "index" we mean that it is somehow causally linked to something else, or associated with it in space or time. A thermometer *indicates* the temperature of water, a weathervane indicates the direction of the wind, and a disagreeable odor might indicate the presence of a skunk. Most forms of animal communication have this quality, from pheromonal odors (that indicate an animal's physiological state or proximity) to alarm calls (that indicate the presence of a dangerous predator). Finally, when we say something is a "symbol," we mean there is some social convention, tacit agreement, or explicit code which establishes the relationship that links one thing to another. A wedding ring symbolizes a marital agreement; the typographical letter "e" symbolizes a particular sound used in words (or sometimes, as in English, what should be done to other sounds); and taken together, the words of this sentence symbolize a particular idea or set of ideas.

No particular objects are intrinsically icons, indices, or symbols. They are interpreted to be so, depending on what is produced in response. In simple terms, the differences between iconic, indexical, and symbolic relationships derive from regarding things either with respect to their form, their correlations with other things, or their involvement in systems of conventional relationships.

When we apply these terms to particular things, for instance, calling a particular sculpture an *icon,* a speedometer an *indicator,* or a coat of arms a *symbol,* we are engaging in a sort of tacit shorthand. What we usually mean is that they were *designed* to be interpreted that way, or are highly likely to be interpreted that way. So, for example, a striking resemblance does not make one thing an icon of another. Only when considering the features of one brings the other to mind because of this resemblance is the relationship iconic. Similarity does not cause iconicity, nor is iconicity the physical relationship of similarity. It is a kind of inferential process that is based on recognizing a similarity. As critics of the concept of iconicity have often pointed out, almost anything could be considered an icon of anything else, depending on the vagueness of the similarity considered.

The same point can be made for each of the other two modes of referential relationship: neither physical connection nor involvement in some conventional activity dictates that something is indexical or symbolic, respectively. Only when these are the basis by which one thing invokes another are we justified in calling their relationship indexical or symbolic. Though this might seem an obvious point, confusion about it has been a source of significant misunderstandings. For example, there was at one time considerable debate over whether hand signs in American Sign Language (ASL) are iconic or symbolic. Many signs seemed to resemble pantomime or appeared graphically to "depict" or point to what was represented, and so some researchers suggested that their meaning was "merely iconic" and by implication, not wordlike. It is now abundantly clear, however, that despite such resemblances, ASL is a language and its elements are both symbolic and wordlike in every regard. Being capable of iconic or indexical interpretation in no way diminishes these signs' capacity of being interpreted symbolically as well. These modes of reference aren't mutually exclusive alternatives; though at any one time only one of these modes may be prominent, the same signs can be icons, indices, and symbols depending on the interpretive process. But the relationships between icons, indices, and symbols are not merely a matter of alternative interpretations. They are to some extent internally related to one another.

This is evident when we consider examples where different interpreters are able to interpret the same signs to a greater or lesser extent. Consider, for example, an archeologist who discovers some elaborate markings on clay tablets. It is natural to assume that these inscriptions were used symbolically by the people who made them, perhaps as a kind of primitive writing. But the archeologist, who as yet has no Rosetta Stone with which to decode them, cannot interpret them symbolically. The archeologist simply infers that to someone in the past these may have been symbolically interpretable, because they resemble symbols seen in other contexts. Being unable to interpret them symbolically, he interprets them iconically. Some of the earliest inscription systems from the ancient Middle Eastern civilizations of the Fertile Cresent were in fact recovered in contexts that provided additional clues to their representations. Small clay objects were marked with repeated imprints, then sealed in vessels that accompanied trade goods sent from one place to another. Their physical association with these other artifacts has provided archeologists with indexical evidence to augment their interpretations. Different marks apparently indicated a corresponding number of items shipped, probably used by the recipient of the shipment to be sure that all items were delivered. No longer merely iconic of other generic writinglike marks, they now can be given indexical and tentative symbolic interpretations, because something more than resemblance is provided.

This can also be seen by an inverse example: a descent down a hierarchy of diminishing interpretive competence, but this time with respect to interpretive competences provided by evolution. Let's consider laughter again. Laughter indicates something about what sort of event just preceded it. As a symptom of a person's response to certain stimuli, it provides considerable information about both the laugher and the object of the laughter, i.e, that it involved something humorous. But laughter alone does not provide sufficient information to reconstruct exactly what was so funny. Chimpanzees also produce a call that is vaguely similar to laughter in certain play situations (e.g., tickling). Consequently, they might also recognize human laughter as indicating certain aspects of the social context (i.e., playful, nonthreatening, not distressing, etc.), but they would likely miss the reference to humor. I suspect that implicit in the notion of humor there is a symbolic element, a requirement for recognizing contradiction or paradox, that the average chimpanzee has not developed.[3] The family cat and dog, however, probably do not even get *this* much information from a human laugh. Not sharing our evolved predisposition to laugh in certain social relationships, they do not possess the mental prerequisites to interpret even the social signaling function of laughter. Experience may only have provided them with the ability to use it as evidence that a human is present and is probably not threatening. Nevertheless, this too is dependent on some level of interpretative competence, perhaps provided by recalling prior occasions when some human made this odd noise. Finally, there are innumerable species of animals from flies to snails to fish that wouldn't even produce this much of a response, and would interpret the laughter as just another vibration of the air or water. The diminishing competences of these species corresponds with interpretations that are progressively less and less specific and progressively more and more concrete. But even at the bottom of this descent there is a possibility of a kind of minimalistic reference.

This demonstrates one of Peirce's most fundamental and original insights about the process of interpretation: the difference between different modes of reference can be understood in terms of *levels* of interpretation. Attending to this hierarchical aspect of reference is essential for understanding the difference between the way words and animal calls are related. It's not just the case that we are able to interpret the same sign in different ways, but more important, these different interpretations can be arranged in a sort of ascending order that reflect a prior competence to identify higher-level associative relationships. In other words, reference itself is hierarchic in structure; more complex forms of reference are built up from simpler forms. But there is more to this than just increasing complexity. This hierarchical structure is a clue to the relationships between these different modes of reference. Though I may fail to grasp the symbolic reference of a sign, I might still be able to interpret it as an index (i.e., as correlated with something else), and if I also fail to recognize any indexical correspondences, I may still be able to interpret it as an icon (i.e., recognize its resemblance to something else). Breakdown of referential competence leads to an ordered descent from symbolic to indexical to iconic, not just from complex icons, indices, or symbols to simpler counterparts. Conversely, increasing the sophistication of interpretive competence reverses the order of this breakdown of reference. For example, as human children become more competent and more experienced with written words, they gradually replace their iconic interpretations of these marks as just more writing with indexical interpretations supported by a recognition of certain regular correspondences to pictures and spoken sounds, and eventually use these as support for learning to interpret their symbolic meanings. In this way they trace a path somewhat like the archeologist learning to decipher an ancient script.

This suggests that indexical reference depends upon iconic reference, and symbolic reference depends upon indexical reference—a hierarchy diagrammatically depicted in Figure 3.1. It sounds pretty straightforward on the surface. But this simplicity is deceiving, because what we really mean is that the competence to interpret something symbolically depends upon already having the competence to interpret many other subordinate relationships indexically, and so forth. It is one kind of competence that grows out of and depends upon a very different kind of competence. What constitutes competence in this sense is the ability to produce an interpretive response that provides the necessary infrastructure of more basic iconic and/or indexical interpretations. To explain the basis of symbolic communication, then, we must describe what constitutes a symbolic interpretant, but to do this we need first to explain the production of iconic and indexical interpretants and then to explain how these are each recoded in turn to produce the higher-order forms.

So, we need to start the explanation of symbolic competence with an explanation of what is required in order to interpret icons and build upward. Usually, people explain icons in terms of some respect or other in which two things are alike. But the resemblance doesn't produce the iconicity. Only *after* we recognize an iconic relationship can we say exactly what we saw in common, and sometimes not even then. The interpretive step that establishes an iconic relationship is essentially prior to this, and it is something negative, something that we don't do. It is, so to speak, the act of *not* making a distinction. Let me illustrate this with a very stripped-down example.

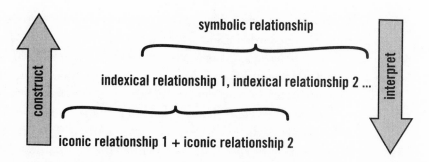

Figure 3.1 *The hierarchic relationships between the three fundamental forms of reference—iconic, indexical, and symbolic. Symbolic relationships are composed of indexical relationships between sets of indices and indexical relationships are composed of iconic relationships between sets of icons (diagrammed more pictorially in Figs. 3.2 and 3.3). This suggests a kind of semiotic reductionism in which more complex forms of representation are analyzable to simpler forms. In fact, this is essentially what occurs as forms are interpreted. Higher-order forms are decomposed into (replaced or represented by) lower-order forms. Inversely, to construct higher representation, one must operate on lower-order forms to replace them (represent them). In C. S. Peirce's terminology, each is an interpretive process, and the new signs substituted for the previous signs at a different level are "interpretants" of those prior signs (see text for details).*

Consider camouflage, as in the case of natural protective coloration. A moth on a tree whose wings resemble the graininess and color of the bark, though not perfectly, can still escape being eaten by a bird if the bird is inattentive and interprets the moth's wings as just more tree. Admittedly, this is not the way we typically use the term *iconic,* but I think it illuminates the most basic sense of the concept. If the moth had been a little less matching, or had moved, or the bird had been a little more attentive, then any of the differences between the moth and the tree made evident by those additional differences would have *indicated* to the bird that there was something else present which wasn't just more tree. If the bird had been in a contemplative mood, it might even have reflected on the slight resemblance of the wing pattern to bark, at least for the fraction of a second before it gobbled the hapless moth. Some features of the moth's wings were iconic of the bark, irrespective of their degree of similarity, merely because under some interpretation (an inattentive bird) they were not distinguished from it.

Now, it might seem awkward to explain iconicity with an example that could be considered to be no representation at all, but I think it helps to clarify the shift in emphasis I want to make from the relationship to the process behind it. What makes the moth wings iconic is an interpretive process produced by the bird, not something about the moth's wings. Their coloration was *taken* to be an icon because of something that the bird *didn't* do. What the bird was doing was actively scanning bark, its brain seeing just more of the same (bark, bark, bark . . .). What it didn't do was alter this process (e.g., bark, bark, not-bark, bark . . .). It applied the same interpretive perceptual process to the moth as it did to the bark. It didn't distinguish between them, and so confused them with one another. This established the iconic relationship between moth and bark. Iconic reference is the default. Even in an imagined moment of reflective reverie in which the bird ponders on their slight resemblance, it is the part of its responding that does not distinguish wing from bark that determines their relationship to be iconic. Iconic resemblance is not based on some prior ground of physical similarity, but in that aspect of the interpretation process that does not differ from some other interpretive process. Thus, although a respect in which two things are similar may influence the ways they tend to be iconically related, it does not determine their iconicity. Iconism is where the referential buck stops when nothing more is added. And at some level, due either to limitations in abilities to produce distinguishing responses or simply a lack of effort to produce them, the production of new interpretants stops. Whether because of boredom or limitations of a minimal nervous system, there are times when almost anything can be iconic of anything else (stuff, stuff, stuff . . .).

What does this have to do with pictures, or other likenesses such as busts or caricatures that we more commonly think of as icons? The explanation is essentially no different. That facet or stage of my interpretive recognition process that is the same for a sketch and the face it portrays is what makes it an icon. I might abstractly reflect on what aspects of the sketch caused this response, and might realize that this was the intention of the artist, but a sketch that is never seen is just paper and charcoal. It could also be interpreted as something that soaked up spilled coffee (and the spilled coffee could be seen as a likeness of Abe Lincoln!). Peirce once characterized an icon as something which upon closer inspection can provide further information about the attributes of its object. Looking at the one is like looking at the other in some respects. Looking at a caricature can, for example, get one to notice for the first time that a well-known politician has a protruding jaw or floppy jowls. The simplification in a diagram or the exaggeration in a cartoon takes advantage of our spontaneous laxness in making distinctions to trick us into making new associations. In this way a caricature resembles a joke, a visual pun, and a diagram can be a source of discovery.

In summary, the interpretive process that generates iconic reference is none other than what in other terms we call *recognition* (mostly perceptual recognition, but not necessarily). Breaking down the term *re-cognition* says it all: to "think [about something] again." Similarly, representation is to pre-sent something again. Iconic relationships are the most basic means by which things can be re-presented. It is the base on which all other forms of representation are built. It is the bottom of the interpretive hierarchy. A sign is interpreted, and thus seen to be a representation, by being reduced (i.e., analyzed to its component representations) to the point of no further re-duceability (due to competence or time limitations, or due to pragmatic con-straints), and thus is ultimately translated into iconic relationships. This does not necessarily require any effort. It is in many cases where interpretive ef-fort ceases. It can merely be the end of new interpretation, that boundary of consciousness where experience fades into redundancy.

Interpreting something as an indexical relationship is this and more. Physical contiguity (nearness or connectedness) or just predictable co-occurrence are the basis for interpreting one thing as an index for another, but as with the case of icons, these physical characteristics are not the cause of the indexical relationship. Almost anything could be physically or tem-porally associated with anything else by virtue of some extension of the ex-perience of nearness in space or time. What makes one an index of another is the interpretive response whereby one seems to "point to" the other. To understand the relationship that indexical interpretations have to iconic in-terpretations, it is necessary to see how the competence to make indexical interpretations arises. In contrast to iconic interpretations, which can often be attributed to interpretive incompetence or the cessation of production of new interpretants, indexical interpretations require something added. In fact, icons arise from a failure to produce critical indices to distinguish things.

Consider the example of a symptom, like the smell of smoke. When I smell smoke, I begin to suspect that something is burning. How did my abil-ity to treat this smell as an indication of fire arise? It likely arose by learn-ing, because I had past experiences in which similar odors were traced to things that were burning. After a few recurrences it became a familiar as-sociation, and the smell of smoke began to indicate to me that a fire might be near. If we consider more closely the learning process that produced the indexical competence, the critical role of icons becomes obvious. The in-dexical competence is constructed from a set of relationships between icons, and the indexical interpretation is accomplished by bringing this as-sembly of iconic relationships to bear in the assessment of new stimuli. The smell of smoke brings to mind past similar experiences (by iconically representing them). Each of these experiences comes to mind because of their similarities to one another and to the present event. But what is more, many of these past experiences also share other similarities. On many of these occasions I also noticed something burning that was the source of the smoke, and in this way those experiences were icons of each other.

There is one important feature added besides all these iconic recognitions. The *repeated correlation* between the smelling of smoke and the presence of flames in each case adds a third higher-order level of iconicity. This is the key ingredient. Because of this I recognize the more general similarity of the entire present situation to these past ones, not just the smoke and not just the fire but also their co-occurrence, and this is what brings to mind the missing element in the present case: the probability that something is burning. What I am suggesting, then, is that the responses we develop as a result of day-to-day associative learning are the basis for all indexical interpretations, and that this is the result of a special relationship that develops among iconic interpretive processes. It's hierarchic. Prior iconic relationships are necessary for indexical reference, but prior indexical relationships are not in the same way necessary for iconic reference. This hierarchic dependency of indices on icons is graphically depicted in Figure 3.2.

Okay, why have I gone to all this trouble to rename these otherwise common, well-established uses of perception and learning? Could we just substitute the word "perception" for "icon" and "learned" association for index? No. Icons and indices are not merely perception and learning, they refer to the *inferential* or *predictive* powers that are implicit in these neural processes. Representational relationships are not just these mechanisms, but a feature of their potential relationship to past, future, distant, or imaginary things. These other things are not physically re-presented but only virtually re-presented by producing perceptual and learned responses like those that would be produced if they were present. In this sense, mental processes are no less representational than external communicative processes, and communicative processes are no less mental in this regard. Mental representation reduces to internal communication.

What, then, is the difference between these uncontroversial cognitive processes underlying icons and indices and the kind of cognitive processes underlying symbols? The same hierarchical logic applies. As indices are constituted by relationships among icons, symbols are constituted by relationships among indices (and therefore also icons). However, what makes this a difficult step is that the added relationship is not mere correlation.

ulus or from context to context occurs as an incidental consequence of learning. These are not really separate forms of learning. Both are based on iconic projection of one stimulus condition onto another. Each arises spontaneously because there is always some ambiguity as to what are the essential parameters of the stimulus that a subject learns to associate with a subsequent desired or undesired result: learning is always an extrapolation from a finite number of examples to future examples, and these seldom provide a basis for choosing between all possible variations of a stimulus. To the extent that new stimuli exhibit features shared by the familiar set of stimuli used for training, and none that are inconsistent with them, these other potential stimuli are also incidentally learned. Often, psychological models of this process are presented as though the subject has learned *rules* for identifying associative relationships. However, since this is based on an iconic relationship, there is no implicit list of criteria that is learned; only a failure to distinguish that which hasn't been explicitly excluded by the training.

Words for kinds of things appear to refer to whole groups of loosely similar objects, such as could be linked by stimulus generalization, and words for qualities and properties of objects refer to the sorts of features that are often the basis for stimulus generalization. Animals can be trained to produce the same sign when presented with different kinds of foods, or trees, or familiar animals, or any other class of objects that share physical attributes in common, even subtle ones (e.g., all hoofed mammals). Similarly, the vervet monkeys' eagle alarm calls might become generalized to other aerial predators if they were introduced into their environment. The grouping of these referents is not by symbolic criteria (though from outside *we* might apply our own symbolic criteria), but by iconic overlap that serves as the basis for their common indexical reference. Stimulus generalization may contribute essential structure to the realms to which words refer, but it is only one subordinate component of the relationship and not what determines their reference.

This same logic applies to the transference of learning sets. For example, learning to choose the odd-shaped object out of three, where two are more similar to each other than the third, might aid in learning a subsequent oddity-discrimination task involving sounds. Rather than just transferring an associated response on the basis of stimulus similarities, the subject recognizes an iconicity between the two learning tasks as wholes. Though this is a hierarchically more sophisticated association than stimulus generalization—learning a *learning pattern*—it is still an indexical association transferred to a novel stimulus via an iconic interpretation. Here the structure of the new training context is seen as iconic of a previous one, allowing the subject to map corresponding elements from the one to the other. This is not often an easy association to make, and most species (including humans) will fail to discover the underlying iconicity when the environment, the training stimuli, the specific responses required, and the reinforcers are all quite different from one context to the next.

There are two things that are critically different about the relationships between a word and its reference when compared to transference of word use to new contexts. First, for an indexical relationship to hold, there must be a correlation in time and place of the word and its object. If the correlation breaks down (for example, the rat no longer gets food by pushing a lever when the sound "food" is played), then the association is eventually forgotten ("extinguished"), and the indexical power of that word to refer is lost. This is true for indices in general. If a smokelike smell becomes common in the absence of anything burning, it will begin to lose its indicative power in that context. For the Boy Who Cried Wolf, in the fable of the same name, the indexical function of his use of the word "wolf" fails because of its lack of association with real wolves, *even though the symbolic reference remains*. Thus, symbolic reference remains stable nearly independent of any such correlations. In fact, the physical association between a word and an appropriate object of reference can be quite rare, or even an impossibility, as with angels, unicorns, and quarks. With so little correlation, an indexical association would not survive.

Second, even if an animal subject is trained to associate a number of words with different foods or states of the box, each of these associations will have little effect upon the others. They are essentially independent. If one of these associations is extinguished or is paired with something new, it will likely make little difference to the other associations, unless there is some slight transference via stimulus generalization. But this is not the case with words. Words also represent other words. In fact, they are incorporated into quite specific individual relationships to *all* other words of a language. Think of the way a dictionary or thesaurus works. They each map one word onto other words. If this shared mapping breaks down between users (as sometimes happens when words are radically reused in slang, such as "bad" for "very good" or "plastered" for "intoxicated"), the reference also will fail.

This second difference is what ultimately explains the first. We do not lose the indexical associations of words, despite a lack of correlation with physical referents, because the possibility of this link is maintained implicitly in the stable associations between words. It is by virtue of this sort of dual reference, to objects and to other words (or at least to other semantic alternatives), that a word conveys the information necessary to pick out objects of reference. This duality of reference is captured in the classic distinction between sense and reference. Words point to objects (reference) and words point to other words (sense), but we use the sense to pick out the reference, not vice versa.

This referential relationship between the words—words systematically indicating other words—forms a system of higher-order relationships that allows words to be *about* indexical relationships, and not just indices in themselves. But this is also why words need to be in context with other words, in phrases and sentences, in order to have any determinate reference. Their indexical power is *distributed*, so to speak, in the relationships between words. Symbolic reference derives from *combinatorial* possibilities and impossibilities, and we therefore depend on combinations both to discover it (during learning) and to make use of it (during communication). Thus the imagined version of a nonhuman animal language that is made up of isolated words, but lacking regularities that govern possible combinations, is ultimately a contradiction in terms.

Even without struggling with the philosophical subtleties of this relationship, we can immediately see the significance for learning. The learning problem associated with symbolic reference is a consequence of the fact that what determines the pairing between a symbol (like a word) and some object or event is not their probability of co-occurrence, but rather some complex function of the relationship that the symbol has to other symbols. This is a separate but linked learning problem, and worse yet, it creates a third, higher-order *unlearning* problem. Learning is, at its base, a function of the probability of correlations between things, from the synaptic level to the behavioral level. Past correlations tend to be predictive of future correlations. This, as we've seen, is the basis for indexical reference. In order to comprehend a symbolic relationship, however, such indexical associations must be subordinated to relationships between different symbols. This is a troublesome shift of emphasis. To learn symbols we begin by learning symbol-object correlations, but once learned, these associations must be treated as no more than clues for determining the more crucial relationships. And these relationships are not highly correlated; in fact, often just the reverse. Words that carry similar referential function are more often used alternatively and not together, and words with very different (complementary) referential functions tend to to be adjacent to one another in sentences. Worst of all, few sentences or phrases are ever repeated exactly, and the frequency with which specific word combinations are repeated is also extremely low. Hardly a recipe for easy indexical learning.

One of the most insightful demonstrations of the learning difficulties associated with the shift from conditioned associations to symbolic associations comes not from a human example, but from a set of experiments that attempted to train chimpanzees to use simple symbols. This study was directed by Sue Savage-Rumbaugh and Duane Rumbaugh,[4] now at the Language Research Center of Georgia State University, and included four chimps, two of which, Sherman and Austin, showed particular facility with the symbols. It is far from the "last word" on how far other species can go in their understanding of languagelike communication, and further studies of another chimpanzee (from a different subspecies) that show more developed abilities will be described subsequently (see Chapter 4),[5] but this work has the virtue of exposing much of what is often hidden in children's comparatively easy entry into symbolic communication, and so provides an accessible step-by-step account of what we usually take for granted in the process. In what follows I will outline these experiments briefly. Only the most relevant highlights will be described and other aspects will be simplified for the sake of my purpose here. Of course, my attempts to "get inside the chimps' heads" during this process are fantasy. Though I will use somewhat different terminology from the experimenters to describe this transition from indexical to symbolic communication, I am reasonably confident that my interpretation is not at odds with theirs. However, the interested reader should refer to the excellent account of these experiments and their significance in Savage-Rumbaugh's book describing them.

The chimps in this study were taught to use a special computer keyboard made up of lexigrams—simple abstract shapes (lacking any apparent iconism to their intended referents) on large illuminated keys on a keyboard mounted in their cage. Duane Rumbaugh's previous experiments (with a chimp named Lana)[6] had shown that chimps have the ability to learn a large number of paired associations between lexigrams (and in fact other kinds of symbol tokens) and objects or activities. But in order to respond to critics and more fully test other features of this ability, Duane and Sue began a new series of experiments with a group of chimps to test both chimp-chimp communication and chimps' ability to use lexigrams in combinations (e.g., syntactic relationships). Not surprisingly, the chimps exhibited some interesting difficulties when they were required to use lexigrams in combinations, but they eventually solved their learning problems and exhibited a use of the lexigrams that was clearly symbolic. In so doing they have provided us with a remarkably explicit record of the process that leads from index to symbol.

In order to test Sherman and Austin's symbolic understanding of the lexigrams, the chimps were trained to chain lexigram pairs in a simple verb-noun relationship (a sequence glossed as meaning "give," which caused a dispenser to deliver a solid food, and "banana" to get a banana).[7] Initially there were only 2 "verb" lexigrams and 4 food or drink lexigrams to choose from, and each pair had to be separately taught. But after successful training of each pairing, the chimps were presented with all the options they had learned independently, and were required to choose which combination was most appropriate on the basis of food availability or preference. Curiously, the solution to this task was not implicit in their previous training. This was evident in the fact that some chimps tended stereotypically to repeat only the most recent single learned combination, whereas others chained together all options, irrespective of the intended meanings and what they knew about the situation. Thus they had learned the individual associations but failed to learn the system of relationships of which these correlations were a part. Although the logic of the combinatorial relationships between lexigrams was implicit in the particular combinations that the chimps learned, the converse exclusive relationships had not been learned. For example, they were not explicitly trained to avoid any number of inappropriate combinations such as "banana juice give." Though these errors are implicit for us, who treat them symbolically from the start, the combinatorial rules that allow pairing in some but not other cases was vastly underdetermined by the training experience (as it is also in a child's experience of others' word use).

It is not immediately obvious exactly how much exclusionary information is implicit, but it turns out to be quite a lot. Think about it from the naive chimpanzee perspective for a moment. Even with this ultra-simple symbol system of six lexigrams and a two-lexigram combinatorial grammar, the chimpanzee is faced with the possibility of sorting among 720 possible ordered sequences ($6*5*4*3*2*1$) or 64 possible ordered pairs. The training has offered only four prototype examples, in isolation. Though each chimp may begin with many guesses about what works, these are unlikely to be in the form of rules about classes of allowed and disallowed combinations, but rather about possible numbers of lexigrams that must be pressed, their positions on the board, their colors or shape cues that might be associated with a reward object, and so on. Recognizing this limitation, the experimenters embarked on a rather interesting course of training. They set out explicitly to train the chimps on which cues were not relevant and which combinations were not meaningful. This poses an interesting problem that every pet trainer has faced. You can't train what *not* to do unless the animal first produces the disallowed behavior. Only then can it be immediately punished or at least explicitly not rewarded (the correlation problem again). So the chimps were first trained to produce incorrect associations (e.g., mistaking keyboard position as the relevant variable) and then these errors were explicitly not rewarded, whereas the remaining appropriate responses were. By a complex hierarchic training design, involving thousands of trials, it was possible to teach them to exclude systematically all inappropriate associative and combinatorial possibilities among the small handful of lexigrams. At the end of this process, the animals were able to produce the correct lexigram strings every time.

Had training out the errors worked? To test this, the researchers introduced a few new food items and corresponding new lexigrams. If the chimps had learned the liquid/solid rule, and got the idea that a new lexigram was for a new item, they might learn more quickly. Indeed they did. Sherman and Austin were able to respond correctly the first time, or with only a few errors, instead of taking hundreds of trials as before. What had happened to produce this difference? What the animals had learned was not only a set of specific associations between lexigrams and objects or events. They had also learned a set of logical relationships *between the lexigrams*, relationships of exclusion and inclusion. More importantly, these lexigram-lexigram relationships formed a complete system in which each allowable or forbidden co-occurrence of lexigrams in the same string (and therefore each allowable or forbidden substitution of one lexigram for another) was defined. They had discovered that the relationship that a lexigram has to an object *is a function of* the relationship it has to other lexigrams, not just a function of the correlated appearance of both lexigram and object. This is the essence of a symbolic relationship.

The subordination of the indexical relationships between lexigrams (symbol tokens) and foods (referents or objects) to the system of indexical relationships between lexigrams is schematically depicted in three stages of development in Figure 3.3. Individual indexical associations are shown as single vertical arrows, mapping each token to a kind of object, because each of these relationships is independent of the others. In contrast, the token-token interrelationships (e.g., between lexigrams or words), shown as horizontal arrows interconnecting symbols, form a closed logical group of combinatorial possibilities. Every combination and exclusion relationship is unambiguously and categorically determined. The indexical reference of each symbol token to an object after symbolic reference is achieved is depicted with arrows reversed to indicate that these are now subordinate to the token-token associations.

In the minimalistic symbol system first learned by Sherman and Austin,

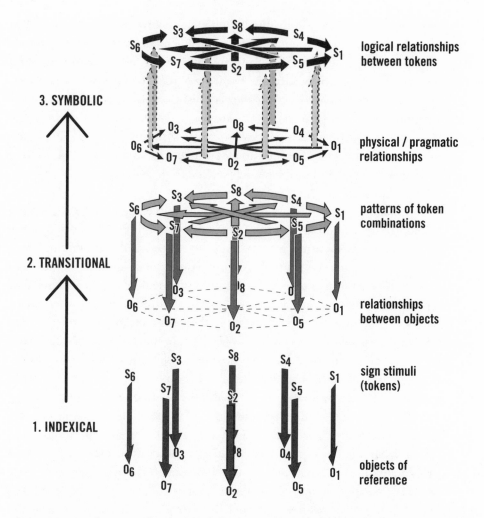

Figure 3.3 *A schematic depiction of the construction of symbolic referential relationships from indexical relationships. This figure builds on the logic depicted in Figure 3.2, but in this case the iconic relationships are only implied and the indexical relationships are condensed into single arrows. Three stages in the construction of symbolic relationships are shown from bottom to top. First, a collection of different indices are individually learned (varying strength indicated by darkness of arrows). Second, systematic relationships between index tokens (indexical stimuli) are recognized and learned as additional indices (gray arrows linking indices). Third, a shift (reversal of indexical arrows) in mnemonic strategy to rely on relationships between tokens (darker arrows above) to pick out objects indirectly via relationships between objects (corresponding lower arrow system). Individual indices can stand on their own in isolation, but symbols must be part of a closed group of transformations that links them in order to refer, otherwise they revert to indices.*

reference to objects is a collective function of relative position within this token-token reference system. No individual lexigram determines its own reference. Reference emerges from the hierarchic relationship *between* these two levels of indexicality, and by virtue of recognizing an abstract correspondence between the system of relationships between objects and the system of relationships between the lexigrams. In a sense, it is the recognition of an iconic relationship between the two systems of indices. Although indexical reference of tokens to objects is maintained in the transition to symbolic reference, it is no longer determined by or dependent on any physical correlation between token and object.

This makes a new kind of generalization possible: logical or categorical generalization, as opposed to stimulus generalization or learning set generalization. It is responsible for Sherman and Austin's ability to acquire new lexigrams and know their reference implicitly, without any trial-and-error learning. The system of lexigram-lexigram interrelationships is a source of implicit knowledge about how novel lexigrams must be incorporated into the system. Adding a new food lexigram, then, does not require the chimp to learn the correlative association of lexigram to object from scratch each time. The referential relationship is no longer solely (or mainly) a function of lexigram-food co-occurrence, but has become a function of the relationship that this new lexigram shares with the existing system of other lexigrams, and these offer a quite limited set of ways to integrate new items. The chimps succeed easily because they have shifted their search for associations from relationships among stimuli to relationships among lexigrams. A new food or drink lexigram must fit into a predetermined slot in this system of relationships. There are not more than a few possible alternatives to sample, and none requires assessing the probability of paired lexigram-food occurrence because lexigrams need no longer be treated as indices of food availability. Like words, the probability of co-occurrences may be quite low. The food lexigrams are in a real sense "nouns," and are defined by their potential combinatorial roles. Testing the chimps' ability to extrapolate to new lexigram-food relationships is a way of demonstrating whether or not they have learned this logical-categorical generalization, which is a crucial defining feature of symbolic reference.

At some point toward the end of the training, the whole set of explicitly presented indexical associations that the chimps had acquired was "recoded" in their minds with respect to an implicit pattern of associations whose evidence was distributed across the whole set of trials. Did this recoding happen as soon as they had learned the full set of combination/exclusion relationships among their lexigram set? I suspect not. Try to imagine yourself in their situation for a moment. You have just come to the point where you are not making errors. What is your strategy? Probably, you are struggling to remember what specific things worked and did not work, still at the level of one-by-one associations. The problem is, it is hard to remember all the details. What you need are aids to help organize what you know, because there are a lot of possibilities. But in the internal search for supports you discover that there is another source of redundancy and regularity that begins to appear, besides just the individual stimulus-response-reward regularities: the relationships between lexigrams! And these redundant patterns are far fewer than the messy set of dozens of individual associations that you are trying to keep track of. These regularities weren't apparent previously, because errors had obscured any underlying systematic relationship. But now that they are apparent, why not use them as added mnemonics to help simplify the memory load? Forced to repeat errorless trials over and over, Sherman and Austin didn't just learn the details well, they also became aware of something they couldn't have noticed otherwise, that there was a system behind it all. And they could use this new information, *information about what they had already learned,* to simplify greatly the mnemonic load created by the many individual rote associations. They could now afford to forget about individual correlations so long as they could keep track of them via the lexigram-lexigram rules.

What I am suggesting here is that the shift from associative predictions to symbolic predictions is initially a change in mnemonic strategy, a recoding. It is a way of offloading redundant details from working memory, by recognizing a higher-order regularity in the mess of associations, a trick that can accomplish the same task without having to hold all the details in mind. Unfortunately, nature seldom offers such nice neat logical systems that can help organize our associations. There are not many chances to use such strategies, so not much selection for this sort of process. We are forced to create artificial systems that have the appropriate properties. The crucial point is that when such a systematic set of tokens becomes available, it allows a shift in mnemonic strategy that results in a radical transformation in the mode of representation. What one knows in one way gets recoded in another way. It gets *re-represented.* We know the same associations, but we know them also in a different way. You might say we know them both from the bottom up, indexically, and from the top down, symbolically. And because this recoding is based on higher-order relationships, not the individual details, it often vastly simplifies the mnemonic problem and vastly augments the representational possibilities. Equally important is the vast amount of implicit knowledge it provides. Because the combinatorial rules encode not objects but ways in which objects can be related, new symbols can immediately be incorporated and combined with others based on independent knowledge about what they symbolize.

The experimenters working with Sherman and Austin provided a further, and in some ways even more definitive, demonstration of the difference between indexical reference of lexigram-object correlations and symbolic reference in a subsequent experiment that compared the performance of the two symboling apes (Sherman and Austin) to a previous subject (Lana), who had been trained with the same lexigram system but not in the same systematic fashion. Lana had learned a much larger corpus of lexigram-object associations, though by simple paired associations. In this new experiment (see Figure 3.4), all three chimps were first tested on their ability to learn to sort food items together in one pan and tool items together in another (Lana learned in far fewer trials than Sherman and Austin). When all three chimps had learned this task, they were presented with new foods or tools to sort and were able to generalize from their prior behavior to sort these new items appropriately as well. This is essentially a test of stimulus generalization, and it is based on some rather abstract qualities of the test items (e.g., edibility). It shows that chimps have a sophisticated ability to conceptualize such abstract relationships irrespective of symbols. Of course, chimpanzees (as well as most other animal species) must be able to distinguish edible from inedible objects and treat each differently. Learning to sort them accordingly takes advantage of this preexisting categorical discrimination in a novel context. In this sense, then, what might be called an indexical concept of food and nonfood precedes the training. Each bin is eventually treated as indexical of this qualitative sensory and behavioral distinction, and so the ability to extend this association to new food and nonfood items involved stimulus generalization (though of an indirectly recognizable stimulus parameter).

This sorting task was followed by a second task in which the chimps were required to associate each of the previously distinguished food items with the same lexigram (glossed as "food" by the experimenters) and each of the tool items with another lexigram ("tool"). Initially, this task simply required the chimps to extend their prior associations with bins to two additional stimuli, the two lexigrams. Although all three chimps learned this task in a similar way, taking many hundreds of trials to make the transference, Sherman and Austin later spontaneously recoded this information in a way that Lana did not. This was demonstrated when, as in the prior task, novel food and novel tool items were introduced. Sherman and Austin found this to be a trivial addition and easily guessed without any additional learning which

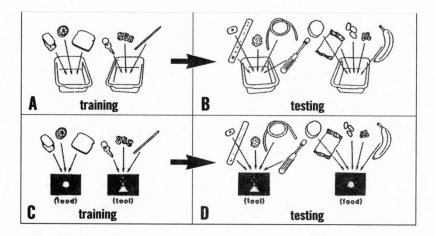

Figure 3.4 *Summary of part of a 1980 test of lexigram reference in chimpanzees by E. Sue Savage-Rumbaugh and her colleagues. This compares three levels of symbolic learning of lexigram reference by the chimps Sherman and Austin to the indexical learning of lexigram reference by another chimp, Lana, who is unable to complete tasks requiring symbolic reference. The panels on the left depict training trials and the panels on the right depict items added in test trials. Test trials introduced new lexigrams and tested to determine generalization to items for which there was no previous experience. The top task was merely a sorting task to determine that all animals understood the distinction between foods and tools (nonfood). The second task required identification with one of two lexigrams ("food," "tool"). Though all three learned it, only Sherman and Austin made the shift to symbolic categorization of reference and were able to generalize to new items (because of past symbol-learning experience). Lana was excluded from the remaining two procedures (not shown), where Sherman and Austin learned first to associate lexigrams to pictures of the foods and tools, and then to associate individual food and tool lexigrams with the appropriate general lexigram for food or tool.*

lexigram was appropriate. Lana not only failed to extend her categorization to the new items, the novelty and errors appeared to produce a kind of counterevidence that caused her to abandon her prior training in a subsequent test. Though on the surface this task resembles the sorting task, these conflicting results demonstrate that there is a critical difference that undermined the rote learning strategy used by Lana and favored the symbolic recoding used by Sherman and Austin. The difference is probably related to the fact that the sorting task involved a physical-spatial association of sign and object, whereas the lexigram "labeling" involved only temporal correspondence. Lana appeared not to be using these underlying qualities to solve the task. For her, each lexigram object association was an independent datum, and so provided no information about other associations.

In contrast Sherman and Austin, as a result of their experience with a previous symbol system, recoded these new lexigram-object associations into two new symbolic categories that superseded the individual associations. It took them hundreds or thousands of trials to learn the first simple one-to-many associations. This was because they began with no systemic relationship in their still small lexigram repertoire for a general reference to "food" or "tool." They had to learn them the hard way, so to speak, indexically. But as soon as they did learn these associations, they were primed to look for another higher-order logic, and once it was discovered, they were able to use this logic to generalize to new associations. Instead of hundreds or even thousands of trials, the availability of a symbolic coding allowed them to bypass further trials altogether, an incredible increase in learning efficiency. The chimps essentially knew something that they had never explicitly learned. They had gained a kind of implicit knowledge as a spontaneous byproduct of symbolic recoding.

I have chosen to recount this ape language study not because it portrays any particularly advanced abilities in chimpanzees, or because I think it is somehow representative. In fact (as noted earlier), more recent studies by these same experimenters, with a pygmy chimpanzee (or bonobo) named Kanzi, have demonstrated far more effortless and sophisticated symbolic abilities.[8] Rather, I have focused on this earlier study because of the clarity with which it portrays the special nature of symbol learning, and because it clearly exemplifies the hierarchic relationship between symbolic and indexical reference. The *reductio ad absurdum* training ploy is particularly instructive, not because it is an essential element but because it provides an explicit constructive demonstration of the index-by-index basis of the eventual symbolic relationship. It also demonstrates how normal associative learning strategies can interfere with symbol learning. Indexical associations are necessary stepping stones to symbolic reference, but they must ultimately be superseded for symbolic reference to work.

Unlearning an Insight

The problem with symbol systems, then, is that there is both a lot of learning *and unlearning* that must take place before even a single symbolic relationship is available. Symbols cannot be acquired one at a time, the way other learned associations can, except *after* a reference symbol system is established. A logically complete system of relationships among the set of symbol tokens must be learned before the symbolic association between any one symbol token and an object can even be determined. The learning step occurs prior to recognizing the symbolic function, and this function only emerges from a system; it is not vested in any individual sign-object pairing. For this reason, it's hard to get started. To learn a first symbolic relationship requires holding a lot of associations in mind at once while at the same time mentally sampling the potential combinatorial patterns hidden in their higher-order relationships. Even with a very small set of symbols the number of possible combinations is immense, and so sorting out which combinations work and which don't requires sampling and remembering a large number of possibilities.

One of the most interesting features of the shift in learning strategy that symbolic reference depends upon is that it essentially takes no time; or rather, no more time than the process of perceptual recognition. Although the prior associations that will eventually be recoded into a symbolic system may take considerable time and effort to learn, the symbolic recoding of these relationships is not *learned* in the same way; it must instead be *discovered* or perceived, in some sense, by reflecting on what is already known. In other words, it is an implicit pattern that must be recognized in the relationships between the indexical associations. Recognition means linking the relationship of something new to something already known. The many interdependent associations that will ultimately provide the nodes in a matrix of symbol-symbol relationships must be in place in order for any one of them to refer symbolically, so they must each be learned *prior to* recognizing their symbolic associative functions. They must be learned as individual indexical referential relationships. The process of discovering the new symbolic association is a restructuring event, in which the previously learned associations are suddenly seen in a new light and must be reorganized with respect to one another. This reorganization requires mental effort to suppress one set of associative responses in favor of another derived from them. Discovering the superordinate symbolic relationship is not some added learning step, it is just noticing the system-level correspondences that are implicitly present between the token-token relationships and the object-object relationships that have been juxtaposed by indexical learning. What we might call a symbolic *insight* takes place the moment we let go of one associative strategy and grab hold of another higher-order one to guide our memory searches.

What I have described as the necessary cognitive steps to create symbolic reference would clearly be considered a species of "insight learning," though my analysis suggests that the phrase is in one sense an oxymoron.

Psychologists and philosophers have long considered the ability to learn by insight to be an important characteristic of human intelligence. Animal behaviorists have also been fascinated with the question, Can other animals learn by insight? The famous Gestalt psychologist Wolfgang Köhler described experiments with chimpanzees in which to reach a fruit they had to "see" the problem in a new way.[9] Köhler set his chimp the problem of retrieving a banana suspended from the roof of the cage and out of reach, given only a couple of wooden boxes that when stacked one upon the other could allow the banana to be reached. He found that these solutions were not intuitively obvious for a chimpanzee, who would often become frustrated and give up for a long period. During this time she would play with the boxes, often piling them up, climbing on them, and then knocking them down. At some point, however, the chimp eventually appeared to have recognized how this fit with the goal of getting at the banana, and would then quite purposefully maneuver the boxes into place and retrieve the prize. Once learned, the trick was remembered. Because of the role played by physical objects as mnemonic place-holders and the random undirected exploration of them, this is not perhaps the sort of insight that appears in cartoons as the turning on of a light bulb, nor is it what is popularly imagined to take place in the mind of an artist or scientist. On the other hand, what goes on "inside the head" during moments of human insight may simply be a more rapid covert version of the same, largely aimless object play. We recognize these as examples of insight solely because they involve a recoding of previously available but unlinked bits of information.

Most insight problems do not involve symbolic recoding, merely sensory recoding: "visualizing" the parts of a relationship in a new way. Transference of a learning set from one context to another is in this way also a kind of insight. Nevertheless, a propensity to search out new "perspectives" might be a significant advantage for discovering symbolic relationships. The shift in mnemonic strategy from indexical to symbolic use of food and food-delivery lexigrams required the chimps both to use the regularities of symbol-token combinations as the solution to correct performance, and to discover that features of the food objects and delivery events correspond to these lexigram combination regularities. In other words, they had to use these combination relationships to separate the abstract features of liquid and solid from their context of indexical associations with the food-delivery events. The symbolic reference that resulted depended on digging into these aspects of the interrelationships between things, as opposed to just mapping lexigrams to things themselves. By virtue of this, even the specific combinations of tokens cannot be seen as indexical, so that it is not just that the ability to combine tokens vastly multiplies referential possibilities, in the way that using two digits instead of one makes it possible to represent more numerical values. Which tokens can and cannot be combined and which can and cannot substitute for one another determines a new level of mapping to what linguists call "semantic features," such as the presence or absence of some property like "solidity." This is what allows a system of symbols to grow. New elements can be added, either by sharing reference with semantic features that the system already defines, or by identifying new features that somehow can be integrated with existing ones. Even separate symbol groups, independently constructed, can in this way become integrated with each other. Once the relationship between their semantic feature sets is recognized, their unification can in one insight create an enormous number of new combinatorial possibilities.

The insight-recoding problem becomes increasingly difficult as additional recoding steps become involved in establishing an association. For this reason, a child's initial discovery of the symbolic relationships underlying language is only the beginning of the demand on this type of learning/unlearning process. Each new level of symbols coding for other symbolic relationships (i.e., more abstract concepts) requires that we engage this process anew. This produces a pattern of learning that tends to exhibit more or less discrete stages. Since the number of combinatorial possibilities that must be sampled in order to discover the underlying symbolic logic increases geometrically with each additional level of recoding, it is almost always necessary to confine rote learning to one level at a time until the symbolic recoding becomes apparent before moving on to the next. This limitation is frustratingly familiar to every student who is forced to engage in seemingly endless rote learning before "getting" the underlying logic of some mathematical operation or scientific concept. It may also contribute to the crudely stagelike pattern of children's cognitive development, which the psychologist Jean Piaget initially noticed.[10] However, this punctuated pattern of symbolic conceptual development is a reflection of symbolic information processing and not an intrinsic feature of developing brains and minds.

The ability of Sherman and Austin to discover the abstract symbolic references for "food" and "tool" provides an additional perspective on the difference between indexical associations and symbolic associations. Consider the potential conflict between the lexigram-object relationships they had previously acquired and this new set of associations. If their prior associations were supported only by the correlations in lexigram-object-reward occurrence, then re-pairing the same objects with a new lexigram would be expected to partially if not totally extinguish the prior association. Although it would be possible to provide additional contextual cues to enable the chimps to decide which of two competing associative strategies to use (e.g., simply run trials without the alternatives available) and thus learn and retain both, there would still be interference effects (i.e., their prior associations might interfere both with relearning the new associations and with shifting between them in different contexts). Unfortunately, data to assess this are not available, but we can infer from Sherman and Austin's learning shifts, and their subsequent maintenance of the prior symbolic associations, that neither extinction nor interference was a significant problem. Though it was not tested explicitly in this series of experiments, we should expect that this should also distinguish Sherman and Austin from Lana. Certainly Lana's rapid decline in performance when new items were added points to such effects.

This ability to remember large numbers of potentially competing associations is an additional power of symbolic reference that derives from the shift in mnemonic strategy to token-token relationships. Competition effects grow with increasing numbers of overlapping associative categories in typical indexical reference relationships. Not only would the choice among alternatives in any use become a source of confusion, but because they were competing for reinforcement, each would weaken the association of the others. Though some of the interference effects also attend symbol use, and often are a cause of word retrieval errors and analysis delays, in terms of associative strength there is an opposite effect. Competing sets of overlapping associative relationships on the indexical level translate into mutually supportive higher-order semantic categories on the symbolic level. These become sources of associative redundancy, each reinforcing the mnemonic trace of the other. So, rather than weaken the strength of the association, they actually reinforce it.

This helps to explain where the additional associative glue between words and their referents comes from. Though token-object correlations are not consistently available to the symbol user, indeed are rare, this loss of associative support is more than compensated by the large number of other associations that are available through symbolically mediated token-token relationships. Individually, these are comparatively weak associations, with a low correlated occurrence of any two tokens in the same context; but they are not just one-to-one associations. They are one-to-many and many-to-one associations that weave symbol tokens together into a systematic network of association relationships, and the pattern has a certain coded isomorphism with relationships between objects and events in the world.

As a result of sharing many weak interpenetrating indexical links, each indexical association gains mnemonic support from a large number of others because they are multiply coded in memory. Together, their combined associative strengths make them far more resistant to extinction due to diminished external correlations with objects than are individual indexical associations. Thus, not only is symbolic reference a distributed relationship, so is its mnemonic support. This is why learning the symbolic reasons behind the bits of information we acquire by rote learning offers such a powerful aid to recall. How else could the many thousands of different words we use every day be retrieved so rapidly and effortlessly during the act of speaking or listening?

Numerous neuropsychological probes of semantic field effects demonstrate this for word meaning. Hearing, memorizing, or using a word can be a source of priming effects for subsequent recall or identification of other words in overlapping categories. For example, hearing the word "cat" might prime later memory tasks involving "dog" or "animal." Even more interesting is the fact that this also transfers to indexical associations involving these words as well. Receiving a mild electric shock every time you hear the word "cat" would cause you to learn to spontaneously produce physiological correlates of stress response (such as change in heart rate or galvanic skin response) upon hearing that word repeated. But a similar but less intense response will also be produced whenever you hear a word like "dog," even though there had never been shocks associated with these sounds. A lesser response will also be produced whenever you hear a word like "meow" or "animal," demonstrating lexical (word-word) associations, and in response to a rhyming word like "mat," demonstrating stimulus generalization effects. All of these distinct associative relationships are brought into relationship to one another by the symbolic relationship. Because each arouses an associative network that overlaps with that of the shock-conditioned word, the shared activation raises an arousal level also associated with shock. The extent of both the symbolic and indexical overlap appears to correlate with the extent of the transference. Though analogous to stimulus generalization, it is clearly different. There are no shared *stimulus* parameters that distinguish "dog" and "cat" from "car," which does not produce a similar priming. The difference is also reflected in the fact that there is an independent transference to words that rhyme, like "flat" or "sat." Rhyme associations are true stimulus generalization effects and also show some transference of physiological responses.

This analogy between effects involving shared stimulus features and shared semantic features shows that the brain stores and retrieves both symbolic and nonsymbolic associations as though they were the same sort of thing. Just as the contingencies of co-occurrence and exclusion in the same context determine the strengths of stimulus associations, so too do these statistics in language affect the strengths of word associations.

With each shift of referential control to a token-token system of relationships, it became possible for Sherman and Austin to add new lexical items to their growing symbol system with a minimum of associative learning, often without any trial-and-error testing. This produces a kind of threshold effect whereby prior associative learning strategies, characterized by an incremental narrowing of stimulus response features, are replaced by categorical guesses among a few alternatives. The result is a qualitative shift in performance. The probabilistic nature of the earlier stage is superseded by alternative testing that has a sort of all-or-none character. This change in behavior can thus be an indication of the subject's shift in mnemonic strategy, and hence the transition from indexical to symbolic reference. The simplest indicator of this shift is probably the rate of acquisition of new lexical items, since this should be highly sensitive to the hundred- to thousand-fold reduction in trial-and-error learning required to reach 100 percent performance.

In young children's learning of language, apparent threshold effects have long been noticed in vocabulary growth and sentence length. Vocabulary and utterance length are of course linked variables in two regards. First, the more words a child knows, the more there are to string together. But this does not simply translate into larger sentences. Creating a larger sentence in a human language cannot just be accomplished by stringing together more and more words. It requires the use of hierarchic grammatical relationships, as well as syntactic tricks for condensing and embedding kernel sentences in one another. Thus, not only does vocabulary need to grow, but the types of words must diversify. In other words, the regular discovery of new grammatical classes must be followed by a rapid filling of these classes with new alternative lexical items.

Each time a new logical group is discovered among a set of tokens, it essentially opens up one or more types of positional slots that can be filled from an open class of symbols. Each slot determines both a semantic and a grammatical category. Recall that although Sherman and Austin could add new food items to their lexigram "vocabulary" with little difficulty, when they had to learn to recode food items in terms of the higher-order semantic category "food," they essentially had to start over. Their prior knowledge of the symbolic designations of distinct foods with respect to food-delivery modes was of no help. It may even have been a source of interference, since the same foods were now being linked with different lexigrams. But again, once this new symbolic association was established, adding new items proved trivial, usually involving no errors.

In the small symbol system initially learned by Sherman and Austin, the semantic features that were implicit in the few combinatorial possibilities available might be specified in terms of solid versus liquid and food versus delivery (of food). Discovering the combinatorial rules was the key to discovering these semantic features, and, conversely, these semantic features provided the basis for adding new symbols without needing to relearn new correlations. All that was necessary was prior knowledge of the object to be represented with respect to one or more of the relevant semantic features in order to know implicitly a token's combinatorial possibilities and reference. Beginning with any initial core, the system can grow rapidly in repeated stages. Each stage represents a further symbolic transition that must begin with incremental indexical learning. But past experience at symbol building and a large system of features can progressively accelerate this process.

In summary, then, symbols cannot be understood as an unstructured collection of tokens that map to a collection of referents because symbols don't just represent things in the world, they also represent each other. Because symbols do not directly refer to things in the world, but indirectly refer to them by virtue of referring to other symbols, they are implicitly combinatorial entities whose referential powers are derived by virtue of occupying determinate positions in an organized system of other symbols. Both their initial acquisition and their later use requires a combinatorial analysis. The structure of the whole system has a definite semantic topology that determines the ways symbols modify each other's referential functions in different combinations. Because of this systematic relational basis of symbolic reference, no collection of signs can function symbolically unless the entire collection conforms to certain overall principles of organization. Symbolic reference emerges from a ground of nonsymbolic referential processes only because the indexical relationships between symbols are organized so as to form a logically closed group of mappings from symbol to symbol. This determinate character allows the higher-order system of associations to supplant the individual (indexical) referential support previously invested in each component symbol. This system of relationships between symbols determines a definite and distinctive topology that all operations involving those symbols must respect in order to retain referential power. The structure implicit in the symbol-symbol mapping is not present before symbolic reference, but comes into being and affects symbol combinations from the moment it is first constructed. The rules of combination that are implicit in this structure are discovered as novel combinations are progressively sampled. As a result, new rules may be discovered to be emergent requirements of encountering novel combinatorial problems, in much the same way as new mathematical laws are discovered to be implicit in novel manipulations of known operations.

Symbols do not, then, get accumulated into unstructured collections that can be arbitrarily shuffled into different combinations. The system of representational relationships, which develops between symbols as symbol systems grow, comprises an ever more complex matrix. In abstract terms, this is a kind of tangled hierarchic network of nodes and connections that defines a vast and constantly changing semantic space. Though semanticists and semiotic theorists have proposed various analogies to explain these underlying topological principles of semantic organization (such as +/- feature lists, dictionary analogies, encyclopedia analogies), we are far from a satisfactory account. Whatever the logic of this network of symbol-symbol relationships, it is inevitable that it will be reflected in the patterns of symbol-symbol combinations in communication.

Abstract theories of language, couched in terms of possible rules for combining unspecified tokens into strings, often implicitly assume that there is no constraint on theoretically possible combinatorial rule systems. Arbitrary strings of uninterpreted tokens have no reference and thus are unconstrained. But the symbolic use of tokens is constrained both by each token's use and by the use of other tokens with respect to which it is defined. Strings of symbols used to communicate and to accomplish certain ends must inherit both the intrinsic constraints of symbol-symbol reference and the constraints imposed by external reference.

Some sort of regimented combinatorial organization is a logical necessity for any system of symbolic reference. Without an explicit syntactic framework and an implicit interpretive mapping, it is possible neither to produce unambiguous symbolic information nor to acquire symbols in the first place. Because symbolic reference is inherently systemic, there can be no symbolization without systematic relationships. Thus syntactic structure is an integral feature of symbolic reference, not something added and separate. It is the higher-order combinatorial logic, grammar, that maintains and regulates symbolic reference; but how a specific grammar is organized is not strongly restricted by this requirement. There need to be precise combinatorial rules, yet a vast number are possible that do not ever appear in natural languages. Many other factors must be taken into account in order to understand why only certain types of syntactic systems are actually employed in natural human languages and how we are able to learn the incredibly complicated rule systems that result.

So, before turning to the difficult problem of determining what it is about human brains that makes the symbolic recoding step so much easier for us than for the chimpanzees Sherman and Austin (and members of all other nonhuman species as well), it is instructive to reflect on the significance of this view of symbolization for theories of grammar and syntax. Not only does this analysis suggest that syntax and semantics are deeply interdependent facets of language—a view at odds with much current linguistic theory—it also forces us entirely to rethink current ideas about the nature of grammatical knowledge and how it comes to be acquired.

Outside the Brain

Nothing that is worth knowing can be taught.

—Oscar Wilde

Chomsky's Handstand

Over the last few decades language researchers seem to have reached a consensus that language is an innate ability, and that only a significant contribution from innate knowledge can explain our ability to learn such a complex communication system. Without question, children enter the world predisposed to learn human languages. All normal children, raised in normal social environments, inevitably learn their local language, whereas other species, even when raised and taught in this same environment, do not. This demonstrates that human brains come into the world specially equipped for this function. Few would argue with this sense of the term *innate*.

But many linguists and psychologists propose a more thoroughly preformationist interpretation of this same phenomenon. They argue that the child's remarkable feat of learning a first language is the result of an innate "language competence." When we say that people are competent at a skill, for example, typing, we generally mean that they possess some proficiency in it, and not just a potential or talent that might be realized under the right conditions. A competence is an available skill, normally one learned or acquired previously. So by analogy an innate language competence is an ability to perform certain language tasks *as though they had previously been learned.* If language competence is innate in this sense, then language knowledge itself, in some form, is already present in the human brain prior to gaining knowledge from any experience with language. But is it really?

There is, without doubt, something special about human brains that enables us to do with ease what no other species can do even minimally without intense effort and remarkably insightful training. We not only have the ability to create and easily learn simple symbol systems such as the chimps Sherman and Austin struggled to learn, but in learning languages we acquire an immensely complex rule system and a rich vocabulary at a time in our lives when it is otherwise very difficult to learn even elementary arithmetic. Many a treatise on grammatical theory has failed to provide an adequate accounting of the implicit knowledge that even a four-year-old appears to possess about her newly acquired language. No wonder so many linguists have thrown up their hands, exclaiming that "language must be unlearnable" and claiming that it is all a magician's trick, that the rabbit (grammatical knowledge) must have been in the hat (the child's brain) from the beginning. But in what form? And how could it have gotten there? Unfortunately, as we have seen, the theory that innate knowledge of grammar is the heritage of all human children simply asserts the answers to these messy questions, and leaves it to evolutionary biology and neuroscience to explain how the answers are to be derived. Before scientists in these fields commit their experimental resources and theoretical modeling efforts to exploring this theory's claims, it may be worth asking if they are biologically plausible and if there really are no alternatives.

The idea that an innate Universal Grammar is the only way to account for language abilities was first argued by the MIT linguist Noam Chomsky.[1] Three insights originally motivated his claim. First, he showed that the logical structure of grammars was more complex and difficult to specify than anyone had previously suspected, and yet that normal speakers of a language seem to know an enormous number of complex grammatical rules and applications without an explicit knowledge of what they know. Second, he argued that although languages appear incredibly variable on the surface, they share a common deep logic, or "deep structure," from which the specific rules used by each can be derived by a sort of deductive logic. This, however, further complicates the discovery of the rules, because what is presented to the language learner is only the surface consequences of applying these rules. The rules have to be inferred from this indirect representation. Third, he argued that to learn a logical system of such subtlety and complexity should require extensive trial-and-error experience with explicit feedback, and yet young children rapidly develop a sophisticated knowledge of grammatical rules and their applications in the absence of these.

Many have taken this last point one step further. They argue that even much more extensive experience of the type that children do not get might still be insufficient to allow one to discover the abstract rules that constitute the grammar of a natural language. To put it another way, no one could learn a language in the same way that one learns the names of the presidents, the letters of the alphabet, or the rules of long division. Grammar is too complex, and the rules exhibited by spoken examples only obliquely reflect its logic. Some researchers have argued that even a scientist or logician would be unable to discover the rules of grammar inductively from texts of a language, except by reference to some previously known grammar.[2] This difficulty of discovering the rules, even in theory, is epitomized by the fact that modern linguists still disagree about what is the most appropriate formal description of natural grammatical rule systems. So how could very young children, who are apparently far less sophisticated in their analytic abilities, nevertheless become rapidly competent language users? In essence, then, the argument supporting the existence of an innate Universal Grammar is an argument from incredulity. How is it possible otherwise?

The conclusions appear inescapable. If grammatical knowledge cannot be acquired by inductive learning in childhood, then knowledge about the rules of grammar (that we all end up possessing after just a few years) must come from somewhere else. And if acquiring this knowledge does not depend on experience, then paradoxically the knowledge must already be present in some form *prior* to experience. It must be available as a corpus of already assumed (innately known) rules that need only to be compared to the available input. Those innate rules which predict the structure of the experienced language are adopted; the remainder are ignored.

This is a compelling argument, but innate Universal Grammar (UG for short) is a cure that is more drastic than the disease. It makes sweeping assumptions about brains and evolution that are no less credible than the claim that children are super-intelligent learners.

The critics of the UG theory mainly take issue with what they claim are straw-man assumptions about the language-learning context: a restricted conception of learning as induction, and the claim that language experience provides no feedback. Children are not just fed a sequence of speech inputs and forced to discover the abstract rules of their production. Children's language experiences are embedded in a rich and intricate social context, which provides them with multiple routes to pragmatic social feedback. Moreover, the language interactions that young children engage in are often simplified by the adults, and certain features are exaggerated to make them more salient. In short, the critics argue that no paradox needs explaining, that a general theory of learning may be enough. It is demonstrably true that children learn language in a rich social context, but the claim that learning alone can account for this ability begs as many questions as the innate knowledge theory. We cannot discount the obvious and immense gap that separates what children accomplish easily from what otherwise highly intelligent species and even sophisticated inductive-learning algorithms cannot. This is especially compelling since young children are quite limited in other aspects of their learning abilities. We need to confront the paradox directly. Some kind of prior preparation for language must be present. But what kind, and where? These are questions that must be answered, not avoided, by either gambit.

There is another possibility that has been almost entirely overlooked, and it is the point of the remainder of this chapter to explore it. I think Chomsky and his followers have articulated a central conundrum about language learning, but they offer an answer that inverts cause and effect. They assert that the source of prior support for language acquisition must originate from *inside* the brain, on the unstated assumption that there is no other possible source. But there is another alternative: that the extra support for language learning is vested neither in the brain of the child nor in the brains of parents or teachers, but outside brains, in language itself. To explain what I mean, let me begin by offering a couple of extended analogies that I think portray a similar problem. I'll start with a computer analogy.

" . . . And you'll see why 1984 won't be like *1984.*" Thus ended an award-winning television commercial seen by millions of Americans watching the Super Bowl in January 1984. It was the commercial for Apple Computer Inc.'s new personal computer, the Macintosh, aired just once during the big game then retired to history, its message sent. Apple depicted a hypnotically transfixed audience of human drones staring, mouths agog, at a huge screen on which the image of George Orwell's "Big Brother" intoned his propaganda about world domination and the good of all mankind. Then, through the high-tech corridors and the numbly nodding masses, chased by uniformed techno-storm troopers, an athletic young woman ran, carrying a large hammer. In the midst of the chanting mass she whirled around and let the hammer fly into the screen, shattering the image of the face of Big Brother, and breaking the hypnotic spell.

What Apple was announcing was not just a new computer but a new way of thinking about computing. The company that brought computers from the tinkerer's garage into the average home was unveiling a computer that was "user-friendly." Before this, computers were machines that you had to figure out. To get them to do what you wanted, you had to use *their* logic—the logic of program codes, I/O functions, and endless varieties of commands and names, some of which were acronyms like "DOS," "bios," and "CD" or truncated words like "txt," "cmd," and "bak," but many of which were just programmers' syntactic operators, such as ". . ." and "\\." And if you didn't type the code line just right, or didn't quite know the right spacing, order, or sequence of commands, you were stuck, left trying to figure out what happened to the previous day's work.

In 1984, everyone knew that computers were very complicated machines that only engineers and technicians really understood, and so in order to use them it was inevitable that you needed training or experience or a lot of time to read stacks of manuals. But the gurus at Apple had a different idea—an idea they borrowed from a group of computer visionaries assembled years earlier in a Xerox research project. The idea was to make interacting with a computer intuitive. Instead of having to figure out how to get a program or data file to do what you wanted, why not represent them with the images of familiar objects, virtual objects on a screen made into a virtual desk top? And instead of using commands and codes to control what happens to programs and data, why not just arrange the system so that by manipulating these virtual objects, as you might think of doing if they had been out there on your real desk top, you were activating programs and modifying your data? This, of course, was the birth of personal computers with "object-oriented" interfaces. Instead of having to page through a stack of unreadable manuals, it was possible to experiment, to learn by trial and error, in an environment where your intuitive guesses were likely to be good ones. Apple may not have banished Big (Blue) Brother, but today few personal computers and work stations lack for icons, windows, and pointing devices. Computer operations have been adapted more to people so that people need to adapt less to computers.

Given the complexity of language and its evolutionary novelty, one might imagine that every two-year-old faces an incredibly complicated set of syntactic rules and thousands of words like a computer novice facing an unforgiving "C>," and without technical manuals to help in sorting them out. This seems to be the implicit assumption behind the claims of the proponents of theories of innate grammatical knowledge. If language were invented and just thrust upon us poor users, we might be justified in thinking of it as not very "user-friendly," in which case a built-in on-line instruction manual might be a necessity. But this inverts the actual relationship. The engineering constraints of computer operations are quite unlike any information processes we normally engage in. Operating systems and programming languages are limited by the requirements of the machine architecture more than by the requirements of learnability. But languages have evolved with respect to human brains, not arbitrary principles or engineering constraints. There need be few compromises. The problem faced by a child learning a first language should not be analogous to the problem faced by a computer neophyte trying to learn to use an unforgiving mechanism. Instead, we should expect that language is a lot more like an intuitive and user-friendly interface. Over countless generations languages should have become better and better adapted to people, so that people need to make minimal adjustments to adapt to them. With this alternative model in mind, let's compare the typical competing explanations of language acquisition.

Chomsky argues that much of the child's knowledge of grammar and syntax is not learned in the way words are. I agree. It is *discovered*, though not by introspection of rules already available in the brain. On the surface, it simply appears that children have an uncanny ability to make "lucky guesses" about grammar and syntax: they spontaneously anticipate the ways words work together. I think this appearance of lucky coincidence accurately captures what happens, though it is not luck that is responsible. The rules underlying language *are* acquired by trial-and-error learning; but there is *a very high proportion of correct trials.* Why are kids such "lucky" guessers? Turning this observation around, we might notice that children do *not* sample the full range of possible alternative ways of organizing words. They are strongly biased in their choices. If language were a random set of associations, children would likely be significantly handicapped by their highly biased guessing. But language regularities are not just any set of associations, and children's guesses are biased in a way that usually turns out to be correct. How else could this happen except by endowing children with some foreknowledge of what constitutes a likely correct guess? I think the object-oriented computer operating system example suggests the alternative. Things could be arranged so that intuitive guesses are more likely to work.

Let me illustrate the logic of this alternative answer to the riddle with another story. Imagine a gambling establishment and the following ingenious scheme to defraud it. A disgruntled casino employee has decided that he wants to get back at the owners by making them lose big. He does not want to leave any way for the owners to get back their losses even if they catch him. So instead of stealing or embezzling the money, he decides to help an unsuspecting gambler break the bank. In order to accomplish this he needs to find a gambler who has characteristic betting habits, or superstitions—for example, someone who always places large bets on numbers containing the digit 5. If the employee can rig the Roulette wheel so that fives appear far more often than chance, he can turn the unsuspecting gambler into a millionaire. The gambler has no special knowledge of the future numbers that will turn up, even though the games have been rigged to conform to his biased guesses. His lucky guesses are not based on innate knowledge about the wheel or about the future. It appears *as though* the gambler has foreknowledge of the outcomes, but in reality he has only a sort of "virtual knowledge," as seen in hindsight. He might even come to believe that he has a clairvoyant ability to see into the future, but irrespective of what he thinks he knows, he is simply being fooled. The point is that one cannot always be sure that apparent knowledge, inferred from successful predictions and guesses, is really what it seems.

Finally, let me offer an analogy that every clever teacher or animal trainer will find familiar. The first time I visited an aquarium where trained dolphins performed, I was astonished by the way they had been taught to jump high out of the water, turn somersaults in midair, and "stand" on their tails. Later, when I had become acquainted with some of the staff, I asked if dolphins could be trained to do this because they were particularly smart. No, they said, dolphins spontaneously produce similar behaviors in the wild and the trainers simply taught them to exaggerate and modify things that they already tended to do naturally. If the trainers had tried to teach them to perform a completely novel behavior, without considering the dolphins' innate predispositions, the trick would have been very difficult if not impossible to train. Successful animal trainers carefully select the tricks they attempt to train their subjects to perform to fit with their subjects' spontaneous behavioral tendencies. What learning psychologists call "shaping" of an operant behavior must begin with a spontaneous behavior to be shaped. Even very unusual and exotic behaviors can be taught, if one begins with what an animal already tends to do, then step by step extrapolates and generalizes to new variations on the theme. The training will be minimized because the animal already"knows" what to do to please its trainer and receive a reward.

Each of these examples offers a twist on the Macintosh (and Windows) strategy. The learning is not so likely to fail if what needs to be learned is "user-friendly," organized in a way that the learner is predisposed to thinking and working. One way to facilitate children's inspired "guessing" about language, and thus spare them countless trials and errors, would be to present them with a specially designed language whose structure anticipated their spontaneous guesses. If we could study children in "the wild" to discover their natural tendencies, we could then design the perfect language that took advantage of what kids do spontaneously. Learning this artificial language would then be more like trying on new clothes and discovering that they just happen to fit, as opposed to going on a diet in order to fit into clothes that don't. Children's minds need not innately embody language structures, if languages embody the predispositions of children's minds!

Okay. Although this is a very interesting fantasy, we are *not* brilliant language trainers with complete insight into our children's minds. We have few enough insights into our own minds. And we don't consciously tailor language for children, except for the rather lame simplifications and exaggerations we call "baby talk" and "motherese." We don't *design* language at all. It "designs" itself. Languages just change spontaneously over the course of many generations. Every effort to design a language has flopped. If no one ever studied children in order to design language for them, then how could any sort of "preestablished harmony" between language and children ever get set up? It seems as if we have to postulate the same kind of miraculous accident for language as has been suggested for explaining the origins of a language organ. Do we gain any more explanatory credibility by postulating infant-friendly languages as opposed to language-savvy infants? How could such a preestablished harmony arise if no one is "throwing the game"?

Someone or something *is* throwing the game, and I don't mean a divine designer. Languages don't just change, they *evolve*. And children themselves are the rigged game. Languages are under powerful selection pressure to fit children's likely guesses, because children are the vehicle by which a language gets reproduced. Languages have had to adapt to children's spontaneous assumptions about communication, learning, social interaction, and even symbolic reference, because children are the only game in town. It turns out that in a curious sort of inversion of our intuitions about this problem, languages need children more than children need languages.

In short, we failed to notice that a flurry of adaptation has been going on *outside* the brain. The reason we haven't considered the relevance of this process before is that we tend to think on a human scale. Focusing on a lifetime scale and the rapid changes that take place in the few years in which a child learns a first language, we naturally think of this as the flexible half of the equation and of language as though it is some fixed entity. But from an evolutionary perspective, the situation is just the reverse. Biological change is vastly more slow and inflexible than language change. Brain evolution takes place on a geological time scale. Even slight changes probably take hundreds of thousands of years to become widely represented in a species, and the basic architecture of brains has been remarkably conserved since the origins of vertebrates. Languages, on the other hand, can become unrecognizably different within a few thousand years. Language evolution is probably thousands of times more rapid than brain evolution. Such a vast difference in evolutionary mobility suggests that we may have assumed that the wrong half of the evolutionary equation contained the critical variables.

The Other Evolution

The world's languages evolved spontaneously. They were not designed. If we conceive of them as though they were invented systems of rules and symbols, intentionally assembled to form logical systems, then we are apt either to assign utility and purpose where there is none, or else to interpret as idiosyncratic or inelegant that for which we cannot recognize a design principle. But languages are far more like living organisms than like mathematical proofs. The most basic principle guiding their design is not communicative utility but reproduction—theirs and ours. So, the proper tool for analyzing language structure may not be to discover how best to model them as axiomatic rule systems but rather to study them the way we study organism structure: in evolutionary terms. Languages are social and cultural entities that have evolved with respect to the forces of selection imposed by human users.

The structure of a language is under intense selection because in its reproduction from generation to generation, it must pass through a narrow bottleneck: children's minds (a relationship depicted in cartoon fashion in Figure 4.1). Language operations that can be learned quickly and easily by children will tend to get passed on to the next generation more effectively and more intact than those that are difficult to learn. So, languages should change through history in ways that tend to conform to children's expectations; those that employ a more kid-friendly logic should come to outnumber and replace those that don't. From this perspective, children don't have to be particularly smart, and parents don't have to be particularly gifted teachers. The limitations of the learners and the teachers are an unavoidable part of the ecology of language transmission. Language structures that are poorly adapted to this niche simply will not persist for long.

If, as linguists often point out, grammars appear illogical and quirky in

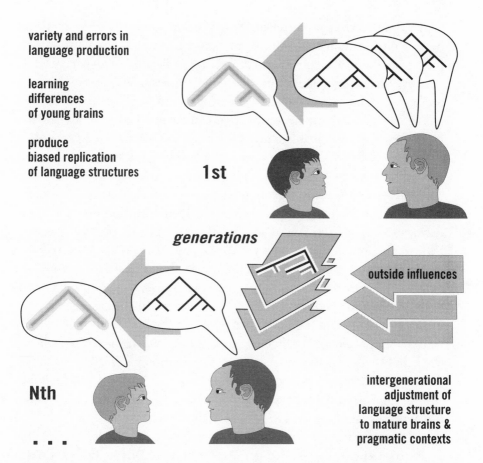

variety and errors in language production

learning differences of young brains

produce biased replication of language structures

1st

generations

outside influences

Nth

· · ·

intergenerational adjustment of language structure to mature brains & pragmatic contexts

Figure 4.1 *Cartoon suggesting the possibility that languages must go through the "filter" of children's reduced associative learning and short-term memory constraints in order to be passed on most effectively from one generation to the next with any degree of fidelity. Children selectively hear some structures and ignore others, and so provide a major selection force for language structure that is "child-friendly."*

their design, it may only be because we are comparing them to inappropriate models and judging their design according to functional criteria that are less critical than we think. Instead of approximating an imaginary ideal of communicative power and efficiency, or following formulae derived from an alleged set of innate mental principles, language structures may simply reflect the selection pressures that have shaped their reproduction.

In some ways it is helpful to imagine language as an independent life form that colonizes and parasitizes human brains, using them to reproduce. Versions of this idea have been independently proposed by many thinkers over the years. My own view is probably closest to that proposed in a recent paper by Morton Christiansen.[3] The key ingredient that unites our approaches is the recognition of a co-evolutionary dynamic between language and its host. By imagining language as a parasitic organism, we can come to appreciate the potential for conflicting reproductive interests, where some language features might occur at the expense of the host's adaptations, and the possibility that many features may have more to do with getting passed on from generation to generation than with conveying information.

Of course, languages are entirely dependent on humans, and are not separate physical organisms, replete with their own metabolic processes and reproductive systems. And yet their very different form obscures deep similarities to living processes. They might better be compared to viruses. Viruses are not quite alive, and yet are intimately a part of the web of living processes. Viruses are on the liminal border between the living and nonliving because they lack organs in any normal sense, including any vestige of metabolic or reproductive systems. They are minimally packaged strings of DNA or RNA that regularly happen to get themselves insinuated into cells that mistake them for their own nucleotides and haphazardly replicate them and transcribe their base sequences into proteins. Though relatively inanimate collections of macromolecules in and of themselves, the information implicit in them nevertheless evolves and adapts with frightening efficiency, as recent epidemics so vividly illustrate.

From this perspective it is not so far-fetched to think of languages a bit as we think of viruses, neglecting the difference in constructive versus destructive effects. Languages are inanimate artifacts, patterns of sounds and scribblings on clay or paper, that happen to get insinuated into the activities of human brains which replicate their parts, assemble them into systems, and pass them on. The fact that the replicated information that constitutes a language is not organized into an animate being in no way excludes it from being an integrated adaptive entity evolving with respect to human hosts.

The parasitic model is almost certainly too extreme, for the relationship between language and people is symbiotic. There are many familiar examples of parasitic symbionts in nature. Two well-known examples are the microorganisms that colonize human and termite digestive systems. Both aid their host's digestive processes and neither parasite nor host could survive without each other. The endosymbiont in termites secretes an enzyme that helps break down cellulose, a trick no eukaryotic organism seems to have evolved on its own. Such useful parasites are not just passively replicated. Because of its critical importance to termite life, termites have even evolved ways to ensure their symbiotic parasite's reproduction and passage into other hosts. When a growing termite molts and sheds its old exoskeleton, this infected shell is immediately eaten either by its owner or by other members of the colony, passing on copies of the parasite in the process. Symbiotic relationships naturally evolve to bring the reproduction of the two organisms into closer alignment, since this is the best way to insure the success of both.

In the case of language and human beings, by analogy, we should not be surprised to find complex human adaptations to language on the one hand, whose purpose is to ensure that language is successfully replicated and passed from host to host, and language adaptations to children on the other, whose purpose is to make languages particularly "infective" as early as possible in human development. Modern humans need the language parasite in order to flourish and reproduce, just as much as it needs humans to reproduce. Consequently, each has evolved with respect to the other. Each has become modified in response to the peculiar constraints and requirements of the other. The adaptation of the parasite to its hosts, particularly children, provides the basis for a theory of prescient language learning. Though this is a caricature, it is no less so than the nativist and empiricist alternatives, and it captures much more accurately the dynamic push and pull of biases that have shaped both languages and the human brain.

Of course, the analogy with an organism treats language as far more autonomous and individuated than is warranted, since any particular language is at most a fuzzy statistical entity and language as a whole is even less well defined. Yet there are a number of useful parallels. For example, the words and rules that constitute a language are not simply members of a collection, they are highly organized and interdependent as are the genes and organs of an organism. From a single individual's perspective, the analogy is even stronger. The logical structure of languages is replicated (acquired) and passed on as a complete system, not just a collection of words. Even though it may be learned word by word and phrase by phrase, what is acquired only becomes a language when the prescribed ways of using these words have been internalized to the point that one is theoretically capable of knowing how to phrase all thoughts for which words are available and able to determine the grammaticality of any novel sentences of known words. Though children may take years to develop an extensive vocabulary, they quickly master the central core of rules, which will give all new words a predefined grammatical function. We might think of this self-sustaining core, consisting of a grammar and syntax and a sufficient number of words to determine all critical word classes, as the completed organ system of an embryonic organism. Additional vocabulary items enlarge the whole, cause it to grow and mature, but do not alter its basic organization.

Languages are abstractions. They are fuzzy collections of behaviors that happen to be reasonably well described by logical rule systems. And although what one person knows about his native language may in some large measure be describable this way, what an entire population of English or Japanese speakers shares in common is only statistically clustered. Like most characteristics of biological species and social groups, the common language that links a social group might be described as a collection of similar but not identical languages. Like a species, then, the language of a whole society is a natural reservoir of variation, with some features becoming less diverse and others becoming more diverse over time.

Variation in language use points to the final and most relevant parallel: languages evolve. As a language is passed from generation to generation, the vocabulary and syntactic rules tend to get modified by transmission errors, by the active creativity of its users, and by influences from other languages. Isolated language communities which begin with a common language will follow divergent patterns of spontaneous change. Eventually words, phraseology, and syntax will diverge so radically that people will find it impossible to mix elements of both without confusion. By analogy to biological evolution, different lineages of a common ancestral language will diverge so far from each other as to become reproductively incompatible. This immiscibility is reflected in the brains of fluent bilingual speakers, who tend to have different neural localization for the two languages they know. This is demonstrated by the fact that in such individuals stroke, electrical stimulation, and other neuropsychological disturbances can affect each language independent of the other. The physical separation is almost certainly dictated by the fact that the two languages would otherwise compete for use of the same neural systems and so interfere with each other (see Chapter 10). At the level of what an individual knows, a language is very much like one's own personal symbiotic organism.

Similar analogies have been suggested as a way to describe the dynamical interdependence between many biological and social evolutionary processes. Although evolutionary models are not new to the social sciences, and have an even longer history in linguistics than in biology, only recently have researchers begun to conceive of this relationship as though artifacts, techniques, customs, and even ideas and beliefs were separate organisms. This is more than just a metaphor. There is an important sense in which artifacts and social practices evolve in parallel with their living hosts, and are not just epiphenomena. They must be reproduced from one generation to the next and replicated as each new person learns them, copies them, emulates them, or is forced to conform to them. But because of this there is a potential for innovations and errors like recombinations and mutations in biological evolution, that introduce new variety into the process over time. Biases can creep into the process and affect what does and does not get reproduced, just as natural selection favors certain genetically specified traits. Those bits of copied cultural information (Richard Dawkins has dubbed them "memes" as a cultural analogue to genes) which happen to lead to an increased probability of being reproduced will persist longer, and spread over time to be used by more individuals, than those that don't potentiate their own reproduction. Sources of selection that determine what does and does not get passed to future generations not only include the utility of these memes and their consequences, but also the biases imposed by their mode of transmission (human minds) and the peculiarities of their cultural ecosystems (systems of other memes).

Language is a social phenomenon. To consider it in purely formal, psychological, or neurobiological terms is to strip away its reason for being. Social phenomena like language cannot be adequately explained without appealing to a social evolutionary dynamic as well as a biological one. The source of information that is used to "grow" a language lies neither in the corpus of texts and corrections presented to the child, nor in the child's brain to begin with. It is highly distributed across myriad interactions between children's learning and the evolution of a language community.

The key to understanding language learnability does not lie in the richly social context of language training, nor in the incredibly prescient guesses of young language learners; rather, it lies in a process that seems otherwise far remote from the microcosm of toddlers and caretakers—language change. Although the rate of social evolutionary change in language structure appears unchanging compared to the time it takes a child to develop language abilities, this process is crucial to understanding how the child can learn a language that on the surface appears impossibly complex and poorly taught. The mechanisms driving language change at the sociocultural level are also responsible for everyday language learning.

Emerging Universals

I believe that recognizing the capacity of languages to evolve and adapt with respect to human hosts is crucial to understanding another long-standing mystery about language that theories of innate knowledge were developed to explain: the source of language universals. Grammatical universals exist, but I want to suggest that their existence does not imply that they are prefigured in the brain like frozen evolutionary accidents. In fact, I suspect that universal rules or implicit axioms of grammar aren't really stored or located anywhere, and in an important sense, they are not *determined* at all. Instead, I want to suggest the radical possibility that they have emerged spontaneously and independently in each evolving language, in response to universal biases in the selection processes affecting language transmission. They are *convergent* features of language evolution in the same way that the dorsal fins of sharks, ichthyosaurs, and dolphins are independent convergent adaptations of aquatic species. Like their biological counterparts, these structural commonalities present in all languages have each arisen in response to the constraints imposed by a common adaptive context. Some of the sources of universal selection on the evolution of language structures include immature learning biases, human mnemonic and perceptual biases, the constraints of human vocal articulation and hearing, and the requirements of symbolic reference, to name a few. Because of these incessant influences, languages independently come to resemble one another, not in detail, but in terms of certain general structural properties, and any disruption that undermines a language's fit with its hosts will be selected against, leading to reconvergence on universal patterns. The forces of selection should be most intense at the level of the logical structure of a language (grammar and sytax), because features at this level are most critical to its successful replication, but less intense converging forces should also influence features as apparently fluid and culture-specific as word reference and sound-meaning relationships.

In fact, probably the best understood example of this co-evolutionary convergence process, though seldom recognized as such, is a case of convergent word reference: the evolution of color terms in different languages and different societies. This is an ideal example because it combines the apparently complete arbitrariness and idiosyncrasies associated with assigning a name to a position on a qualitative continuum, on the one hand, with a well-understood human neurobiological universal, on the other. The assignment of color term reference is essentially unconstrained because: (a) the name of a color can be any combination of human vocal sounds; (b) the human eye can see every gradation of color between certain limits of wavelength; and (c) people can assign any term to any point on the visible light spectrum. So any association between utterable sound and perceivable light frequency is possible, in principle. Surprisingly, in the real world's languages, the mappings of color terms to light frequencies are not only limited, they are essentially universal in many respects. In this regard, they are a lot like other language universals. But does this mean color terms are built into the brain as an intrinsically modular and categorical form? Or have all the color terms in the world derived by diffusion from a single parent language? Though these answers are logically possible, the evidence does not support either. The answer is far more interesting and sheds important light on other language universals.

Interest in this problem was sparked by the work of Brent Berlin and Paul Kay, who in a series of cross-cultural comparisons demonstrated an interesting regularity in the presence or absence of terms for different colors in different societies.[4] What they showed was a hierarchic sequence in the order in which color terms were included or not in different languages. In societies where they found the fewest terms for colors, there were at least three terms: the equivalents of black (dark), white (light), and red. In languages where there were only two color terms in addition to black and white, these always were red and green. In languages where there were three or four color terms, either yellow or blue or both were added. The pattern became progressively less predictable as languages with more terms were considered, but in general, additional color terms were acquired in complementary pairs fitting between prior color terms, with the "primary" colors (e.g., red, green, blue) leading the pack (see Figure 4.2). Of course, the fewer the terms, the more broadly they are applied, so that societies lacking a term for "blue" might refer to blue objects as "dark green." This, however, brings up an additional finding discovered subsequently by Elenor Rosch and her colleagues.[5] The best physical exemplars of particular color terms, irrespective of how many are present in the language, are essentially the same when chosen from an arbitrary set of color samples. In other words, though the boundaries between one color and the next for which one has a term tend to be graded and fuzzy, i.e., not categorical, color terms do apparently have something like a category center; a best red or best green. Surprisingly, the best red and the best green, whatever the terms used, are essentially agreed upon by people from around the world. Though the words themselves are arbitrary and the colors continually grade into one another, words are not arbitrarily mapped to points on the color spectrum. They are universally constrained.

To understand why this might be so, we need to consider two things: the way the brain transforms light frequency into the subjective experience of color, and the way this might influence the nongenetic evolution of color-word/color-perception associations in a society. The important fact about color perception is that color is in a sense a feature that is created by the brain as a means of maximizing distinctive experiences of photons striking

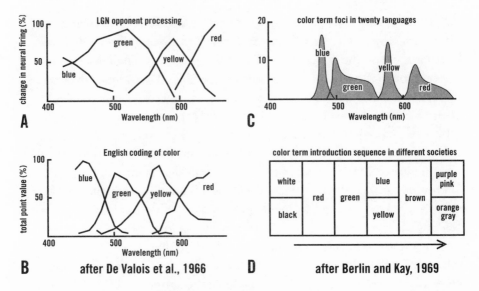

Figure 4.2 *The factors underlying the evolution of near universality of color naming. Correspondences between neural and linguistic coding of color. (A) Response characteristics of neurons in the lateral geniculate nucleus of the thalamus (LGN), the first central nervous system relay for visual inputs, in response to different wavelengths of light. Different neurons respond preferentially to different wavelengths with essentially no overlap in response to opposite colors (e.g., red vs. green, blue vs. yellow). (B) When presented with color samples that span the visual spectrum and asked to rank them according to closeness to four primary colors, English speakers also show nearly categorical separation of opposite colors and close correspondence to neural response patterns. (C) Wavelengths of samples chosen by informants that "best" correspond with each of twenty languages' primary color terms. The exemplars of these category centers for each primary color term map across languages, are categorically distinct from one another both across and between languages, and correspond with neural category centers. Their categorical distinctness is more pronounced than is the neural distinction that underlies it. (D) Sequence from left to right in which languages tend to add new color terms.*

the cones of the retina in blended streams of different wavelengths. The neural computational logic of this enhancement process is called *opponent processing* because it pits the signals from each of the three different types of cones (photoreceptors in the retina with different wavelength sensitivities) against one another, to derive a difference signal. This second-order signal best corresponds with the color discriminations we actually see (which often do *not* correspond to specific matching frequencies of incoming light).

It is this oppositional sampling process that determines color complementarity (i.e., that red is opposite green and yellow is opposite blue on a color wheel representing all the colors).

It is also the source for biases that underlie the universality of color naming. Not that we see these colors as discrete, or even that we are biased only to name certain points on the spectrum (the advertising and paint industries have proven that thousands of color distinctions can be given individual names), but rather that this will bias the co-evolution of color names. The patterns of errors in use and transmission of color terms will be biased like a loaded die, so that over time the linguistic references will converge to match the neuropsychological foci of perceptual experience. It is a case of neurological bias acting as a relentless selective force in social evolution.

To show how this biasing of language change might work, consider the thought experiment of coining a new name for the color of a specific object, say the color of a kind of moss. Let's call this color "grue."[6] Assume also that there is no color term other than red preexisting in the language. Since the moss is unlikely to be exactly the same green as is perceptually opposite to the category center for red, this new term will begin with a reference that is idiosyncratic and specific to that object. But it is unlikely to stay that way if it persists for very long in the language. This is because both the memory of the color of the moss and the ability to communicate exactly what color is meant by "grue," particularly in the absence of the original moss, are prone to error and imprecision. In the absence of any biases these errors would be random, and we would expect that over time the reference of grue would drift away at random from the original reference. But since the perceptual spectrum has high and low points, so to speak, the drift will *not* be entirely random. Errors in remembering and communicating the original reference of grue will tend to be biased toward the perceptually most salient and the pragmatically most contrasting opposite color to red: green. These errors in transmitting the precise color reference are the social and psychological analogues to mutations. The biases our brains introduce in the way we remember and reproduce the reference are the analogues to natural selection favoring some variants, but not others. As a result, the reference will tend to evolve in adaptation to human nervous systems and to the pragmatic constraints of human uses. Grue will eventually evolve to refer to green if there is no term for green, or blue if there is, or to refer to the most salient intermediate if both already exist.

In summary, the universality of color term reference is an expression of shared neurological biases, but—and this is a crucial point—the translation of this biological constraint into a social universal is brought about through the action of *nongenetic* evolutionary forces. In the absence of a social evolution process, color term references would remain forever idiosyncratic to individual users and brief historical epochs in each society. The mediation of this relationship by evolutionary tendencies amplifies the effects of slight biases in much the same way as a statistical analysis amplifies our ability to identify small biases affecting large samples. Because both errors of indicating exemplary objects and ambiguities of describing what is meant are biased in similar ways, they each amplify the other in tens of thousands of repeating sampling errors from generation to generation.

In many ways, primary color term reference relationships are as invariant and universal in character as is any aspect of the deep logic of human grammars, and yet it is clear that color terms themselves are not built into the brain, and that choice of what color to designate with a particular word is not dictated by any mental compulsion. What is universally built into the brain is a rather subtle set of perceptual biases that have none of the categorical and symbolic properties of words. Now, this may at first appear to be a comparatively trivial example of some minor aspect of language, but the implications for other aspects of language evolution are truly staggering. It demonstrates that even weak biases, *if constantly and invariably present*, can produce social evolutionary consequences that appear so ubiquitous as to be considered completely determinate. Consider the subtlety of the differences in perceptual salience of colors and how little constrained we are in assigning names to colors. Such slight biases in the introduction and retention of naming errors can hardly be a source of intense selection pressures. Yet consider the effect: nearly universal consistency of color reference. Phenomenologically, this qualifies as a true linguistic universal, even though it is neither a necessary design feature of language nor even an innate linguistic category.

If tiny biases can produce near-universal language features, then imagine what strong biases might do! How much more powerful might the selection pressures be for some of the more limiting constraints on human attention, working memory, sound production, and automation of functions? The effects of such fundamental and universal cognitive boundary conditions on the social selection of syntactic habits must be profound. And what about biases from our visual-manual-dominated primate heritage? Is it any wonder that so many descriptive and explanatory devices in languages throughout the world borrow heavily from visual and manipulative analogies? Some have even suggested that these ubiquitous patterns of expression, such as using spatial metaphors to refer to semantic dimensions (e.g., "higher truth," "further developed," "distantly related") are the result of innate cognitive concepts.[7] But once we recognize this evolutionary process as the primary source behind the universality of linguistic features, and abandon the assumption that to be universal a feature must be hard-wired into the brain, it becomes evident that we may have vastly underestimated the range and variety of language universals, or near universals. Benjamin Lee Whorf suggested that different languages might be so radically divergent that many basic modes of thinking expressed in one might be found to be utterly untranslatable. I think, in contrast, we should not be surprised by the extent to which even high-level conceptual patterns of linguistic representation and discourse share near-universal features in most languages, simply because we are all members of the same species, sharing many common perceptual, behavioral, and emotional biases.

Whether or not we can draw an exact parallel for all aspects of the universal features of grammars, it is clear that no matter which formal, cognitive, or phonological universals of language we consider, they probably arose as a result of biases affecting nongenetic evolutionary processes of language in a manner that is similar to the way that the designation of colors in different languages tends to converge on the same referents and multiply color terms along parallel lines. There is nothing necessary or predetermined about such a process. It is just exceedingly unlikely that, even if introduced, any exceptions to these tendencies will survive for more than a very brief period of time. Language universals are, in this interpretation, only statistical universals, but supported by the astronomical statistics of millions of speakers over tens of thousands of years. They are, despite their almost epiphenomenal origin, for all practical purposes categorically universal.

The theory that there are innate rules for grammar commits the fallacy of collapsing this irreducible *social evolutionary* process into a static formal structure. It not only ignores the effect of forces that could modify word formation and syntax over time, and of factors promoting converging or parallel trends, it also ignores the forces that stabilize language structure and are thus responsible for maintaining concordant use among contemporary individuals. All this is "given" without explanation if we assume universally fixed instructions for deriving language structures. My point is not that these instructions are located outside the child in the language community, but that there are no instructions (except in the imaginations of those who write texts on innate or "proper" grammar and syntax). A set of basic design rules or principles is simply superfluous. Once we accept the fact that the link from psychological universals to linguistic universals is exceedingly indirect at best, it becomes evident that the explanation for language structure will require that we broaden the search to include a wide variety of biasing factors. Language structures at all levels are the products of powerful multilevel evolutionary processes, to which innate mental tendencies contribute only one subtle source of the Darwinian selection biases.

Thus, even if all the paradoxical problems posed by Universal Grammarians—the theoretical unlearnability of grammars, the apparently miraculous acquisition abilities of children, and the universality of often illogical abstract rules of grammar and syntax—are real, we nevertheless do not need to turn to a theory of innate linguistic knowledge to explain them. They can be understood as the products of convergent social evolutionary trends, as spontaneous parallel adaptations of language structure to the unavoidable and ubiquitous limitations and biases provided by human brains, especially children's brains. Languages have adapted to human brains and human brains have adapted to languages, but the rate of language change is hundreds or thousands of times more rapid than biological change. The underlying architecture of all languages remains highly conserved across great distances in space and time, and despite modifications and corruptions remains learnable, because the less learnable variants—less well adapted to human minds—are continually selected against. Even if the syntax of some natural language was once as "unhuman" as the syntax of low-level computer languages, it would not have persisted long. Slow learning rates and high error rates during use would have slated it for rapid extinction or selected for variants in its structure that eased the burdens. Human children appear preadapted to guess the rules of syntax correctly, precisely because languages evolve so as to embody in their syntax the most frequently guessed patterns. The brain has co-evolved with respect to language, but languages have done most of the adapting.

Better Learning Through Amnesia

The paradoxical ease with which young children acquire their knowledge of language is not the only aspect of children's behavior that has seemed to offer support for the concept of a language instinct. An apparently equally compelling reason for an innate, highly specialized, and modular acquisition mechanism is that children seem to do a lot better than adults at acquiring language. They appear to learn it in a very different way. In fact, even before Chomsky, researchers had suggested that there was a "critical period" for acquiring language. This concept has old roots. Sigmund Freud's theories of the formative influence of very early sexual and socialization experiences on adult personality patterns and disorders have long provided a precedent for theories about the special nature of early experiences. Students of natural animal behaviors (ethologists) have provided numerous examples of special learning patterns in very young animals. One of the founding fathers of the field, Konrad Lorenz, demonstrated, for example, that newly hatched geese exhibited a very specialized rapid learning of the features of their caretaker, whether goose or human. This was called "imprinting," and after they imprinted on a caretaker shortly after hatching, the birds exhibited an intense emotional attachment and would stay as close to the imprinted animal, person, or object as possible at all times. A more relevant example, discovered subsequently and still the subject of intense study, is the acquisition of local "dialects" of birdsong. Young songbirds learn the sound of their local variant of the species' song shortly after hatching, and appear to use this as a sort of perceptual "template" against which to compare and match their own singing when they finally begin singing at puberty. Like imprinting, the acquisition of this perceptual exemplar of the local song occurs rapidly and during a specific brief stage of development. If no appropriate inputs are presented during this period in their lives, songbirds lose the ability fully to supplement what was not acquired, and their later behavior is significantly affected. They are sensitive to a very restricted class of inputs during a critical period when their ability to learn those details is greatly amplified. For this reason it is often called "prepared learning" or even "superlearning."

The similarities to how children acquire language are obvious. In the context of otherwise poorly developed learning abilities, children show one glaring, time-dependent, island of competence for learning language. The ages that children tend to pass through corresponding stages of pre-language and language development are remarkably similar from individual to individual, despite very different rearing contexts, and young children's acquisition of a second language tends to be more rapid and more complete than what most adults achieve. Children's perceptual responses to speech sounds appear to become "committed" to features that are present in their first language by as early as the first year of life, and this tends to produce a permanent difficulty in distinguishing phonemes (individual sound elements of syllables) that they did not hear or produce during their childhood. Children who for some reason were deprived of the opportunity to learn a language when young, such as so-called feral or wild children raised in contexts with minimal human contact, appear to demonstrate considerable difficulty acquiring a normal language facility at older ages and tend to exhibit permanent language deficits. Finally, there have been numerous reports which demonstrate the difference in the effects of brain damage on the language abilities of younger versus adult patients. Even surgical removal of most of the left hemisphere, including regions that would later have become classic language areas of the brain, if done early in childhood, may not preclude reaching nearly normal levels of adult language comprehension and production. In contrast, even very small localized lesions of the left hemisphere language regions in older children and adults can produce massive and irrecoverable deficits. All these facts seem to provide powerful arguments for the existence of a specialized language acquisition device (LAD) deployed during a critical period in children's development and later shut off. This seems like the obvious human counterpart to a songbird's instinct to sing a particular type of song and rapidly acquire information about how to modify it during a special window of opportunity in early development.

However, the significance of these patterns may not be quite what it seems. We have seen how the grammatical and syntactic structures most commonly represented in languages may have been selected with respect to the constraint of children's brains. Could a parallel inversion of the classic interpretations of these patterns of children's development shed new light on this apparently critical period? A new clue to an alternative approach has again been provided by efforts to teach a languagelike communication system to chimpanzees.

The most advanced symbolic capabilities demonstrated by any nonhuman species to date appear to have developed almost by accident in a very young chimp named Kanzi.[8] Kanzi, who is a *bonobo* or pygmy chimpanzee, as opposed to a common chimpanzee, joined the chimp language studies of Sue Savage-Rumbaugh and Duane Rumbaugh a few years after the studies involving the common chimps Sherman, Austin, and Lana summarized in Chapter 3. Kanzi developed the ability to communicate using the same visual lexigram system that Sherman and Austin were trained to use. But Kanzi has far outshined his predecessors. He is now capable not only of communicating symbolically with a lexigram keyboard but has also demonstrated sophisticated comprehension of normal spoken English, including the ability to analyze a variety of grammatical constructions. For example, Kanzi has demonstrated nearly 90 percent correct responses to novel, pragmatically anomalous, but syntactically correct requests to manipulate foods, object, or tools ("Put the soap on the apple"). This demonstrates that he is not merely using semantic analysis to guess at the meaning of the sentences. He even appears to have spontaneously regularized lexigram combinations to produce a minimal but consistent syntactic order in his output.[9] These abilities alone pose a significant challenge for theories which argue that language abilities are unattainable without some innate knowledge to jump-start the analysis of grammar. But this is not what I want to focus on and is not the most devastating challenge Kanzi offers to the nativist, Chomskyan perspective. Although we might suspect that Kanzi's more sophisticated abilities are a result of continual improvements in training methods as the experimenters have become more savvy at training chimps, in fact, Kanzi's success cannot be attributed to experimental sophistication. Kanzi learned to understand speech and use lexigrams symbolically without explicit training.

Kanzi was raised by a foster mother, Matata. It was Matata who was the subject of the experimenters' training efforts, while Kanzi was clinging to her and climbing on her during the process. In the end, Matata was a good foster mother but a very poor language learner. Kanzi, on the other hand, surprised everyone by demonstrating that all the while his mother was struggling in vain to figure out what these experimenters were on about, he was spontaneously, even vicariously, picking up on the game while playing and crawling about his mother and generally being more of a distraction than a participant. When the experimenters decided to turn their attention to him, at a point when they assumed he might now be old enough to be trainable, he showed them that he already knew most of what they were planning to teach him and more. The incidental student was far more adept than any previous chimpanzee at the language tasks explicitly posed to him for the first time, and he has continued to develop and use these abilities in the years since. So, whereas Sherman and Austin required very explicit and carefully structured training to overcome natural learning predispositions that worked against their discovery of the symbolic reference associations of the lexigrams they were taught, Kanzi avoided these cul-de-sacs and crossed the same cognitive threshold supported mostly by his own spontaneous structuring of the learning process.

There are two possible interpretations of Kanzi's success. One is that it derives from a species (or subspecies) difference. Perhaps bonobos are just innately better at language-type tasks. Sue Savage-Rumbaugh has even suggested that bonobos in the wild might engage in spontaneous symbolic communication of a sort, though nothing quite so sophisticated and languagelike has yet been demonstrated. There is, however, a somewhat more conservative, easier to defend, and to my mind far more intriguing possibility. Kanzi learned better simply because he was so immature at the time. This possibility is more intriguing because of its implications for human language development and evolution. It forces us to turn our attention away from an essentialist perspective, focused on the contribution of something intrinsic to the species (i.e., an innate language competence or predisposition), and to pay attention to the relevance of maturational factors. If Kanzi benefited in symbol acquisition and in the development of speech comprehension as a result of his youth at the time of exposure, then he appears to exhibit a critical period effect, as though some special language-learning mechanism is activated at this time of life.

It is not surprising that the other chimps who have been trained in languagelike communication were not quite so immature. Compared to human infants, chimps are relatively mature at birth, roughly equivalent to a human child of one year of age, so the comparable age of peak acquisition in children would be just shortly after birth for a chimp. But like a human toddler, young chimps need to be constantly in the care of their mothers (or a very good substitute). There is another pragmatic reason why slightly older chimps were preferred. Previously, chimpanzee infants have been found to be rather poor at learning to perform complicated tasks, and so were assumed to be poor candidates for language training as well. Just like young children, they have poor memories for details, are extremely distractible and difficult to motivate, and they get bored easily. Of course, for children, this parallels an apparent advantage when it comes to language learning, and this seems to be the case for Kanzi as well. There are other ways in which Kanzi's abilities resemble the patterns that children exhibit and which are characteristic of what we would consider to be critical period effects. Sue Savage-Rumbaugh notes (in personal communication) that Kanzi seems to have a far better sense of what is and is not relevant to symbolic and linguistic communication. He attends to the appropriate stimulus and context parameters whereas other chimps, trained at older ages, seem to need a lot of help just recognizing what to pay attention to. Like a child attuned to the phonemes of the local language, Kanzi's whole orientation seems to have been biased by this early experience.

But observing an apparent critical period effect in Kanzi and then inferring that some special language-learning mechanism is activated at this time of life ignores a troubling contradiction. The glaring problem with this explanation as applied to Kanzi is that chimps in the wild do not learn a language. So, there is no reason why Kanzi's brain should have evolved a language-specific critical period adaptation. If the critical period effect is evidence of a language acquisition device, then why should an ape whose ancestors never spoke (and who himself can't speak) demonstrate a critical period for language learning? Kanzi's example throws a monkey wrench into the whole critical period argument as it applies to language acquisition in children. If early exposure to language is even part of the explanation for

Kanzi's comparatively exceptional language acquisition, then it must be attributable to something about infancy in general, *irrespective of language*. And, if some nonspecific feature of immaturity accounts for Kanzi's remarkable success, then it must at least in part also account for human children's abilities as well. But how?

In a famous paper produced shortly after Chomsky suggested that grammars might be unlearnable, a linguist-philosopher named Gold tried to provide a more rigorous proof of Chomsky's argument.[10] He provided a logical proof which concluded that, without explicit error correction, the rules of a logical system with the structural complexity of a natural language grammar could not be inductively discovered, even in theory. What makes them unlearnable, according to this argument, is not just their complexity but the fact that the rules are not directly mapped to the surface forms of sentences. They are instead embodied in widely distributed word relationships, and are applied to them recursively (i.e., rules are applied again and again to the results of their application). The result is that sentences exhibit hierarchic syntactic structures, in which layers of transformations become buried and implicit in the final product, and in which structural relationships between different levels can often produce word-sequence relationships that violate relationships that are appropriate within levels. From the point of view of someone trying to analyze sentence structure (such as a linguist or a young language learner), this has the effect of geometrically multiplying the possible hypothetical rules that must be tested before discovering the "correct" ones for the language. Without explicit error correction, an astronomical number of alternative mappings of word relationships to potential rules cannot be excluded, especially in the limited time that children appear to take to master their first language. Few contemporary theories rely on this argument alone to support claims of innate grammatical knowledge, but it has provided one of the clearest formulations of why natural language grammar should be exceedingly difficult to learn.

A critical factor in this argument is the way that learning is understood. Learning is construed in its most generic sense as logical induction. An induction process begins with a set of positive instances (a set of grammatical sentences) and derives the general rules that produce them by comparing their similarities. The hypothesized rules are thus only constrained by the statistical relationships implicit in the input strings.[11] This logic has also been adapted to the language-learning problem in an intriguing discussion by the philosophers William Ramsey and Stephen Stich, who argue that even an empirical scientist with all the reasoning and analysis tools of modern science would be unable inductively to determine the rules of a language by analyzing a finite set of grammatical utterances.[12] The number of possible variants of rule systems that could be induced from this sampling of utterances, even under a variety of simplifying conditions, makes it hopeless to test each in a limited time frame. They conclude that only by comparing against some independent and limited set of grammatical rules (such as the scientist's own knowledge of Universal Grammar) would the scientist be able to succeed. But this logical problem may not quite match the pragmatic problem faced by real language learners. Grammatical rules allow one to predict which word combinations are and are not likely to convey unambiguous symbolic propositions in a language. To be able to do this is essential to successful communication. But inductively deriving grammatical rules isn't the only way to arrive at this competence without relying on preformed hypotheses. Despite the statement that it applies to any empirical induction method, this way of posing the problem of what must be learned and how it may be learned is still overly restrictive.

Learning is not one process, but the outcome of many. The efficiency with which one learns something depends in part on the match between the learning process and the structure of the patterns to be learned. What provides efficient learning in one situation may be very inefficient in another. In the last chapter I argued that symbolic relationships themselves are difficult to learn because their structure violates expectations that are appropriate for more typical learning paradigms. Individual symbols are markers for points in a global pattern of relationships that are only quite indirectly reflected by individual instances. Consequently, attending to the surface details can actually be a disadvantage. The problem is similar for learning the logic of grammar and syntax, in large measure because these facets of language are also surface expressions of the deep web of symbolic relationships.

Recently, researchers from a number of cognitive science labs have taken a novel approach to the language-learnability problem, by exploring our assumptions about the kind of learning processes that are best suited for developing a skill as complex as using syntactically structured speech. The most fruitful approach to this seems to have come again from a somewhat counterintuitive inversion of common sense thinking about learning. This approach is summed up by the phrase "less is more," suggested by Elissa Newport as a possible clue to an explanation of the paradoxical nature of child language learning.[13] Like the innate grammar theorists, she was struck by this savantlike ability to learn something that seemed on the surface to be far more difficult than other things that children are unable to learn. For example, the younger the child, the more difficulty he or she has with tasks that require conscious memorization of novel associations. Not only is memory consolidation itself apparently less efficient in younger compared to older children, but both their distractibility and the brief span of their working memory contribute additional handicaps to explicit learning. This is reflected by our common sense practice of keeping children out of school until they are at least four to six years of age, because they are just not ready for explicit classroom instruction. Rather than concluding that this paradoxical ability proved that learning couldn't account for children's language acquisition ability, Newport wondered instead whether children's learning limitations might actually be advantageous when it comes to language.

In this regard, Newport's theory shares a similar logic to the language evolution arguments sketched above. Just as constraints affecting the sorts of errors in transmitting and reproducing bits of language can powerfully shape the patterns of language structure, so can constraints and biases in learning shape what is and is not learnable. In designing a learning device capable of learning the widest possible range of arbitrary associations, there might be an advantage in making it unbiased. But language is not just any system of association. As we have seen, its deep logic of associations, which derives from the indirect systemic logic of symbolic reference, is highly distributed and nonlocal, and the syntactic implementation of these relationships tends to form complicated hierarchic patterns. Computer and cognitive scientists have both come to recognize that this sort of distributed pattern recognition problem is easily capable of defeating "brute force" learning approaches, and requires, instead, some special tricks to focus learning at the right level of analysis and avoid cul-de-sacs created by false leads. Starting small and simple, with a learning process incapable of tackling the whole problem, might offer this sort of learning constraint. Following these clues, Newport and others have attempted to find independent support for the less-is-more intuition about language learning.

In a series of efforts to model the effect of different learning biases on language learning, many researchers have turned to neural network simulations. To understand these approaches, it is necessary to know at least superficially the way that neural networks "learn," as well as a few details about the design of the specific type of neural network architectures that are most successful at language-learning simulations. Calling these model systems *neural* networks (neural nets, for short) is at best a very loose analogy. As the term implies, these computer models borrow some design features that we imagine to be crucial to information processes in brains. In fact, however, the vast majority of neural nets are programs that are "run" on the

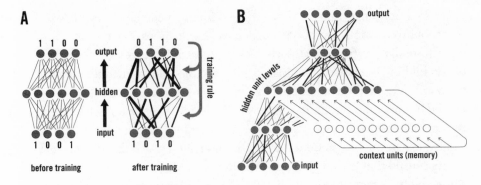

Figure 4.3 *The basic logic of neural network simulations. A neural network can be a real electronic circuit made of physical nodes and connections, or, more often, a computer simulation of the behavior of such a device. A basic neural network consists of three "layers" of nodes (input, output, and hidden units) and the connections between them.*

A. In this diagram circles represent nodes, lines represent connections between nodes, and the thickness of the lines represents differing connection strengths. Input values on input nodes (1s and 0s) cause them to send an "on" or "off" signal to all hidden unit nodes (nodes with no direct links to input or output) they are connected to, which affect the values on these units according to the strength of the connection. Training a network requires using output consequences to strengthen or weaken connections (shown by different thickness lines) between nodes in such a way that it correlates with success or failure to predict "correct" associations.

B. In order to produce predictions of future sequences, a recurrent network design can be used which reenters previous values of the hidden units into subsequent stages of processing. This was the sort of network design used by Jeff Elman for a grammar-predicting network (see text). Elman's "simple recurrent net" design includes multiple levels of hidden units, as well as feedback of past hidden unit information which provides a sort of "memory" input to modulate the immediate input.

usual sorts of large, fast digital computers. It is the "virtual architecture" of the way these programs handle information that gives them their name and their interesting characteristics.

Neural nets are composed of simple elements (nodes) that respond in simple ways (e.g., turning on or off) to their inputs, and connections between nodes that convey signals indicating the states of other nodes (see Figure 4.3). The nodes are the analogues of neurons and the connections are the analogues of the axons and dendrites through which neurons are linked. The operation of a neural net is determined by each node's "reading" the signals on its input connections from other nodes, and, using some very simple input-output transformation rule, producing a signal that is sent out through its output connections to other nodes. To create a network capable of interesting behaviors, a large number of nodes are connected together, often in semi-randomly specified and highly interconnected patterns. In addition, some nodes are connected to external input signals and others to external output registers. Those not connected directly to either inputs or outputs are called "hidden units." The function of the net is determined by the global patterning of signals from output nodes with respect to the patterns presented to input nodes. These input-pattern to output-pattern relationships are thus mediated via the patterning of signals distributed through the web of interconnections that link output to input nodes, by way of the intervening hidden nodes, and not by the state or activity of any individual node.

What makes the behavior of such nets interesting, and similar to their biological counterparts, is that they can be set up so that all connections between nodes can be modified with respect to their correlation to certain input-output patterns. If individual "connection strengths" can be adjusted to weaken or strengthen the effect of one node on another, the net's behavior can be progressively adapted to fit a given rule linking input patterns to output patterns. This is the analogue to training and learning. There is an almost unlimited range of possible strategies for organizing and training nets. All share the common logic of modifying the local connection with respect to some index of global behavior. Over the course of many trials with many inputs, the performance of a net can thus be trained to converge on a given target set of input-output relationships. The network becomes progressively adapted to produce a given set of behaviors as a result of an incremental and indirect selection of certain minimal design elements.

Trained neural networks exhibit remarkable pattern recognition abilities, something that is difficult to program digital computers to do using top-down approaches with step-by-step instructions for analysis. Networks trained to "categorize" one set of stimuli (i.e., respond similarly to them) will often also produce this response when presented with novel stimuli similar to those on which they were trained. In other words, they are capable of (susceptible of) something like stimulus generalization. This property is also reflected in their response to "damage." If one were to cut a handful of connections or pull a chip out of a digital computer at random, or even just insert nonsense strings into a program running on it (as periodically happens when disk information gets corrupted), the inevitable outcome is catastrophic failure: a "crash," as computer vernacular describes it. In contrast, if a trained network is damaged by randomly eliminating nodes or connections, its behavior seldom fails in an all-or-none fashion. Instead, the performance of a damaged neural net degrades incrementally as the extent of the damage is increased. This, too, is reminiscent of the way nervous systems appear to behave in response to damage, and demonstrates that the analogy is not entirely superficial.

The essence of these behaviors lies in the way these nets distribute the information they embody within all parts of the whole. Such a global representation of information has often been compared to a hologram (a laser-light interference "photograph" that produces a three-dimensional image when viewed from different angles, examples of which can now be found on most credit cards). Holograms "contain" multiple views of a three-dimensional image. Which perspective is visible depends on the angle at which the hologram is viewed. Even when only a fragment of the original remains, there is some angle at which it can be viewed to see the imaged object, though the smaller the fragment, the more limited the perspectives viewable.[14] In a neural network, the relationship that maps input to output is decomposed into minute facets of the whole that are distributed throughout the net and are literally embodied in its connectional logic. Since the input-output relationship is in effect computed by the whole network, it is called parallel distributed processing (PDP) of information.

One of the pioneers in applying this approach to language learning problems is Jeff Elman of the University of California at San Diego.[15] He and his colleagues used a modification of neural net design in order to create a net that was able to learn to predict sequentially presented patterns, not just categorize static ones. For this, his nets needed something analogous to short-term memory in order to represent the immediate past and future states to the present state of the net. He accomplished this using an architecture called a *recurrent* net, in which the past states of the hidden units were reentered as additional inputs to these same units at subsequent stages of processing. This allowed him to translate the syntax-learning problem into a predictive mapping of past sequences to future sequences of inputs. Given an incomplete series of inputs, the net would be required to predict which output should most likely follow; more specifically, given a partial sentence, it would predict which word should most likely follow according to English grammar and syntax. The inputs he fed into the net consisted of a corpus of simple sentences, in which different words were encoded as distinct strings of 0s and 1s (in other words, meanings were treated as irrelevant). The training consisted in comparing predicted next "words" with actual next words, and then modifying network connection strengths according to how each contributed to correct predictions.

A fully trained net that could make correct predictions for novel sentences (analogous to stimulus generalization) would necessarily embody aspects of the statistical structure of English grammar and syntax in its architecture, even though it would not contain any semantic information.[16] If a net could be trained to do this as well as a person, it would indicate two things: (1) that the statistics of the relationships between grammatical classes of words in the training strings contained sufficient structure from which to recover grammatical regularities; and (2) that these regularities were learnable in some form without explicit (rule-based) error correction.

Elman showed that recurrent nets were indeed capable of extrapolating what they learned from a set of kernel sentences used in training to novel kernel sentences composed of the same words.[17] However, when trained with sentences that were only a bit more complicated, the recurrent nets failed to learn. As soon as recursive structures were introduced (e.g., embedded clauses or structural transformations), which had the effect of separating and inverting simple contiguity relationships between words as found in kernel sentences, no amount of training seemed to converge towards predictability. Up to this point, the experiment exemplified Gold's theorem.

Elman realized that this problem was a variant of something commonly observed in other neural net experiments. Neural nets often tend to converge on suboptimal solutions to complex mapping problems that provide only weak predictability, because they get attracted by local patterns that "obscure" more global ones. They can get stuck in suboptimal response "basins" (sort of like statistical potholes in the learning landscape) because training produces only incremental changes, which can only converge on a solution if closer approximations provide ever-increasing predictive accuracy. Learnability, then, depends on a certain match between the learning algorithm and the structure of the problem, so that optimal solutions are found in the "vicinity" of local solutions. Apparently, natural language syntax and simple conditional learning processes don't match in this way. Indeed, a variety of learning problems show this same nonlocality feature, and thereby defeat many otherwise powerful learning paradigms. In general, solving them requires recoding the inputs in such a way as to reduce the distributed character of the associations.[18] Preventing neural nets from getting trapped in learning potholes can sometimes be accomplished by introducing "noise" into the net to disrupt convergence toward weakly predictive states, and forcing a wider "sampling" of possible solutions, or else by introducing biases against features that are common only to suboptimal solutions, if there are any.

Elman employed both strategies to aid his nets in overcoming this difficulty. First, he found that staged training, starting with an initial set of simple sentences followed later by training with sets of more complex sentences, could produce nets eventually able to predict complex syntax. The early restricted training essentially provided enough of a biased sample of within-clause relationships to minimize the training influence of exceptions produced by higher-order between-clause relationships. Second, he found that by randomly disrupting recurrent inputs (essentially degrading the short-term memory) during early training, then gradually decreasing this "noise," the net could learn complex syntax from a corpus of complex sentences presented right from the beginning. At early stages of learning, the net would essentially be unable to retain a trace of local regularities long enough to produce any persistent changes in structure; but more global regularities of the input would be more redundantly represented throughout the net, and so would produce weak but consistent changes in structure. When the net eventually was made capable of retaining a trace of local associative relationships, it was already biased against incorporating information from those that violated the large-scale statistics of word associations. Though Elman describes this pattern of learning as "starting small," it is probably more accurate to describe it as "starting crude"; it is more like interpreting a visual image by starting with a very blurred version of the whole and then progressively resolving individual details, than like trying to assemble many independent close-up views of the image at fine detail into a complete picture. Both of Elman's incremental learning strategies effectively partitioned the learning processes so that learning to predict patterns at one level of structure was partially decoupled from learning to predict patterns at other levels. By structuring the learning process so that certain aspects of the problem could not be learned from the outset, he prevented the learning of relationships that are employed at different levels of language structure from interfering with one another.

This simulation thus demonstrated that it was possible to design a device that could learn to predict grammatically correct sentence structure in a purely inductive fashion, given nothing more than a corpus of positive examples of allowable texts—exactly what the UG theorists had said was impossible. The key to accomplishing what many had "proved" was impossible was to structure the learning process differently at different stages of learning. What was available for learning at early stages was "filtered" (either by the training sequence or by the competence of the net), so that only some aspects of the input were available at any time. This set up learning biases that just happened to correspond with crucial structural features of the problem space. Although the information embodied in this incremental learning strategy was extrinsic to the language data presented to the net, it was far less explicit than a Universal Grammar, or indeed any specific grammatical information. There is no difficulty imagining how such generic constraints could happen to be available to support the learning of grammatical speech. Although critics might argue that such a simulation hardly demonstrates the ability consistently to generalize this finite statistical "knowledge" of grammar to the theoretically infinite variety of possible grammatical sentences that would be specifiable by a set of grammatical rules, and indeed that it doesn't demonstrate induction of *rules of grammar* at all, it does show how important the structure of the learning process is to what can and cannot be learned. More important, it suggests that the structure of language and the way that it must be learned are linked. What may be essentially inaccessible under quite general learning conditions may become more accessible under more limited conditions.

This provides some novel clues about why children might be more facile language learners than adults and why early language experience might have aided Kanzi's access to language as well. Being unable to remember the details of specific word associations, being slow to map words to objects that tend to co-occur in the same context, remembering only the most global structure-function relationships of utterances, and finding it difficult to hold more than a few words of an utterance in short-term memory at a time may all be advantages for language learning. This is the proposal that Elman and Newport each offers to counter the strong nativist alternative. Precisely because of children's learning constraints, the relevant large-scale logic of language "pops out" of a background of other details too variable for them to follow, and paradoxically gives them a biased head start. Children cannot tell the trees apart at first, but they can see the forest and eventually the patterns of growth within it emerge.

The co-evolutionary argument that maps languages onto children's learning constraints can be generalized one step further to connect to the most basic problem of language acquisition: decoding symbolic reference. Symbolic associations are preeminent examples of highly distributed relationships that are only very indirectly reflected in the correlative relationships between symbols and objects. As was demonstrated in the last chapter, this is because symbolic reference is indirect, based on a system of relationships among symbol tokens that recodes the referential regularities between their indexical links to objects. Symbols are easily utilized when the system-to-system coding is known, because at least superficial analysis can be reduced to a simple mapping problem, but it is essentially impossible to discover the coding given only the regularities of word-object associations. As in other distributed pattern-learning problems, the problem in symbol learning is to avoid getting attracted to learning potholes—tricked into focusing on the probabilities of individual sign-object associations and thereby missing the nonlocal marginal probabilities of symbol-symbol regularities. Learning even a simple symbol system demands an approach that postpones commitment to the most immediately obvious associations until after some of the less obvious distributed relationships are acquired. Only by shifting attention away from the details of word-object relationships is one likely to notice the existence of superordinate patterns of combinatorial relationships between symbols, and only if these are sufficiently salient is one likely to recognize the buried logic of indirect correlations and shift from a direct indexical mnemonic strategy to an indirect symbolic one.

In this way, symbol learning in general has many features that are similar to the problem of learning the complex and indirect statistical architecture of syntax. This parallel is hardly a coincidence, because grammar and syntax inherit the constraints implicit in the logic of symbol-symbol relationships. These are not, in fact, separate learning problems, because systematic syntactic regularities are essential to ease the discovery of the combinatorial logic underlying symbols. The initial stages of the symbolic shift in mnemonic strategies almost certainly would be more counterintuitive for a quick learner, who learns the details easily, than for a somewhat impaired learner, who gets the big picture but seems to lose track of the details. In general, then, the initial shift to reliance on symbolic relationships, especially in a species lacking other symbol-learning supports, would be most likely to succeed if the process could be shifted to as young an age as possible. The evolution of symbolic communication systems has therefore probably been under selection for early acquisition from the beginning of their appearance in hominid communication. So it is no surprise that the optimal time for beginning to discover grammatical and syntactic regularities in language is also when symbolic reference is first discovered. However, the very advantages that immature brains enjoy in their ability to make the shift from indexical to symbolic referential strategies also limit the detail and complexity of what can be learned. Learning the details becomes possible with a maturing brain, but one that is less spontaneously open to such "insights." This poses problems for brain-language co-evolution that will occupy much of the rest of the book in one form or other. How do symbolic systems evolve structures that are both capable of being learned and yet capable of being highly complex? And how have human learning and language-use predispositions evolved to support these two apparently contrary demands?

Elissa Newport was one of the first to suggest that we should not necessarily think of children's learning proficiency in terms of the function of a special language-learning system. She suggests that the relationship might be reversed. Language structures may have preferentially adapted to children's learning biases and limitations because languages that are more easily acquired at an early age will tend to replicate more rapidly and with greater fidelity from generation to generation than those that take more time or neurological maturity to be mastered. As anyone who has tried to learn a second language for the first time as an adult can attest, one's first language tends to monopolize neural-cognitive resources in ways that make it more difficult for other languages to "move in" and ever be quite as efficient. Consequently, strong social selection forces will act on language regularities to reduce the age at which they can begin to be learned. Under constant selection pressure to be acquirable at ever earlier and earlier stages in development, the world's surviving languages have all evolved to be learnable at the earliest age possible. Languages may thus be more difficult to learn later in life only because they evolved to be easier to learn when immature. The critical period for language learning may not be critical or time limited at all, but a mere "spandrel,"[19] or incidental feature of maturation, that just happened to be coopted in languages' race to colonize ever younger brains.

So immaturity itself may provide part of the answer to the paradoxical time-limited advantage for language learning that Kanzi demonstrates. Kanzi's immaturity made it easier to make the shift from indexical to symbolic reference and to learn at least the global grammatical logic hidden behind the surface structure of spoken English. But equally important is the fact that both the lexigram training paradigms used with his mother and the structure of English syntax itself had evolved in response to the difficulties this imposes, and so had spontaneously become more conducive to the learning patterns of immature minds. The implications of Kanzi's advantages are relevant to human language acquisition as well, because if *his* prodigious abilities are not the result of engaging some special time-limited language acquisition module in his nonhuman brain, then such a critical period mechanism is unlikely to provide the explanation for the language prescience of human children either. The existence of a critical period for language learning is instead the expression of the advantageous limitations of an immature nervous system for the kind of learning problem that language poses.

And language poses the problem this way because it has specifically evolved to take advantage of what immaturity provides naturally. Not being exposed to language while young deprives one of these learning advantages and makes both symbolic and syntactic learning far more difficult. Though older animals and children may be more cooperative, more attentive, have better memories, and in general may be better learners of many things than are toddlers, they gain these advantages at the expense of symbolic and syntactic learning predispositions. This is demonstrated by many celebrated "feral" children who, over the years, have been discovered after they grew up isolated from normal human discourse. Their persistent language limitations attest not to the turning off of a special language instinct but to the waning of a nonspecific language-learning bias.

This may also be responsible for another curiously biased language phenomenon that linguists have taken as evidence for innate knowledge of language structure: the transition from pidgin languages to creole languages. Pidgins are forced hybrids of languages that have arisen in response to the "collision" of languages, typically arising as a result of colonization or trade relationships. They are not the primary language of anyone and often have enjoyed a very transient history, typically disappearing within a generation or so. They are like makeshift collections of language fragments from both languages that are used as a common translation bridge that is mutually understandable. But during recorded history, a number of populations have also evolved new languages directly from pidgins, which differ from either "parent" language. Such "creole" languages appear to be able to originate quite rapidly—within as little as one or two generations—especially when a population finds itself transported and isolated in a new context, such as periodically occurred as a result of the slave trade of the past few centuries. What is even more remarkable, according to the linguist Derek Bickerton, is that the syntactic structures of different creole languages often appear to be more similar to one another than to the languages that spawned them, despite being isolated in different parts of the world.[20] For example, they all tend to be minimally inflected, use particles to replace tense markers, use repetition in place of adjectives and adverbs, and have restricted word order.

Historically, linguists dismissed such similarities as a result of comparing simplified languages, like comparing the syntax of newspaper headlines, or as a result of coincidental similarities between the languages involved, or just coincidences resulting from the limited numbers of major syntactic alternatives available. But most contemporary comparative linguists have come to recognize that the coincidences are real and too numerous to be explained by spontaneous parallelisms. This has suggested to some that these similarities are evidence for a single historical origin (possibly from the early Mediterranean pidgin, Sabir, carried by the first Portuguese explorers). This explanation seems no less incredible to other linguists, however, because of the separate eras and great distances involved, the lack of Portuguese contact in many cases, and the fact that the local pidgins and creoles exhibit specific roots directly traceable to the colonial English, French, Spanish, Portuguese, etc., spoken by colonists in these areas. Moreover, pidgins and creoles that have developed without European influence also appear to share these features.[21] In response to the implausibility of both alternatives, Bickerton interprets such similarities as evidence for an innate grammatical template reasserting itself in each case. But I think there is an intermediate alternative suggested by the co-evolutionary approach, which parallels Bickerton's approach in the same way that the co-evolutionary theory of child language acquisition parallels the UG approach.

Creolization occurs when children are exposed to a pidgin as their first and only language and so reflects the ways that these young minds reinterpret a partial symbolic system *as though it were complete*. In this regard, I follow Bickerton in suggesting that the similarities between creoles reflect how children spontaneously fill in what is missing and recategorize what is given. As such, these similarities provide a special opportunity to view the distorted mirror through which children view language, and so offer an insight into what children normally bring to the problem. Instead of reflecting a built-in knowledge of grammar, however, I think the similarities between create generic learning constraints that nonetheless strongly bias what is learned and what is spontaneously "invented."

Given children's constrained learning abilities, which force them to employ a top-down global-to-specific reconstruction of the language presented to them, they are initially forced to ignore many details present in the input in their effort to decode the basis of its symbolic reference—the first and most basic demand they face. This may help to explain certain biased choices that seem to be consistently made in creolization, as well as in normal language development. For example, early recognition of global representational relationships, necessary to discover the symbolic reference of words and phrases, and an inability to track local word-association patterns, lead young children to initially treat many phrases as unanalyzed wholes or else to ignore all but the most salient elements relevant to their symbolic analysis of meaning and reference. Normally, subsequent maturation of learning abilities enables children to later pick up the word-association patterns buried within phrases that previously escaped their notice, and this forces a recoding with respect to within-phrase structure. This is constrained, however, by their previous symbolic commitment to global phrase relationships, and biases them to seek out subordinate symbolic functions for these later combinatorial patterns. This may be the bias that favors the hierarchic phrase structure reflected in both acquisition and evolution of languages. So, how might this also help to explain commonalities in creolization? The constraints of the acquisition process are similar irrespective of whether the input is a full-blown language or a pidgin, except that in the latter case there will be insufficient variety of phrase structures to force within-phrase dissection of elements. As a result, whole phrase units may become crystallized as wordlike units, modifiers will be minimally dissected from phrases, and the lack of grammatical morphemes that ensues will tend to be compensated by syntactic operations instead. These limitations would naturally lead to a minimally inflected, constrained word-order language, as is characteristic of most creoles and well-developed pidgins. So we have come full circle. Not only does it appear that languages have evolved to take advantage of the learning biases of children, but to whatever extent prior language input is impoverished, it appears that these learning biases tend to reshape language to fit. Universal (i.e., convergent) trends emerge either way in the absence of a specific preexisting plan, whether in the input or in the mind.

This strategy of employing learning handicaps as a means of overcoming certain learning difficulties also has considerable generality beyond symbols, grammar, and syntax. It may, for example, offer an insight into the remarkable sparing of certain special learning abilities exhibited by so-called idiot savants. Individuals who may be mentally handicapped in other ways sometimes show geniuslike abilities in very restricted domains of cognition—often quite specific "talents," such as lightning calculation abilities, prodigious musical talent, or artistic-spatial skills. This has led some to speculate that these individuals possess some special modular "organ" or "instinct" for such talents. Brains have evolved to be able to employ many different learning strategies at different times and in different circumstances and these strategies are often competitive with one another in their recruitment of neural resources. This is why a severe impairment affecting one learning strategy can inadvertently provide a release of resources to some other complementary, or mutually exclusive, learning strategy.

In summary, I have argued that the underlying source of each of the apparent paradoxes of language learning is a misleading assumption that learning is a one-dimensional process, in which a collection of individual memories is built up bit by bit, like adding items to a list, and in which general rules can only be derived by inductive generalizations from a finite set of instances. This blinkered view of learning has limited both our understanding of the nature of symbolic reference and our analysis of how children acquire competence in producing a symbolic system that is structured like a hierarchic rule-governed logical system. Both what is learned and the context in which something is learned may have characteristic invariant features that can be incorporated into learning strategies to increase the compatibility between the learning and the learned. Specific learning handicaps are examples from this more general class of learning biases. Biases affecting how often and when different learning strategies are employed or withheld can radically change what can be learned and how easy or difficult it is to learn it.

Immaturity of the brain is a learning handicap that greatly aids language acquisition. Despite its counterintuitive symbolic basis, however, immaturity is not the whole explanation for the human language capacity. It is only, in a way, an evolutionary afterthought that has arisen as languages have adapted to take advantage of their host organisms' natural biases. The simple fact that other species experience nearly insurmountable difficulties with even simple language learning, and even when immature, demonstrates that significant modifications of the human brain during our evolution additionally enable us to overcome these same difficulties. One might be tempted to call such neural predispositions a "language instinct," as Steven Pinker suggests, because these predispositions are both innate and universal, and because they determine that we alone find linguistic communication natural. But this tends to be interpreted in terms of a false dichotomy that has deeply confused research into the basis for language. It is misleading to imagine that what is innate in our language abilities is anything like foreknowledge of language or its structures. Rather than a language organ or some instinctual grammatical knowledge, what sets human beings apart is an innate *bias* for learning in a way that minimizes the cognitive interference that other species encounter when attempting to discover the logic behind symbolic reference—a bias that is far more intense and ubiquitous than mere immaturity.

Thinking of our mental difference in these terms provides a crucial clue to the mystery of human brain evolution. Just as incomplete differentiation of neural circuitry can be responsible for the helpful learning biases of young children, and congenital brain abnormalities are likely responsible for the learning biases of idiot savants, differences in the global organization of human brains are likewise responsible for the more extensive learning biases that help human children alone among species surmount the symbolic threshold and learn an immensely complex language. So instead of looking for neural language modules, or postulating some global increase in general learning abilities, we need to begin to think of human brain evolution in terms of changes that could have produced certain biases in how we tend to learn. But the relevant biases must be unlike those of any other species, and exaggerated in peculiar ways, given the unusual nature of symbolic learning. Such unprecedented differences in human brain function must be supported by equally unprecedented differences in human brain structure. Discovering how these two radical departures from the general pattern correspond to one another will likely provide important insights concerning the design principles behind global brain functions.

2

Brain

The Size of Intelligence

When an idea is wanting, a word can always be found to take its place.

—Goethe

A Gross Misunderstanding

Having identified a previously unrecognized cognitive problem that is the cause of the language-learning barrier, we can now turn our attention back to the mystery of human evolution charged with a new aim: to determine what happened to human brains to make it possible for our ancestors to break through this barrier. If the symbol-learning problem is the threshold that separates us from other species, then there must be something unusual about human brains that helps to surmount it. This one-of-a-kind difference ought to be readily apparent, but only if we know what to look for. Part of the problem is that something else which appears more obviously relevant has captured our attention, and like the too obviously guilty suspect in a murder mystery, this one fact about human brains is the perfect distraction causing us to miss the pattern of more subtle clues.

Despite the glaring fact that the relationships between brain structure and brain function are still poorly understood, many feel that the explanation for the exceptional nature of human mental abilities is known—in fact, that it has been known for a century! The answer can be found in almost every textbook on human origins and in most popular books and magazine articles about the mind and brain. According to these sources, it is a well-established fact that the human brain is a better, more powerful computing device than any other species' brain because it is proportionately bigger and can process and retain more information. As a result of this greater mental capacity, humans can build more sophisticated mental models of the world, solve more complicated social and survival problems, and learn to communicate with more complicated signal systems (e.g., language). The idea that human beings' encephalization—the extent to which we have large brains for a primate our size—is so much a part of our culture that it has essentially achieved the status of unquestioned fact. It has become synonymous with the definition of what it means to have humanlike mental abilities to the extent that even science fiction caricatures of "advanced" alien species are portrayed with oversize brains.

The accepted view can be reduced to a simple phrase: "Bigger is smarter." Well, almost. Researchers do not all agree on how best to measure the *effective* size increase—whether absolute size or some measure of relative size is most relevant (a problem we will return to shortly)—but there is broad agreement that *some* measure of human brain expansion will be found to be the primary correlate of our increased mental abilities. All that seems left to determine is exactly how the increase in the one translates into increase in the other. Most researchers believe that they know what the answer should be; it's just that they are not quite sure how to derive it from the evidence.

Though there is undoubtedly some correspondence between the size of a brain and its capabilities, I suspect that it is far less obvious than we tend to imagine. There are two critical errors hidden in these flattering assumptions. First is the notion that either brain size or intelligence can be usefully treated as a unitary linear trait, and second that the relative quantity of brain tissue compared to nonbrain tissue in the body somehow correlates with something like the relative amount of uncommitted computational power. Together, these presumptions have blinded us to innumerable other features of brain evolution and differences in brain function related to issues of scale. An understanding of the relationship between human brain size and language abilities has been one of the major casualties.

The presumed relationship between brain size and intelligence is an intuitively compelling idea. It has produced a century of ideologically motivated studies of the relationship to genius and criminality, ignited acrimonious debates about how best to measure these variables, and generated hundreds of papers examining the potential behavioral and ecological correlates of brain size from every conceivable aspect. Still, after more than a century of rethinking the possibilities, theories about *the function* of brain size remain in a state of continuous flux. This is not because we haven't yet discovered what intelligence is good for. Everyone agrees that more intelligence should be good for any number of things, and almost every conceivable use for more intelligence has been suggested in some theory to explain why human brains are large. But that's just the trouble. With such a vague and general answer already assumed, it is hard to phrase the research question clearly enough to be able to rule out spurious correlations. What if brain size is not a single trait, but a reflection of many possible complex internal changes in brain organization, each of which has quite different functional consequences? And what if mental functions reflect a delicate balance of many complementary and competing learning, perceiving, and behavioral biases, not just a single "capacity"?

Indeed, we shouldn't be surprised to find out that nature has not been straightforward in its design of brains. The brain is by far the most complex organ of the body, and one whose outward appearance provides little hint of its functional organization. Brains don't do only one thing, and different brain functions are not uniformly distributed throughout the brain. If human brain size is a complex trait, enfolding numerous deeper changes in brain structure and function, then all the correlative studies looking for what might have caused brains to grow may be pointless mathematical exercises, and the differences in neural architecture that make the difference in human cognitive abilities will continue to elude us. We can't afford to assume that the human brain is *just* bigger, and put all our efforts toward discovering what selected for its size. We need to look deeper.

My own perspective has also been well stated by Ralph Holloway of Columbia University, one of the pioneering figures in the analysis of fossil hominid brain casts. He wonders in dismay at the forest of comparative brain size studies that seem oblivious to the likely possibility that neither intelligence nor brain size is a simple trait. How can we imagine, he asks, that the brain is some "unitary organ with a simple behavioral task to accomplish such as 'intelligence,' 'language,' 'adaptive behavior,' or any other such ped-agogical fig leaf to cover our ignorance about how the brain evolved?"[1] If a unique function (language), global assessment of function capacity (intel-ligence), and the crudest possible measure of brain structure (size) have any-thing to do with each other, it will probably be discovered in the neurobiological details and not in global extrapolations from these most su-perficial features.

In the preceding chapters we encountered a series of paradoxical results suggesting that the language-learning problem is not just difficult, it is counterintuitive, in the deepest sense of that word. Both the hierarchic-recursive logic of sentence structure and the distributed multilevel asso-ciative relationships that support symbolic reference are essentially unlearnable by "brute force" information-processing approaches. This ex-plains why even simple languages are nearly impossible for nonhuman brains to learn—not because of other species' limited learning abilities, but because their intrinsic learning biases undermine the process before it can even get started. The point was driven home by the fact that the learning handicaps of immature brains may actually be beneficial for some aspects of language learning. But these odd results do not fit easily into a bigger-is-smarter interpretation of the crucial human mental difference. Evolving a more powerful learning device is not the solution to the language-learning problem. This suggests that some critical distinctions may have been missed by assessing the evolution of human intelligence in global terms.

And yet there is no escaping the fact that human brains are unusually large, in both absolute and relative terms. Any story about human brain evo-lution that does not incorporate this fact ignores a major clue to the mys-tery of the human difference. The question is not whether brain size is an important correlate of human brain evolution. It is. It is not whether our unusually large brains and the differences in our cognitive abilities are somehow linked. They undoubtedly are. The question is: What other changes in brain organization correlate with this global change in brain size, and what are their functional consequences? We can be certain that some-thing having to do with the sizes of brain structures is central to the origins of the human mind, but what? To answer this, we need to get some gen-eral idea of how quantitative changes affect brain functions.

There are two aspects of the size question that need to be addressed. First, exactly how is this change in size distributed within the human brain? Is it global, like enlarging a photograph; is it an extrapolation from a more complicated trend exhibited in many other species as well; does it only in-clude certain limited parts of the brain? Determining which of these options is the case is more difficult than one might suspect. Second, what are the possible ways that differences in the size of the brain or its parts can influence its functions? For this we have little more than crude analogies to go on. Few neuroscientists have seriously investigated this question, except in the most general way. There are many dimensions of size-correlated effects that we might need to consider before we determine that we understand the significance of the human example. Larger brain structures may mean more storage capacity or discriminatory ability, but they may also mean changes in rate of processing, changes in relative excitatory or inhibitory influence over other linked systems, or a difference in intrinsic signal production tendencies (like longer periodicity of cyclic activities). How can we be sure we know which is most relevant? But there is an even more basic problem: size is not so simple as it appears. The question is always "large or small with respect to what?"

As large-brained apes on the top rung of the ladder of biological progress and perfectibility—or so we like to think—we habitually rank people's mental abilities for various purposes, in work, school, and casual conversations. We naturally feel comfortable ranking other species this way as well. Assessing intelligence has become a widely used tool for deciding who should and should not be able to do all manner of things, from play quarterback to enter medical school. We assume that we all have a certain amount of intelligence, like height, and that it can be measured and compared from person to person. Since intelligence seems to change little if at all over one's lifetime, we assume that it has a fixed value determined very early in life. We have devised tests to measure this computational power, IQ tests, and we encourage the gatekeepers in our society—school administrators and employers, among others—to wield them freely. We cannot begin to measure what we gain and lose thereby, just as we cannot isolate exactly what intelligence is. But the comparatively large size of human brains seems to provide a satisfying confirmation that the amount of mental ability one has is related to the amount of tissue one devotes to producing it.

Assessing the human brain size difference is not just a matter of comparing weights, volumes, or even neuron numbers. The question is, what do we want to know about this difference? Just size? Our brains are not the biggest, nor do they have the most neurons or connections. Elephants and whales are vying for that honor. We do have large brains for our bodies, but in simple ratio terms mice are more brainy. Exactly how most usefully to assess brain size is not obvious, and it turns out that the same problems surface in this analysis whether we are considering the brain as a whole or the sizes of its parts. Perhaps we can gain a clearer idea of the problem by considering some less complex organ system as an analogy. A good candidate is body musculature.

Larger muscles are capable of generating more force, a larger heart is capable of pumping more blood per minute, and a larger gland is able to synthesize more hormone in a given period of time. By analogy, a larger brain should be capable of greater computing power, of processing more information per second, and of producing more complex mental representations and communications than smaller brains. If the brain secreted thought as a gland secretes hormones, then this might make sense, but it doesn't. There are some ways in which such quantitative analogies can help make sense of differences in brain functions, but also many ways in which they can be misleading.

Even though a large animal with greater total muscle mass can exert more force with its limbs than a smaller one, it may not be able to jump proportionately higher or run proportionately faster as a result. Larger bodies need more muscle mass to move them with the same facility as smaller ones. This is essentially the difference between gross strength and net strength. Gross strength probably correlates well with total muscle mass, but net strength depends on many other factors. The amount of weight one can lift is a fair predictor of gross muscle mass. Larger bodies with larger muscle mass can lift heavier weights. Certain athletic performances and exercises— the standing broad jump, pull-ups, and push-ups, for example—provide indices of net strength of certain muscles with respect to body mass as a whole. The number of sit-ups or pull-ups one can perform does not favor large over small individuals, but rather those whose abdominal or arm muscles, respectively, are most developed in comparison to the rest of their bodies.

This emphasizes one crucial difference between gross and net strength. There is only one measure of gross strength, but there are innumerable different measures of net strength, even for a single individual. This is because net strength is a comparison of part to whole. Any particular measure of net strength depends entirely on what part is being compared to the whole. Having greater net leg strength for body mass might produce a longer long jump, while having greater arm strength might produce a greater number of pull-ups. But these two net results are to some extent mutually exclusive. Increasing the one decreases the other. And even though by combining results from a large number of diverse physical tests, one might be able to arrive at a single value with which to compare individual athletes (which might provide an assessment of muscle to fat content), it would miss most of the relevant and interesting aspects of net strength and would have only modest predictive power for sports abilities.

Analogous arguments apply to size-function relationships of the brain. The amount of brain tissue an animal posesses is probably proportional to something like gross information storage and processing capacity, but assessing net brain power is far more problematic, and there may be no one best measure. Psychologists have long argued over this implicit problem with respect to intelligence tests. But the problem reappears in only slightly modified form for brain size issues or brain structure size comparisons. Something about human brain size almost certainly is the clue to the human cognitive difference, so to solve this mystery we must first sort out this gross/net problem, even if in the end we conclude that, like muscle strength, there is no single answer possible.

Whether gross or net strength is important depends on the context. Often both are important in different aspects of an activity and both may interact to determine performance in complicated ways. For most behavioral purposes net strength is probably more significant than gross strength, since it correlates with locomotor capabilities and postural support. Nevertheless, there will be some adaptations for which gross strength will be far more important. It is crucial for certain kinds of foraging where some threshold amount of force is necessary to gain access to food. Some nuts and seeds, for example, are protected by strong shells; only species with jaw muscles of sufficient gross strength (and appropriate teeth) will be capable of cracking them. Physical combat can also select for increasing gross strength with respect to net strength. Highly sexually dimorphic species, in which (typically) males physically compete with one another for territories or mates, are good examples. The more intense the physical combat, the greater the disparity in physical size between males and females. Selection for greater gross strength can be achieved by increased total size, if the costs in net strength are of less importance. So extending this analogy to the problem of brain size, we first need to understand the difference between gross and net information processing.

The idea that natural selection is driven by use or disuse of an organ has a long and influential history in evolutionary thinking. Nineteenth-century evolutionists took it for granted that habitual use of an organ, over many generations, could bring about evolution to develop and enlarge it, and that organs that were not used would eventually become reduced or vestigial. The presumed correlation of use with the evolutionary enlargement or reduction of organs is, even now, seldom questioned. Large size means more used; small size means less used. Obvious examples spring to mind in the shapes of animal bodies. A kangaroo's large hind limbs reflect selection for their specialized form of locomotion, whereas its small forelimbs reflect their evolutionary disuse as a means of support. The analogy between the effects of exercise on muscle size during the lifetime and during evolutionary time appears to be straightforward. Indeed, natural selection theory borrowed this conception of use and disuse from Lamarckian theories, based on the inheritance of acquired characteristics, where exercise was thought directly to alter genetic transmission. The Darwinian reinterpretation of the role of use and disuse in evolution replaced exercise as the cause of variation in organ size, with differential selection for spontaneous variations in organ size. Nevertheless, the size-use correlation has become an almost axiomatic rule for analyzing morphological evolution.

In terms of the musculo-skeletal system, a Darwinian interpretation of this rule might be phrased as follows: "Larger, more robustly reinforced bones can endure greater stresses and larger muscles can generate greater forces. Therefore, in an environment in which a particular part of the skeletal or muscular system is habitually subjected to unusual stresses, generation after generation, those individuals that tend from birth to grow more robust muscles and bones will tend to thrive and reproduce better than those who do not. These individuals will tend to pass on this propensity to a greater percentage of offspring, resulting in an increasing prevalence of increased muscle and bone mass in future generations." Parallel arguments can be made for any number of organs that appear subject to size variation in response to the stresses or demands of use, including glands and digestive organs. But it is important to understand how size contributes to the function under consideration, and to know whether this factor is free to vary independently of other features of organism design, before assuming that this logically reasonable argument applies across the board to all organs.

In the case of muscles, bones, and glands, we have a relatively unambiguous correlation between functional demand and size as a result of physiological changes during a lifetime. Physical exercise can cause muscle mass to increase and inactivity can cause muscles to atrophy. Changes in hormonal demand are often associated with increase or decrease in the size of endocrine glands, as in gonad enlargement at puberty or adrenal hypertrophy under conditions of chronic stress. The analogy between changes in a lifetime and changes in evolution appears to work in this case. The size-use rule seems naturally to extend from physiological arguments to evolutionary arguments. In evolutionary theories, however, the order of the argument tends to be reversed from consequence to antecedence. When we infer an evolutionary cause for large muscular limbs, it is like guessing that a person with large arm muscles must have been doing a lot of weightlifting. An animal with an atypically enlarged or reduced organ is automatically suspected of having evolved under conditions where more or less was required of that organ, respectively. We tend to feel quite confident of many such extrapolations into the evolutionary past because of the correlations we observe in living species.

The physiological analogy can be both informative and misleading. Evolution has equipped vertebrate bodies so that they are adapted to certain functional demands, but it has also built in some flexibility as well. This allows room for fine-tuning. These mechanisms for physiological fine-tuning can parallel evolutionary mechanisms in some ways but not others. For example, there is more than one way that muscles or glands can appear altered in size. On the analogy of weightlifters versus couch potatoes, one might expect to find comparatively muscle-bound species and comparatively weak species in the world. Surprisingly, differences in the proportions of muscle mass in mammals do not tend to parallel this difference. Larger species have correspondingly more total muscle mass supporting them, but species that are roughly the same size do not run the spectrum from muscle-bound to weakling. The proportion of muscle to body weight in mammals exhibits a remarkably predictable relationship from species to species across the full size range. So the analogy is only provisionally useful, and clearly fails to take some crucial factors into account.

This caveat is particularly relevant for brain evolution. Unlike muscle mass, brain size does not increase or decrease with use during the lifetime, independent of normal growth, aging, and pathological atrophy.[2] There is no way to appeal to a physiological adaptation where habitual brain use is associated with enlargement or reduction in brain size. And, whereas we seem to be on pretty solid theoretical footing when we extrapolate changes in force and structural strength from changes in size of bones, teeth, or muscles (because we can appeal to the physics of forces and lever arms), it is not at all clear that we have a corresponding theory that is adequate for predicting changes in information-processing functions from changes in size of the brain. The first step, then, is to see if we can determine what kind of relationship this might be.

Brains in Bodies

To begin this investigation, let's start with whole brains and bodies and progressively narrow down our search to find exactly what has changed in human brains. In the long history of comparative psychology, a failure adequately to disentangle net and gross brain function has been a consistent source of confusion. Even the most basic questions remain unsettled. At present, there appears to be wide agreement among those who analyze the statistics of comparative brain size that a version of net brain function fits best with our data and intuitions about brain size and intelligence. The basic intuition underlying this interpretation is that some fraction of brain function must always be devoted to handling the information-processing demands of the body, and is therefore unavailable for other cognitive uses. Thus, gross brain function might be partitioned into a visceral and a cognitive fraction as a first step toward analyzing net brain function. If larger brains also have to service the information-processing demands of larger bodies, they will not necessarily offer their possessors any net increase in cognitive power with their greater size. All other things being equal, we might expect that brains will at least keep pace with the demands of the body as bodies get larger, so that degree of neural control over hormonal, digestive, and basic somato-sensory and muscular functions should not distinguish species with different body sizes. But how do we figure out how much information-processing capacity must be dedicated to maintaining somatic functions?

A simple and obvious possibility is that the fraction of brain size that must be dedicated to body maintenance is directly proportional to body size itself. If this were true, then the ratio of brain to body size would reflect the ratio between cognitive and somatic brain functions. Individuals and species with a higher ratio of brain to body size should have more of their brains free for nonsomatic functions—a higher net cognitive capacity.[3] Brain weight is a much smaller fraction of body weight in fish, reptiles, and amphibians than in mammals and birds, and we tend to think of these cold-blooded vertebrates as less mentally sophisticated than our warm-blooded compatriots. A low ratio of brain to body size has also been used to argue for low intelligence in large dinosaurs and to explain the apparently greater intelligence of humans compared to larger-brained whales and elephants. Unfortunately, a simple ratio approach fails to make sense of the high ratio of brain to body size in mice (as much as twice the ratio in humans) and other small mammals, since no one seems ready to claim that mice have a slight intellectual edge over humans and a large advantage over other great apes. Moreover, the difference in ratio is not compensated by differences in neuron densities. Small mammals' brains are more densely packed with neurons because density increases with decreasing size. As a result, the ratio of neuronal number to body size favors small mammals even more than does the brain/body ratio. Neither an appeal to absolutely large size nor to greater size ratio adequately fits with intuitions about comparative intelligence.

Mice are not alone in exceeding the human brain/body ratio. In fact, most very small mammals have comparable brain/body ratios to humans. This reflects the fact that the ratio steadily decreases with increasing body size in most groups of animals. Common experience and intuitions about the mental abilities of animals do not suggest that such abilities decline with increasing size, but by the late nineteenth century, a number of scientists recognized that within this relationship was a hint of a way to salvage the intuition that intelligence increases with increasing brain size. As a result of a somewhat obscure study of an extinct sea cow, which possessed a brain that was absolutely larger than human brains, Alexander Brandt in 1867 speculated that a simple fractional comparison was insufficient for an accurate assessment of intelligence. Instead, he argued that the brain's size might rather have something to do with its special relationship to metabolism and to the body surface. Brandt suggested that brain size might scale with respect to metabolism (which researchers at the time thought was directly related to heat dissipation by the body surface—this turns out not to be true, but more of that later). He also suggested that the mass of the brain's sensory and motor systems should be related to the body's surfaces, not its total size.

The modern era of research on brain size effectively began in 1892, when a German physician named Otto Snell took this analysis to its logical conclusion.[4] Snell showed that mammalian brains and bodies were enlarged with respect to one another according to the power function 2/3 (see Figures 5.1 and 5.2). He derived this value by taking the average of log-transformed brain and body weight differences from pairwise comparisons of a number of different species across a range of body sizes. Snell provided compelling, if not systematic, evidence that the relationship of brain to body size in mammals was essentially comparable to a surface-to-volume relationship. The idea that brain and body weights were related by a systematic function begged the question of what this regularity had to do with brain function. But a metabolic interpretation of this pattern, initially promoted by both men, was soon supplanted by alternative interpretations and analyses that attempted to link it with intelligence

Snell's finding stimulated a burst of interest in the scaling of brain size and body size that has lasted to the present day. In the century that has followed, this basic insight has been the starting point for dozens of alternative statistical approaches and many hundreds of theoretical analyses of its significance. Differences in numerical methods have suggested alternative scaling exponents. Differences in the pattern among different taxonomic groupings have suggested that there is more than one regularity to explain. But the vast majority of interpretations all share two assumptions in common. First, that comparative intelligence is a function of how much of the brain is left when that fraction which is dedicated to basic body functions is analytically subtracted. Second, that this proportion is somehow directly reflected in the comparative scaling of brains to bodies. The regularity of this relationship offered a frame of reference. It determined a means of predicting what a typical mammal brain or body weight would be, and thereby indicated the outliers. It was natural also to assume that "average mammals," as predicted by this trend, should also be of average, and therefore comparable, intelligence, and that outliers were of respectively greater or less than average intelligence. In this view, intelligence is construed as the net or surplus brain function left over after the bodily needs of brain processes are subtracted.

The logic behind this conception found its clearest explanation in a widely cited book on the evolution of intelligence by Harry Jerison.[5] Jerison based his assessment of brain evolution on a concept he called "proper mass." On the assumption that evolution is miserly, and will tend to provide only as much of each tissue as is optimal in the context of the other systems of the body, one might expect that a larger and more costly organ should reflect a greater need for its function, and vice versa. The size of each organ should thus tend toward a proportion that is scaled to its relative importance to the rest of the body. In these terms it seems almost like a tau-

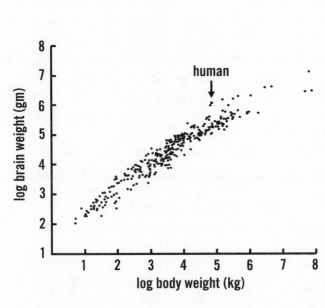

Figure 5.1 *Plot of brain and body size in a wide selection of mammals showing the almost linear distribution in log coordinates and the relative position of the human value with respect to the others. Humans show a greater divergence from predicted brain size for our body size than any other species. In general, monkeys, apes, and dolphins occupy the upper left edge of the distribution along its length.*

tologous expression of natural selection theory, a sort of metabolic economy argument. Brain size should, on average, tend to reflect the balance between the evolutionary advantages and costs of neural information processing, so that the average mammalian brain size for a given body weight should provide a useful index of the "proper mass" of brain for a given body. From this it should follow that the mammalian trend in brain versus body size should reflect an optimum balance of the costs and benefits conferred by the size of the brain, at least when all other things are equal.

Jerison argued that the fraction of total brain computation that is dedicated to serving the demands of body maintenance and overhead functions should scale to the 2/3 power of body size, because most neural representations of the body reflect maps from the sensory surfaces of the body.[6] Basing the assessment of proper mass on this empirical relationship is a powerful simplifying assumption, but is it as straightforward as the trend in mammal brain and body sizes suggests? One problem is that a whole organism economy of proper mass cannot be assessed solely in terms of information-processing capacity; other costs and benefits may scale according to independent factors. For example, one major cost that will scale with respect to the size of the brain is its metabolic demand. The brain is the most metabolically expensive organ at rest, consuming up to ten times more glucose and oxygen for its mass than any other organ.[7] Since basal metabolism scales to the 3/4 power of body mass in mammals (not 2/3 as Brandt and Snell and others had predicted from surface-to-volume heat dissipation assumptions), this cost must also be factored into the equation of proper mass. But precisely because such noncognitive costs must be considered, we should not expect that the average scaling of brain size to body size reflects anything like a line of isometric intellectual capacity. Proper mass, assessed in morphological terms, cannot be equated to net computational power, and so equivalent encephalization at different scales is not a reliable index of *equivalent* cognitive capacity. Indeed, there are many reasons to suspect that it misses some of the most important effects of scale on neural functions.

An assessment of net brain function is further complicated by its nonlinearity. A nonlinear mathematical relationship is one in which some fraction of the result must be figured into the calculation itself. Compounded interest on a debt has this characteristic (a painfully familiar example for many of us). Unpaid interest on a debt adds to the total debt, which increases the debt and so results in higher interest payments, and so on. This kind of relationship exists in estimates of net strength because muscles have to move themselves as well as other structures. Though having greater muscle mass increases the amount of force that can be generated, it also increases the amount of mass that must be moved. This means that when weightlifters put on more muscle mass, they do not get an increase in net strength that keeps pace with their increase in gross strength. This is one reason why most athletes, especially those who depend in significant ways on net strength (e.g., gymnasts), do not merely add muscle bulk as they train. There are clearly diminishing net strength gains with increasing muscle proportion beyond a certain level. (When that level is passed, we say that someone is muscle-bound.)

In terms of brain function, there are probably many counterparts to this sort of nonlinear relationship. Consider an analogy between the information-processing demands of a business and those of a brain. Growth of a business not only increases the need for more workers but also the need for increasing levels of bureaucracy. As the number of workers increases, so must the numbers of managers, administrators, and secretaries. These mid-level workers are necessary to keep track of other workers' hours, paychecks, duties, and so on. Though managers and secretaries comprise only a modest fraction of the total number of employees, this fraction tends to increase with the size of the business, because such middle-level employees themselves must be managed and paid, and so further contribute to the managerial overhead. As organizations grow, they tend therefore to increase the proportion of employees who handle employees, as compared to employees who directly handle the product or service being offered. Expansion of gross output inevitably demands a slight increase in the net proportion of management to production work in order to break even.

Undoubtedly analogous information-management demands come with increased brain size. Like an expanding business, brains may have to devote an increasingly larger proportion of their information-processing capacity to managementlike functions, just to maintain equivalent levels of functional integration and control in the face of increased size and complexity. The proportion of neural computations that can be devoted to input-output functions may consequently decline with size, leaving larger brains progressively less efficient in terms of the number of computations performed per functional output. Larger brains will thus tend to gain a greater gross computational capacity, but with a decreasing net computational efficiency. A larger body will demand a larger proportional commitment of managerial functions in order to break even. We can consider this managerial work to be analogous to the "higher functions" of a brain, those functions that direct, coordinate, and monitor input-output functions. From this perspective, larger brains will inevitably need to be more and more top-heavy, so to speak. More and more will need to be devoted to "middle management," and the executive hierarchy will need to grow new levels just to deliver comparable performance. With larger size, there will be more computational overhead that must be subtracted with respect to input-output capacity. This is an important figure-background shift in the analysis. It is not at all clear that what we generally mean by net intelligence is something like a surplus capacity, or that any simple analytical "subtraction" will be able to take this into account. More important, it provides a reason to suspect that larger brains will need to be differently organized from smaller brains, with different proper masses required for different component functions, further complicating comparisons across ranges of size.

Thinking Your Own Size

Intuition suggests that small-brained creatures like bats, moles, and mice are less intelligent, or at least less "cognizant" of the many possible options and consequences of their activities, than are large-brained creatures such as horses, lions, and elephants. The differences seem even more obvious when we compare large mammals to lizards, amphibians, and fish whose brains are usually smaller than the smallest mammal brains. And what about the tiny brains of insects? It takes a very big leap of imagination to extend to them even a fraction of the mental representational abilities of the lowliest vertebrates. Very large differences in absolute brain size seem to have an undeniable correlation with some aspects of mental ability that we recognize as intelligence. But is the number of neurons the only relevant factor, or is there more to it?

One facet of the brain size/intelligence problem that has been almost universally ignored is the fact that different-sized animals live in very different worlds. Since the pioneering work of D'Arcy Thompson on the effects of scale on organism design,[8] it has been clear that natural forms are subject to quite different types of forces and physical constraints if they differ significantly in size. Change in geometric relationships of body parts in response to size (e.g., surfaces with respect to volumes, forces with respect to support structures) is only one such expression of scale. The old adage about "fleas on the backs of fleas on the backs of fleas . . . ad infinitum" is not strictly possible, because at each level of scale totally different chemical and structural principles apply. Fleas must be designed very differently from the dogs that they parasitize. Dog-size fleas and flea-size dogs are impossible. Likewise, the bacterial microorganisms that parasitize the fleas are totally different in design from either fleas or dogs. Although the range in sizes is not so extreme among different vertebrates as to require radically different body structure, there is still a range of many orders of magnitude in scale—enough to make a difference. Within mammals we find a range of sizes that is nearly as extreme as that between dogs and fleas, and yet the body plans of all mammals, including their brains, are surprisingly similar. Nevertheless, the information-processing demands of being large as compared to small almost certainly make very different sorts of demands on large and small species' brains.

A telescoping of time correlates with size. Small animals' reflexes must be quicker in order to control much smaller limbs and respond to rapid locomotor feedback. Further, decision making must be streamlined in small species because their high metabolic rates and minimal energy reserves offer little leeway in foraging activities, defending against predators, or mating behaviors. And, perhaps most important, a short lifetime offers little time for learning from experience. As a result, being short-lived puts a premium on the effectiveness of preprogrammed behavior patterns that require little in the way of environmental priming or fine-tuning. Large animals, in comparison, can get by with rather slower reflexes, can afford to vary their sexual and foraging behaviors in an effort to better optimize their behaviors, and may have a considerable opportunity to learn by observation and trial and error. Being longer-lived puts a greater premium on learning and memory, and less on automatic preprogrammed behaviors. In addition, living a long time or having the capacity to travel for long distances (which are often though not always linked) is more likely to expose an animal to significant changes in the environment. Consequently, large, long-lived animals must be able to assess the effects of and adapt their responses to changing environmental conditions, whereas small, short-lived species don't face such changes within a single lifetime. There will be correlated differences in strategies of intergenerational information transfer. Large species will tend to do better by transferring learned information from parent to offspring, and by focusing effort on just a few malleable offspring, whereas small species will tend to do better to sample alternative adaptational strategies by producing large numbers of offspring with different variants of preprogrammed behavior patterns, and leaving the rest to natural selection. Many of these cognitive correlates of scale are schematically summarized in Figure 5.2.

Scaling brain functions up or down with size and life span, then, is not simply a matter of more or less computing power. Size changes have inverse consequences in a number of information-processing domains. A small species' brain simply scaled up for a large body or a large species' brain sim-

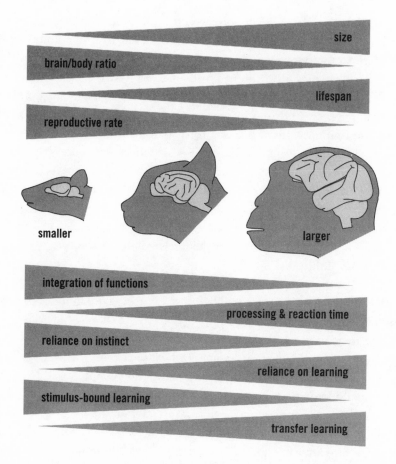

Figure 5.2 *A summary of some of the cognitive differences that correlate with differences in scale of brain, body, and lifespan (left to right = increasing scale; vertical thickness of triangular bars indicates extent or importance). Other general body features (shown above the depiction of three different-sized species' brains) include a reduction in brain/body ratio and reproductive rate with increasing size. A few cognitive correlates are depicted by the triangles below. These include a reduction of integration of functions, due to a more subdivided and less interconnected brain, and a corresponding increase in processing time and reaction times; a reduced reliance on inbuilt responses and an increased reliance on learning, because of a greater tolerance for less efficient but more flexible trial-and-error adaptation; and a shift in learning strategies from highly stimulus-bound learning to learning that is more open to generalization and transfer of information between diverse contexts. For example, transfer of training to conditions where associative relationships are exactly reversed has been shown to correlate fairly well with increased total brain size, but not with encephalization alone (see for example Rumbaugh and Pate, 1984; Rumbaugh et al. 1996), suggesting that this is a correlate of "gross" as opposed to "net" information processing capacity.*

ply scaled down for a small body would both be maladapted. Even in the domain of learning, though most species benefit from the ability to learn to adapt to temporary conditions, the emphasis may be on very different learning strategies at different levels of scale. For example, though both large and small species will likely be selected for learning to respond to immediately present noxious or rewarding stimuli, there is likely to be an increasing emphasis on responses to distant stimuli, and those that have merely predictive value, as the species' size increases. Larger species may also display an increase in the value of the ability to modify learned responses from one situation to fit another, and even a value to learning that is slowed by playful curiosity and exploration. To the extent that size-correlated differences in information processing are reflected in brain structure, we might expect a correlated allometry of brain structure relationships in animals of different size.

Perhaps the most serious problem faced by large brains is due to an inevitable geometric feature of network structure: as the number of neurons increases, the number of connections between them must increase in geometric proportions in order to maintain a constant level of connectional integration. This relationship is generalizable to many information and control processes. In brains that differ by millions or billions of neurons, maintaining a comparable degree of functional connectivity would require astronomical increases in connections, well beyond any reasonable hope of housing in one body. In addition, the metabolic constraints on the size of individual neurons also limit the number of synaptic connections any one neuron can support. So, inevitably, it is impossible to meet this scaling-up demand in any real brain, and a progressive reduction in many dimensions of connectivity is required with increasing size.

In real brains, the connections per neuron increase slightly with size, but the proportion of the brain's neurons to which any one is linked drops rapidly, decreasing the relative "fan-out" and "fan-in" of connections to and from different areas. Consequently, increasing size means an increasing fragmentation of function. It also means loss of speed, both because of the increased distances and because of the increased numbers of nodes that must be traversed by a signal to reach comparably removed sites in the overall network. Compared to electronic computers, the conduction of impulses along axons and across synapses is very slow. Though long, projecting axons incorporate design features (myelination) that speed and protect the propagation of action potentials, propagating signals over greater distances between brain structures inevitably takes more time in larger brains. This time penalty on larger size will also be amplified by the geometric increase in connections and indirectness of connectivity. Finally, a vastly larger and less integrated network will also tend to be far more susceptible to local perturbations, making the neural activities far more "noisy." This, too, will slow down recognition and decision-making processes and impede processing efficiency.

Thus, even if size confers greater information-carrying capacity, these gains may be balanced by significant costs in other areas of function. The inevitable information-processing geometry of size determines that bigger brains cannot be just scaled-up small brains, and this makes the brain size/intelligence equation all the more messy and complicated. But reduced processing speeds and loss of integration of function may not be prohibitive prices to pay for increased discrimination and storage capabilities, so long as its larger size also shields the organism from the need to produce rapid learning and responses. Again, it appears that bigger is different as far as cognition goes.

A few functional features do not change much with scale. These tend to be the functions that are more directly determined by molecular and cellular processes. Paradoxically, basic learning may be one such feature. The basic mechanisms that allow experiences to leave their trace in brain structure changes are highly conserved cellular and molecular mechanisms that are shared by most animals, from snails to simians. For example, a cellular-biochemical process known as *long-term potentiation* allows synaptic connections to be progressively strengthened or weakened in their effective transmission of information from cell to cell. And there appear to be parallel means by which the signals it carries can change a neuron's capacity to propagate signals through the different sectors of its extensive arbor of input branches (dendrites) and modulate signals down its one output line (axon).

Such processes as these are the basis for transforming patterns of experience to patterns of neural behavior—i.e., learning and memory—and it appears that even very simple nervous systems are built around essentially the same basic learning principles as the most complex ones. In a challenging book on brain size and intelligence, Euan MacPhail has criticized the idea that brain size is correlated with intelligence in any comparative sense, because he finds little evidence that basic learning abilities differ significantly across species of vastly different brain size.[9] He documents remarkably sophisticated learning abilities in a wide range of vertebrates not generally assumed to exhibit well-developed cognitive abilities. This could also be expanded to include a wide range of invertebrate learning studies as well. Sophisticated learning in honeybees, for example, has many remarkable parallels with vertebrate learning, in both the rate and complexity of learning.[10] At the very least, MacPhail's analysis offers the caveat that for basic neuronal processes, such as general patterns of simple associative learning, we should not expect brain size to have a very significant effect. Where size should matter most is in terms of differences in reliance on alternative learning strategies and perhaps extent or organization of mnemonic storage. One of the clearest demonstrations of this has come from studies of different species' abilities to transfer learned information across tasks and stimulus conditions. This second-order learning capacity, which enables animals to produce novel "emergent" responses by reusing information in new contexts, has been shown to correlate with brain size but not encephalization.[11]

This pattern of size-correlated differences in the relative importance of different learning strategies is of particular relevance to the special learning problems posed by language. Relatively flexible and indirect learning strategies can only be of use if there is sufficient time to employ them. This is even more of a problem for multistage learning-recoding-unlearning processes. Such learning strategies would have little utility for a short-lived, smaller animal. Smaller brains should instead be biased against such learning strategies, since there will not be sufficient time for them to pay off. This may help explain why symbolic communication didn't evolve until fairly large-brained, long-lived, ecological generalists like apes evolved. Since language learning is an extreme example of a highly distributed learning problem, smaller-brained species would likely be far more biased against the appropriate learning strategies than larger-brained species. Thus, absolute brain size might have played an important limiting role in language evolution irrespective of any increase in computing power.

Growing Apart

. . . nothing is great and little otherwise than by comparison.

—Jonathan Swift, *Gulliver's Travels*

The Chihuahua Fallacy

Differences in relative brain and body proportions are not always what they seem. Encephalization—the degree to which an animal's actual brain size exceeds what would be predicted for a typical animal of its size—is a relationship *between* the brain and body, and so it can be affected by changes in either. This problem is dramatically demonstrated by a familiar example: the differences in brain and body sizes in domestic dog breeds. An average-sized dog has a brain approximately the size that would be predicted for an average mammal of that size (a bit lower than for feral canines such as wolves), but small dogs have comparatively high ratios of brain to body size and large dogs have comparatively low ratios of brain to body size, because brain size is far less variable than body size between dog breeds. From the point of view of encephalization, very small dogs are among the most encephalized of all mammals, and very large dogs are rather poorly encephalized. Do we generally think of small dogs as being unusually intelligent and large dogs as unusually dull? There isn't any support for such a pattern of comparative dog intelligence,[1] and anecdotal evidence does not indicate that higher relative intelligence is found in smaller breeds.

Body size is far more variable than brain size within a species. Plotting log brain size against log body size for *intra*species samples produces trend lines of a low slope when compared to *inter*species brain/body proportions.[2] Smaller individuals within the same species have only slightly smaller brains than much larger individuals. Since middle-sized dogs are near the general mammalian prediction, the effect of breeding for large or small size shifts individuals up or down this shallow slope, exaggerating the deviation from the general mammalian trend. Thus, small dogs appear hyper-encephalized and large dogs appear hypo-encephalized compared to typical mammals (see Figure 6.1).

What if anything can we deduce from this about the mental abilities of dogs as a result of breeding? Stanley Coren, who has written a wonderfully informative book on dog intelligence, does not rank chihuahuas or pekinese among the smartest of dogs, even though they are among the most highly encephalized.[3] Although there may be variations in the mental abilities and predispositions of various breeds—some of which were intentionally selected by breeders—no one imagines that breeding for smartness would inevitably produce miniaturization or that breeding for dullness would produce giantism. We assume that chihuahua brains are fairly typical dog brains (possibly even deformed by miniaturization), with fairly typical dog abilities. They are just in very small dog bodies. Differences in dog encephalization are the result of breeding for body size effects, as well as for a number of other body proportions (relative leg length, head shape, etc.). There are numerous ways to alter the ratio between brain and body size in dogs that have nothing to do with selection for cognitive traits per se. Comparing breeds only in terms of encephalization obscures all these other possible influences on the statistic of encephalization. Dogs bred for short legs only and dogs bred for generally reduced trunk dimensions could have the same encephalization for very different reasons, and in neither case would it reflect selection operating on brain traits. Natural or artificial selection may favor a variety of combinations of body-segment proportions, and in different cases it is not easy to determine which are the independent and which are the dependent variables—which features constitute the fig-

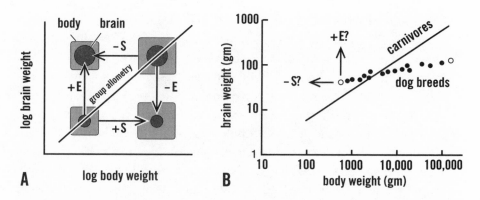

Figure 6.1 *The problem with comparative brain size as an index of brain evolution is well demonstrated by the different ways large or small brain size can be achieved.*

A. Graphic depiction of the four possible mechanisms that can cause an animal to diverge from the typical scaling of brain and body size of the group to which it belongs (e.g., all mammals, all primates, etc.). These deviations have usually been explained only as relative encephalization (±E), but here it is pointed out that evolutionary change in body size without concomitant change in brain size (somatization; ±S) is at least as likely to be involved.

B. Simplified graph of the relationship between brain and body scaling in domestic dog breeds compared to the trend for carnivores in general. Notice that the smallest dog breeds are above the carnivore trend and the largest dog breeds are below it because of the flatter within-species allometry. The smallest breeds could either be said to be more encephalized or less somatized than the average carnivore. The relative position with respect to the trend does not provide sufficient information to distinguish which, but information about other breeds and dog-breeding history indicates that the apparent encephalization of small dogs was rather an indirect product of selection on small comparative body size irrespective of brain size. The domestic dogs' trend line is actually below wild canids like wolves, but breeding has produced much greater variation of body size.

ure and which constitute the stable background. The problems arise from our desire to collapse a multivariate relationship into a bivariate relationship. Since the one variable, body size, is so much more inclusive than the other, it will likely contain the source of much more of the variance in the relationship, all of which will be obscured in an analysis of encephalization. Independent variations in all manner of body proportions are as likely to be evident in nature as in dog shows.

Curiously, though we might imagine people having an interest in breeding dogs for superior intelligence, it doesn't appear that this breeding criterion is the *sine qua non* of any particular breed. In fact, this should cause us to wonder if any domestic animal is bred for increased general intelligence. This was clearly not the guiding principle in the breeding of small dogs. In fact, it appears that domestic breeds have, on average, smaller comparative brain sizes than their feral cousins. This turns out to be the case for domesticated species of all sorts: horses, cows, pigs, goats, and so on.[4] Taking the encephalization argument literally, some brain size researchers have argued that in the process of domestication, we inevitably select breeding stocks for their dullness, not intelligence. But this, too, suffers from the same cephalic bias. Breeding species of farm animals for gross strength, meat production, milk production, and so forth, probably has been a major factor in this trend as well.

The problem with focusing on the brain/body relationship, then, is that it is usually not the brain that makes the difference. How this might affect our interpretation of comparative brain size can be dramatized by a simple thought experiment. Imagine that all other breeds of dogs but the smallest ones had perished at some point in recent prehistory along with supporting fossil evidence. Scientists coming upon one of these miniature dogs for the first time might not see them as miniatures of some larger type, but simply as small carnivores with large brains. Brain and body size statistics would confirm that this is a remarkably encephalized species. Researchers might even be tempted to theorize about the possible cognitive causes for their braininess—complex cooperative hunting, large social groups, and so forth. In the imaginary scenario of the single surviving extreme dog breed, we know that the predisposition to assume brain evolution is unjustified; but, I submit, scientists studying brain evolution have consistently taken this same leap of faith with no more solid justification. Why? What makes selection on brain size a more attractive hypothesis than selection on body size? Is brain evolution a more parsimonious assumption? Is there independent evidence that brain size evolution is far more prevalent than body size evolution? As far as I can tell, there is not. I am afraid that our predisposition to see brain size evolution where there is none mostly grows out of extra-scientific prejudices having to do with the comparatively large size of our own brains.

If the link between encephalization and intelligence is measured in "extra neurons," then the chihuahua example is hard to explain away. We might try to salvage the theory by introducing arbitrary special conditions with respect to types of encephalization. Chihuahuas and human dwarves are encephalized because their body growth has been stunted. Perhaps we can simply exclude such cases. I think this is on the right track, but a different neural logic must be invoked to explain why this should matter. We could not simply frame the argument in terms of the proportions of neurons to body mass, since by this measure it doesn't matter whether brain or body proportions are the source of the shift. So why aren't chihuahuas the darlings of comparative psychology, and why isn't human dwarfism a source of great genius? The answer must be that, just as encephalization doesn't carve the body into the appropriate units of analysis, the extra neuron hypothesis doesn't carve the brain appropriately either.

How can we sort out figure from ground when analyzing body-part growth? When we see a dog like a dachshund, we immediately recognize that it has been bred for short legs with respect to trunk length because we have a frame of reference—an average dog body—to compare it to. Proportional differences between a part of the body and the whole body tend to reflect change in the part with respect to a stable whole, all else being equal, but this is not always a safe assumption. Using dogs as our model, we can see that there are at least two kinds of morphological information that can prevent us from jumping to mistaken conclusions. Selecting a single body part, such as the brain, to compare to the size of the rest of the body introduces an unwarranted bias because it considers this one part out of context. To control for this, the rest of the body needs also to be analytically divided into segmental units that are roughly at the same level of the anatomical hierarchy (e.g., limbs / trunk / head or muscles / digestive organs / cardiopulminary organs / brain). Only in this way can we hope to find unbiased evidence of which is figure and which is ground. In general, I suspect that most of those mammals we determine to be poorly encephalized are analogues of great danes, selected for larger overall body size, and most of those mammals we determine to be highly encephalized are analogues of chihuahuas, selected for small body size. Unfortunately, the multivariate data necessary to pursue this question are unavailable for most species. Fortunately, there is another source of evidence that may help resolve this question for primate and human brain evolution.

One of the least controversial claims about brain size evolution is that primates tend to have larger brains for their size than most other mammals. When seen in terms of encephalization, primates have nearly twice as much brain for their body size as other typical mammals, and humans have almost three times more brain than a typical primate. It would appear that during the primate radiation, there was an increase in brain size over other mammals, and that this trend culminated in human brain size evolution. A great deal has been made of the apparent importance of brain evolution in the primate order. Most researchers take this to mean that primates are on average smarter than other mammals and that humans are the smartest of the primates. Because of the tendency to see evolution from an anthropocentric perspective, and human brain evolution as the culmination of a much larger trend, we tend to view other primates (beginning with our closest relatives, the great apes) as having achieved an intermediate level of "higher intelligence" over and above other mammals. But are the larger proprtions of primate brains with respect to their bodies a reflection of more rapid brain growth or reduced body growth? We can determine whether brains increased with respect to bodies or bodies reduced to brains in primate evolution, as we did with dog breeds, by using information about patterns of growth. Since there is a limited set of patterns by which the masses of brains and bodies grow within young mammals, a comparison of growth patterns can help determine which of these two variables is figure and which is background.

One might expect there to be different brain/body growth trajectories with diverse slopes and ratios for each species, owing to their unique adaptational specializations. But, remarkably, the curves that describe brain/body growth for all mammal fetuses tend to cluster along just two parallel trajectories during fetal development: one that includes primates, cetaceans, and elephants, and a second that includes the remainder of the mammals (see Figure 6.2A). During this early phase of growth, individuals' brains and bodies grow almost in unison, so that enlargement is essentially isometric for all mammal fetuses (like an expanding picture on a balloon being inflated). This suggests that growth rates are likely the same throughout the whole body. Growth is multiplicative: tissue duplicating itself at the same rate will produce progressively more rapid growth. As development proceeds, bodies and brains are growing absolutely faster, but at the same weight most fetal mammals grow at the same rates: a 5-gram cat fetus and a 5-gram cow fetus will add the same amount of new tissue in the same period of time, even though they are at different points along their respective developmental trajectories and headed toward very different end points.

So what accounts for the difference between primates and most other nonprimate mammals? Surprisingly, primate brains grow at essentially the same rate as other mammal brains. Brains that reach the same size at birth take approximately the same amount of time to get there, whether in a primate or nonprimate species (see Figure 6.2B). Primates do not grow their brains faster than other mammals, but they grow their bodies slower. From an early embryonic stage, primates have smaller bodies than expected for their age. The apparent increase of encephalization in primates is then, more accurately, a decrease in somatization. The locus of primate encephalization is not in the head!

This poses a serious challenge to the traditional view of primate evolu-

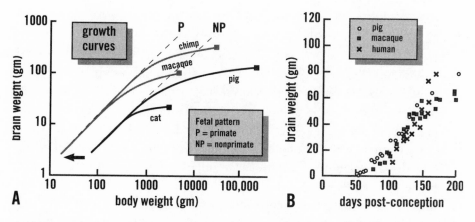

Figure 6.2 *The common shapes but different intercepts of developmental brain/body growth curves distinguishing primates and nonprimate mammals are shown in a slightly idealized comparison of mammalian growth. The left graph schematically depicts the general pattern exhibited by brain/body growth throughout life for four mammals, starting from early embryonic stages. Both curves demonstrate a two-phase pattern of brain/body growth caused by the fact that the brain ceases growth earlier than the rest of the body. P stands for the typical primate fetal brain/body growth pattern and NP for the nonprimate pattern. Notice that the primate line is considerably shifted to the left. Comparison of brain-body growth in mammals shows that all anthropoid primates follow the same growth trajectory prenatally and most other mammals follow one other right-shifted parallel trajectory. Adult allometries within these two groupings of mammals are the result of a scalar expansion of the whole growth curve. The leftward shift of the primate growth curves (producing larger brain per body) is evident from the youngest embryos. The graph on the right shows total brain growth rates for two of these species (pig and macaque) as well as for humans. This demonstrates that the left-shifted primate growth is not the result of faster brain growth but reduced body growth. Human brains follow this pattern also.*

tion as characterized by selection for increased intelligence. If primates were selected for increased intelligence, why should this produce a change in body growth but not brain growth? Shouldn't we instead consider primates to be examples of phyletic dwarfism, rather than brain hypertrophy? We are faced once again with the chihuahua question: Are we justified in inferring anything about brain evolution from a reduction in body growth with respect to brain growth?

Perhaps the answer can be approached by determining how primate bodies have been reduced. Is the primate transformation of fetal growth relationships what we should expect for dwarfism? In other words, does primate brain and body growth differ from other mammal patterns the way chihuahua brain and body growth differs from the growth patterns of other dogs? In fact, these two paths to reduced body size without a corresponding reduction in brain are different. Dwarfed animals exhibit slowed body growth in the later phases (late fetal and postnatal), but they track along with more normal members of their species (and order) for most of fetal development. Primates, however, follow a shifted, parallel gestational growth curve compared to that of other mammals from the start. Fetal brain/body growth in primates is isometric, as in other mammals. Within each fetus, brains and bodies are growing at nearly identical rates, yet the whole primate body is smaller at all corresponding stages of gestation, even though primate brain growth keeps pace with other mammals. It is as though a significant fraction of the rest of the body is missing from the very first moment that head and somatic divisions of the body become evident in the embryo. So primates are like chihuahuas only in a superficial sense. Both have relatively big heads and brains due to reduced body growth, but primates start out with small bodies while chihuahuas only end up with small bodies.

What about humans? According to the traditional view of our place among other mammals, primates evolved bigger brains than other mammals as a response to a more cognitively demanding niche, and humans simply carried this trend further than any other primate. But now that it appears that primates did not evolve bigger brains, just smaller bodies, we need to reconsider the view that human brain evolution was simply an extrapolation of the primate trend. A comparison of human and nonhuman primate brain and body growth patterns (see Figure 6.3) shows us deviating from a typical primate trajectory only postnatally. Thus, we have not shifted away from other primates the way primates have shifted away from other mammals, and so have not simply extrapolated the primate deviation. Instead, like dwarves, the difference between human and monkey-ape growth patterns appears to be the result of a truncation of the growth curve. And yet we are not dwarfed primates. We are among the very largest of primates, and our adult brain size is greater in both absolute and relative proportions than in any other primate. Although our growth curve appears to be cut off early, as in dwarfism, our body growth rates are not suddenly slowed immediately after birth as would be the case in dwarfism. We start out growing according to the standard fetal primate plan, and our brains continue to grow for longer than expected. Further proof that hominid brain size did not arise by somatic miniaturization is provided by fossil evidence. We are

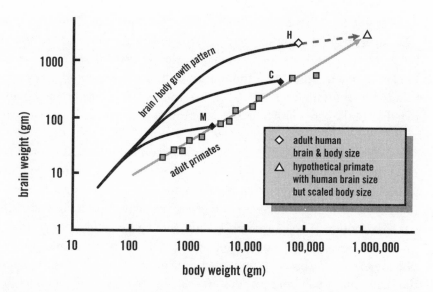

Figure 6.3 *A somewhat idealized graph showing the human brain/body growth curve as compared to other species of primates (H = human, C = chimpanzee, M = macaque monkey). Squares and diamonds indicate average adult values of brain and body weight for selected primates. The gray line is the average adult brain and body size allometric trend for primates. Human fetal growth follows the fetal primate trajectory, but the overall shape of the curve is altered so that the early phase is comparatively extended and the postnatal phase is comparatively truncated. If our body growth followed the pattern indicated by our brain growth, we would grow to be a very large ape (indicated by the dashed line and the upper right triangle). This also contributes to the necessity of giving birth at an earlier stage in the maturation of our offspring. The common trajectory we share with other primate during the fetal phase indicates that increased human encephalization compared to other primates is not an extrapolation of the process that caused primates to diverge from other mammals. Ours is a different mode of encephalization.*

among the largest members of our order (the primates), and the fossil record leading up to us suggests that there was an absolute increase—not decrease—of body size, with an even more extensive increase in brain size at a somewhat slower pace. In humans, the locus of the change was indeed the brain.

Nevertheless, the rate of human brain growth does not distinguish us from other mammals. Differences in adult mammal brain sizes are a function of duration of growth, and our brains reach their large size in about the same amount of time as we should expect to find in a large dolphin or a very large ungulate with a brain this size. Our brains grow as though in a primate with an adult body size that well exceeds 1,000 pounds, and yet our body growth pattern is quite similar to that of a chimpanzee. Both the human brain and the rest of the human body grow according to expected trends for their target adult sizes. These trends just don't normally belong together in the same individual. It's as though the "designer" got the parts mixed up.

Our preoccupation with brain size and encephalization has led us to ignore the differences and confuse these three distinct processes: dwarfism in small mammal breeds, embryonic reduction of body growth but not brain growth in primates, and brain growth prolongation in humans without an extension of body growth. To treat them all as reflections of a single progressive evolutionary trend not only equates very different biological phenomena on the basis of superficial appearance, it also leads us to ignore the likely important role of developmental processes on functional outcomes. These three very different modifications of brain and body growth almost certainly have significantly different implications for brain organization, cognitive function, and the selection pressures that produced them. What we need to figure out is just how these differences in means effect differences in ends. The first problem, of course, is to understand how such differences come about, and this is just beginning to be understood.

Using Fly Genes to Make Human Brains

What determines which fractions of a developing embryo will become brain and nonbrain structures of the body and which will become the different brain structures? We owe thanks to the lowly fruit fly for providing the initial clues that are helping to solve this mystery. The position within an embryo where the brain, heart, stomach, limbs, and other organs will develop is controlled by a constellation of genes known as *homeotic genes*, so named because their expression patterns correlate with the repetition of similarly organized segments along the body axis (*homeo* meaning "similar;" sharing a common underlying plan). What is most significant about these body-plan genes is their evolutionary conservatism. Corresponding homeotic genes appear to be critical for determining segmental organization in the developing bodies of worms, vertebrates, and insects. A homeotic gene important for the development of the human head and brain has even been transferred to mutant fly embryos whose development is abnormal because the corresponding genes have been inactivated, and this human gene has been able to compensate partially for the fly abnormality. This attests to the miserly nature of evolution, not just in genes but in their functions as well, especially when it comes to the events of early embryogenesis.[5]

Homeotic genes all tend to all have one thing in common: they each encode a DNA-binding region that enables the protein molecule they produce to bind to other sites on a chromosome and so regulate the expression of other genes. The most common version of this DNA-binding region is termed the *homeobox*, because of its homeotic functions. Homeobox-containing genes appear to have the ability to control whole suites of other genes, including other homeotic genes. It is probably this hierarchic feedback among different families of homeotic genes that enables them to orchestrate the same sequence of genetic events, with slight variations on the same theme, at different locations along the developing body, and at different stages in development.

Homeotic genes initiate pattern formation in the embryo by virtue of different genes switching on in spatially separated groups of cells in a series of progressively restricted subdivisions of the embryo. Beginning with genes that are differentially activated in the head and tail halves, and the front (belly) and back halves, and progressing to the formation of regular bands of expression down the length of the body, to more complicated combinations of these patterns created by orthogonal and overlapping expression patterns, the cells in the developing embryo inherit developmental fates from their precursors to become distinct adult cell types. At the early stages there is considerable similarity between very diverse species, so that the expression patterns of different homeotic genes in the developing vertebrate embryo produce a series of bands that resemble the segments of a worm or caterpillar. As in such more obviously segmental animals, these vertebrate body segments are each essentially replicas of one another to some degree, containing corresponding parts in corresponding positions. Vertebrae, ribs, and limbs resemble each other because they are corresponding elements in segmental units. Unlike a segmented worm or insect larva, however, the segments in vertebrates become obscured as the fetus develops, both because individual segments grow to differ and because they do not all remain in simple serial alignment. This is particularly the case with the development of the head.

The great German eighteenth–nineteenth-century writer and philosopher of science Johann Wolfgang von Goethe was one of the first to suggest that the serial repetition of vertebrate body segments might be continued in cryptic form in the construction of the head. Later nineteenth-century biologists took up Goethe's hypothesis, including the influential anti-Darwinist Richard Owen. But determining whether there was such a correspondence was complicated by the large number of structures that are unique to heads (e.g., special sense organs like the eyes and ears). Only after almost two centuries have developmental geneticists uncovered evidence that partially vindicates this insight. In the body of the embryonic fruit fly, where the effects of homeobox genes were first understood, it is easier to find structures of the head that appear homologues to those of the rest of the body. For example, the segmented structure of the antennae resembles the segmented structure of the limbs. Mutations affecting homeobox genes expressed in the head have shown that these similarities are deep. A homeotic mutation called *antennapodia* actually causes antennae to develop into limbs. This suggests that the same basic program of genetic expression is slightly modified by homeotic controller genes to make both limbs and antennae.

In mammals, too, homeotic genes are expressed in the developing head and brain, and most of these have fly counterparts. But in the vertebrate brain, there is an apparent breakdown of strict serial segmentation of gene expression (and later structure) as compared to the patterns in the brain stem and spinal cord. The early nervous system begins as a long tube running down the back from the head to the tail end of a wormlike, undifferentiated body. A set of homeobox genes, called Hox genes in the mouse (and which correspond to a nearly identical series of HOM genes in flies), are expressed in serial order along this neural tube axis in partially overlapping patterns like tubes in a telescope. This pattern is depicted in Figure 6.4, and can be seen to become active in a nested pattern during development, with the first Hox genes to be expressed occupying the full length of the brain stem and spinal cord, and later Hox genes turning on within progressively more restricted and posterior subsets of the tube. The Hox genes produce a segmental pattern that is nearly as regular as their counterparts in flies, and they are arrayed in roughly the same front-to-back order.[6]

In the early stages of brain development, the front end of the neural tube enlarges to either side, producing two small, balloonlike bulges. These eventually form the *telencephalon*, made up of the cerebral cortex, limbic system, and basal ganglia, on either side, which will expand to dominate the cranial cavity. In the middle between them is the front end of the tube, which will become the *diencephalon* (composed of the thalamus and hypothalamus); behind it is a bent section of the tube that will become the *mesencephalon*, or midbrain (so named because it is midway between the hindbrain and spinal cord below and forebrain above); and further back is a regular set of bulges that will become the cerebellum, pons, and brain stem. From the midbrain forward, the Hox genes are not expressed, and this corresponds to a transition from relatively simple seriality of organization to a more complicated geometry. This transition occurs roughly at the same point as the front end of the notochord (the mesodermal rod that foreshadows the developing vertebral column and influences the formation of the neural tube), at which point the neural tube appears to curl around the end of the notochord, as though it has lost its linear organizing influence. It is an interesting coincidence that this transition zone roughly corresponds to the point where neural tube formation is initially induced, where major pathways into and out of the brain cross over to opposite sides, and where maturational processes tend to begin and then propagate forward and backward.

Although strict segmentation breaks down forward of this point, a sort of modified segmental gene expression continues to subdivide the head end of the neural tube. The forebrain homeobox genes, whose expression patterns are most similar to the segmental expression patterns of Hox genes, derive from two families of two genes each. These are designated by the acronyms "Emx" and "Otx," named for their associated mutation effects in the developing fly head (Emx = Empty Spiracle and Otx = Orthodenticle) where their homologues were first identified. Like the Hox genes, these genes are expressed in a nested pattern in the vertebrate brain (Otx2>Otx1>Emx2>Emx1, shown in Figure 6.4), with the largest domain (Otx2) extending from the front of the brain to the back of the midbrain, and the smallest domain centered in the dorsal telencephalon roughly corresponding to the entire isocortex (Emx1).[7] The timing of gene expression from largest to smallest expression region is like that of Hox genes, except that the direction of these expression trends is reversed. Hox genes are turned on in a front-to-back order and Otx-Emx genes are turned on in a roughly back-to-front order. One other major difference, however, is that the expression of the Otx and Emx genes is mostly confined to the upper or dorsal part of the forebrain neural tube, whereas Hox genes are expressed in complete cross-sectional segments of the tube. These are only a few of the homeotic genes that are expressed in the developing forebrain and that play major roles in determining its organization, but the segregation of dorsal from ventral gene expression and the back-to-front segmentation appear to be common to most other forebrain genes' expression patterns.

The discovery of homeotic genes has initiated a revolution in the study of brain development and brain evolution. Because they determine the boundaries of the major classes of cell lineages and growth fields within the developing brain, they provide an important new source of data about pre-

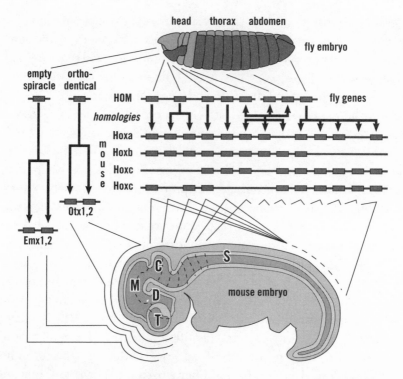

Figure 6.4 *Expression in the developing fly and mouse of three families of homeotic genes (which are presumed to be the same in humans). These include the HOM, orthodenticle, and empty spiracle gene families in flies; and the corresponding Hox, Otx, and Emx gene families in mammals. Each small rectangle on a line depicts the position of a gene on a string of DNA in a chromosome. Vertical arrows indicate probable homologies; also there are many gene duplications in mammals. Note the spatial correspondence of expression patterns along the body axes of both species. The Hox genes are expressed in the brain stem and spinal cord in spatially overlapping segments, with the earliest to be expressed covering the largest portion of the neural tube (indicated by the innermost line curving around the top of the embryonic mouse brain). Each successive Hox gene to be expressed starts a little further down the brain stem (indicated by lines overlapping), creating a segmental expression pattern that plays an important role in determining the segmental positions of the various sensory and motor nuclei of the brain stem and also organizes the segmentation of major body parts (e.g., limbs). The Otx and Emx genes are expressed in the midbrain and forebrain of mammals and are mostly confined to dorsal structures (i.e., on the outside of the curve of the forebrain). The earliest to be expressed is Otx2, which covers the largest sector of the dorsal forebrain after becoming successively restricted to this region over time from an initial distribution that includes most of the undifferentiated embryo. The last of these to be expressed, Emx1, is almost entirely confined to the region that will develop into cerebral cortex. S = spinal cord, C = cerebellum and brain stem, M = midbrain, D = diencephalon, and T = telencephalon.*

viously cryptic sources of information used to design brains. These correspondences are particularly relevant for understanding how evolution might be able to adjust proportions in bodies and brains, as it has in the human case. Indeed, transgenic embryos of mice, chicks, frogs, and flies in which the expression of homeotic genes has been experimentally modified often exhibit significant changes in body plan, including even the addition or deletion of segmental divisions of the brain. So, the sorts of changes in body plan that are characteristic of both the primate shift and the human shift in brain/body proportions may well be traceable to homeotic gene effects. Unfortunately, although the expression domains of these genes have been well mapped out in developing mouse brains and bodies, there is almost no corresponding information for comparing human and nonhuman primate brains. Though it is almost certain that highly similar patterns of homeotic gene expression will be present, we are not in a position to make the appropriate direct comparisons to decide whether subtle expression differences determine either the primate shift in body development or the human shift in brain development. Nevertheless, because of the incredible conservatism of these developmental genetic mechanisms, some important clues to the genetic basis for these primate and human differences can be extrapolated from experiments involving other species' embryos, even those of nonmammals.

How might changes in homeotic gene expression affect such global neural proportions? An extreme example is provided by a recently described gene called Lim1 (the initial version of this gene type was identified in roundworms). In mice, the corresponding gene appears to be critical for initiating head formation. Mouse embryos lacking a working copy of this gene fail to develop heads altogether, though most of the postcranial structures develop relatively normally.[8] Though head deletion is hardly a useful adaptation, such independent determination of head and body development suggests that whole suites of other developmental events may also be able to affect heads and bodies differentially. This may help explain the relative ease with which body growth and proportions can be readily affected by selective breeding, while heads and especially brains change comparatively little (as in small and large dog breeds).

An even more relevant experimental manipulation of brain and body proportion has been demonstrated in frogs (see Figure 6.5). The expression of the frog version of the Otx2 gene (X-Otx2; X- for *Xenopus*, a genus of frog), has been followed from the fertilized egg to the late stage embryo, and has been shown to become progressively restricted over development to the head end of the body, where it is eventually only expressed in the forebrain, as in mice. This narrowing and focusing of expression appears partly controlled by interactions with other genes and gene products in the embryo. The extent to which X-Otx2 becomes restricted to the head end of the neural tube can be experimentally manipulated by bathing the embryo in another differentiation factor (retinoic acid). This decreases X-Otx2 expression levels all over, and causes the head and brain to recruit a much-reduced fraction of the embryonic body as a result. An inverse effect is induced if X-Otx2 expression is artificially increased (e.g., by injecting extra X-Otx2 RNA into the zygote). This causes overall X-Otx2 expression to be higher throughout the embryo, and results in a much-enlarged fraction of the embryonic body becoming recruited to head and brain development. Such experiments suggest that concentration threshold effects may play a crucial role in the global proportioning of these large segmental distinctions during development. Relative timing or rate of gene product production might thus be important factors distinguishing species where these proportions have become modified.

One likely means of increasing levels of expression that seems to be utilized by homeotic gene systems is gene duplication. Gene duplication is a common evolutionary phenomenon that has produced most families of related genes, such as the many variants of hemoglobin genes that are active at different times of mammal development corresponding to different oxygen-transfer requirements (e.g., in the womb as opposed to out of it). Gene duplication is also quite prevalent among homeotic genes, as in the Hox, Otx, and Emx gene families discussed above. There is clearly some degree of redundancy of developmental effect in many of these genes, because it is not uncommon for knockout mutants (missing working copies of a particular gene) to develop relatively normally or with only minor structural modifications, even where ectopic expression (turned on indiscriminately in inappropriate body regions) of the same gene may cause significant structural modifications. Variants of the same gene active at nearly the same time in overlapping domains may contribute to a selective expansion of corresponding segmental expression domains within the neural tube by shifting the relative concentration gradients, where threshold levels produce borders between regions determining different cell fates. Thus, a homeotic gene duplication in evolution might be the genetic equivalent of the experimental introduction of extra gene product in the frog embryos described above.

Though we are still a ways from proving any clear association between a known homeotic gene difference and any naturally occurring species brain structure difference, the field is still very young. We at least know where

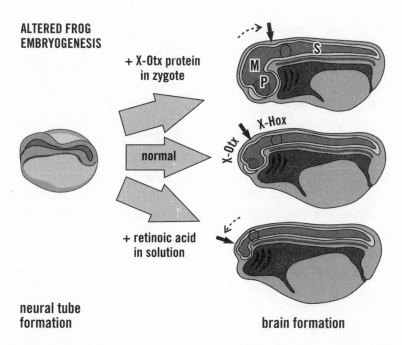

ALTERED FROG EMBRYOGENESIS

+ X-Otx protein in zygote

X-Otx   X-Hox

normal

+ retinoic acid in solution

neural tube formation

brain formation

Figure 6.5 *Schematic depiction of the effects on embryonic brain and body propor-*
tions of modifying X-Otx2 gene expression in developing frog embryos (X-Otx and X-
Hox refer to Xenopus homologues of mammalian Otx and Hox-cluster genes). The
normal growth pattern (middle right) is depicted from the stage of initial neural tube
formation to the point at which the major forebrain structures are formed. The top
right image depicts the effects of increasing levels of X-Otx2 in the cells (added to em-
bryonic stem cells). The results are a posterior shift of the transition between the ex-
pression of X-Otx and X-Hox genes (demarcated by the black arrow), a corresponding
increase in the size of the head and brain (beginning at the level of the midbrain), and
a corresponding decrease in the postcranial body. The bottom right image shows the
reverse effect produced by blocking X-Otx expression. When X-Otx is present in low
levels, it becomes restricted to only a very small part of the front of the neural tube,
resulting in a greatly reduced midbrain and forebrain. The images graphically sum-
marize experiments described by Boncinelli and Mallamaci (1995). P = prosen-
cephalon (early stage telencephalon + diencephalon); M = midbrain; S = spinal cord.
Recently the development of a mutant mouse lacking Otx2 has been described by
Ang et al. (1996). These mouse embryos are abnormal from the gastrula stage and
fail to develop the most forward part of the brain.

we should be looking for the appropriate developmental genetic links. Consider in this context the reduction of the primate body with respect to the
bodies of other mammal embryos. The nature of the shift in proportions
and the timing of its appearance are important clues about possible devel-

opmental genetic correlates. Since this shift in proportions is evident at the earliest point in embryogenesis at which primate and nonprimate brains are distinguishable, it suggests that there has been some change in the expression of some of the very early acting genes responsible for establishing this segmental distinction. The Otx experiments offer a good model for processes that would affect proportions at this stage of development. Comparative neuroanatomical data should be able to guide us to the appropriate juncture in development, to the extent that it can specify exactly where segmental shifts in proportions have occurred. In fact, information about the deviation of human brain growth from other primates may actually be of sufficient resolution to enable us to make a more precise prediction about its genetic correlates.

Large sets of quantitative data comparing human to other primate brain structures have been available for decades, and these have been subject to numerous statistical analyses. To date, none of these analyses has utilized information about the patterns of brain development analytically to carve the brain into the appropriate growth fields and to determine the nature of the underlying segmental shifts in proportions, because until recently such information was unavailable. And this is critical. To measure things appropriately, it helps to know the locations of the natural developmental divisions. If one wanted to study the growth of the human body during childhood, it would not be useful to measure arbitrary divisions (such as the distance from the neck to the middle of the forearm compared to the distance from the middle of the forearm to the first knuckle). It would make better sense to measure body segments that grow separately, for example, long bones as measured from joint to joint. Since the homeobox gene expression boundaries determine the "joints" between brain growth fields, they should be the most useful guides in determining which measurements will be most informative.

The analysis of brain structure allometries, like the analysis of brain/body allometries, has been led down many blind alleys because of this uncertainty about the appropriate units for analysis, and a failure to pay attention to developmental patterns. In addition to a lack of developmental clues, two other analytical problems have complicated the interpretation of quantitative brain data and have led to confusing and conflicting assessments. These have mostly derived from a failure adequately to control for size effects in part/whole comparisons, similar to the brain/body problems we encountered earlier. But unlike the brain and body comparisons, where brains make up a small fraction of the body at best, some brain parts constitute a large fraction of the brain, and so proportional comparisons can misrepresent growth relationships by failing to treat the parts separately.[9] This is a particularly thorny problem for brain development and evolution. The size of a body structure (especially a subdivision of the brain, such as a thalamic nucleus or a cortical area) is determined both by the number of cells generated (proliferation), and by nonproliferative processes that carve this field into functional subdivisions (parcellation). Development proceeds via the interactions of both functions at different stages (these processes are discussed in the next chapter). The complication for quantitative analyses is that whereas cell multiplication has a scalar influence on all structures involved, parcellation is a zero-sum process in which enlargement of one structure only occurs at the expense of another. So, failure to distinguish comparisons that are within from those that are between growth fields can produce confusing and misleading results.[10]

The shift in human brain/body and brain structure proportions inevitably involves both effects. Since proliferative effects are primary, and occur in the context of homeotic segmentation, we should initially try to match the quantitative neuroanatomical analyses to comparisons that roughly correspond to the hierarchy of homeotic segmentation processes in brain development. In general, more global proliferative effects will be expressed earlier and local parcellation effects will be determined later in development. So let's begin with large-scale size relationships and progressively focus in on more localized effects in the search for what is different about human brains.

The discovery of growth deviations is aided by the fact that in general the relative proportions of different brain structures are predictably scaled with respect to overall brain size. When the volumes of mammal brain structures are plotted, the points marking the volumes of pairs of structures in different species tend to fall neatly along a smooth curve (or a straight line in logarithmic coordinates). These highly predictable relationships between the sizes of major brain structures are a reflection of the fact that brains of vastly different sizes derive from a similar highly conserved homeotic starting point chiefly by extrapolating global proliferative processes. The numerical predictability of brain structures differs somewhat from one mammalian order to another; but within orders, families, and genera there is progressively less individual divergence from the trends.

We have established that the fetal brain/body growth pattern for human development is essentially the same as for any "generic" monkey or ape, so we need to look one level down (and later in development) for what might be the human difference. When comparing such major brain "segments" as the telencephalon, diencephalon, and so on, patterns of deviation begin to emerge, and a clear, regionally correlated shift away from typical primate proportions is evident when we begin to consider major subdivisions of each. When the values for structures in the human brain are added to plots of the corresponding structures of other primate species, the curves extrapolated from other primate data predict the sizes of many, but not all human brain structures. Those major structures that are predictable from typical primate size trends turn out to be located within contiguous segmental divisions of the human brain. Thus, components of the basal ganglia (such as the striatum and the pallidum) are matched in size to one another according to the pattern typical of primates, and these are also matched in size to major components of the diencephalon (thalamus and hypothalamus). But the human volumetric relationships most consistently deviate from the primate extrapolations only in certain comparisons between separate brain structures. For example, when we compare any of the basal or subcortical forebrain structures (basal ganglia, thalamus, hypothalamus) to the cerebral cortex, or compare the brain stem and spinal cord to the adjacent cerebellum, this predictability breaks down (see Figure 6.6). The nonhuman primate data underestimate how these structures have scaled up in size with respect to one another in the human brain, so that, for example, the cerebral cortex is roughly twice as big as predicted for many other forebrain structures, and three times as big as predicted for the brain stem, spinal cord, and the rest of the body (see Figures 6.6 and 6.7).

The patterning of these deviations is interesting. Two of the most deviant structures, the cerebellum and cerebral cortex, originate from the dorsal (or back) side of the developing neural tube (as does the dorsal midbrain, which also appears enlarged compared to ventral—front oriented—structures). This deviation from primate patterns suggests that the growth relationship *between* major dorsal and ventral segments of the developing forebrain has been somehow altered, whereas the many relationships *within* each of these broad divisions have remained relatively constant. Such a general pattern is an important clue. It indicates that the extensive functional interdependence of these dorsally and ventrally derived brain regions is not crucial for determining their respective growth (though one might have predicted that the cerebral cortex would need to be strongly coupled in size with its major sources of inputs and major targets for outputs). On the other hand, the fact that there is growth linkage between major groups of structures that are not so extensively interconnected or functionally dependent suggests that the

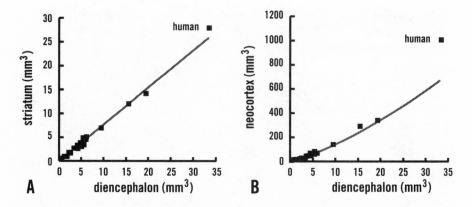

Figure 6.6 *Two examples of human brain structure proportions showing both proportional and disproportional relationships.*

A. *Proportional expansion (i.e., predicted by the primate allometric growth pattern for these structures) of the diencephalon and striatum in the human brain (both mostly confined to the ventral forebrain.*

B. *Deviation from the predicted relationship between the size of the cerebral cortex of the diencephalon, showing increased cortex despite the intimate functional and connectional interdependence between these structures (data from Stephan, Frahm and Baron, 1981).*

---

shift must be determined at a fairly early stage in neuroembryogenesis, when these segmental divisions are just forming. At this stage, only crude divisions of the brain are specified, neural stem cells have not yet given rise to the differentiated neurons and *glia* (additional support cells around neurons) that will comprise these structures.

When we trace back the developmental history of those structures in human brains that are significantly enlarged versus those that are only slightly enlarged, they divide up the neural tube in a pattern that corresponds to the expression domains of distinct groups of homeotic genes. The regions of the embryo that will give rise to enlarged structures in the human brain tend to be on the dorsal surface of the neural tube, and they form an essentially continuous sheet from the cerebellum to the dorsal telencephalon (see Figure 6.7B). This is paralleled by a mostly dorsally restricted sequence of gene expression domains of the Otx and Emx homeobox genes, and matches the general segregation of homeotic gene expression for this dorsal/ventral forebrain distinction. Though we should not simply ascribe the human neuroanatomical divergence to the actions of Emx and Otx genes, their expression matches the entire pattern particularly well, and suggests

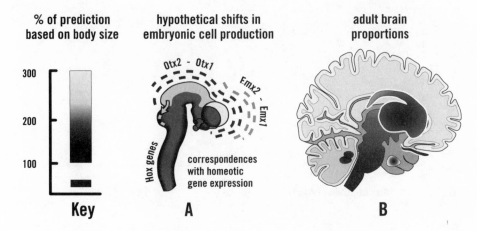

| % of prediction based on body size | hypothetical shifts in embryonic cell production | adult brain proportions |

Key   **A**   **B**

Figure 6.7 *Figure summarizing the size deviation in major structural divisions of the human brain compared to what would be predicted for a "typical" ape brain for human body proportions (i.e., slightly larger than a chimpanzee).*

Key. Graphic indicator of brain proportions with lighter shades of gray indicating greater relative expansion.

A. Corresponding regions of the developing human embryonic brain that must have undergone additional early stem cell production to produce the adult pattern of proportions. The expression domains of Otx and Emx genes are superimposed to show the correspondence with expanded regions. The arrow indicates that cerebellar granule cells originate from the midbrain and migrate to the cerebellum so that this structure is a sort of mosaic made up from the combination of developmentally distinct lineages of cells.

B. Schematic depiction of relative proportions in adult human brains compared to what would be appropriate for an ape with our body size (e.g., a large chimpanzee), with proportional deviations indicated by gray levels corresponding to the key.

that the human difference is correlated with their expression in this region.

What can we conclude from this parallel between gene expression domains and growth field differences in the human brain? First, the difference is not the result of a segmental shift at the level of the whole brain and body, as it appears to be in primate evolution, because the whole brain is not uniformly enlarged. Second, the embryological patterning of proliferative differences that produces this proportional shift in structures corresponds with the segmental pattern of certain gene expression domains in the brain. Whether or not these homeotic genes directly contribute to the human shift in cell production in these regions, it seems clear that the effect is restricted to the cell lineages specified by them. Indeed, as we have seen, there are experimental precedents for changes in the expression of one of these gene groups (Otx) affecting brain proportions by a fairly simple concentration effect, so it does not take a great leap of imagination to imagine a mechanism (such as additional gene duplication) whereby the human proportional shift might be produced.

The Developmental Clock

The cells that comprise large and small animals differ only modestly in size. What determines an organism's size is mostly how many cells are produced. But the signals that tell embryonic mouse cells to undergo a smaller number of cell divisions than embryonic elephant cells has yet to be identified. The number of cell divisions that ensue after fertilization of an ovum is not just the crucial determinant of the size of the whole body; the same mechanism also indirectly specifies the size of each major organ and region of the body. Though there are many tantalizing clues about how this developmental clock is linked to other developmental mechanisms, what we don't know about this clock has kept one of the most important variables of organism design shrouded in mystery.

The decision about how many cell divisions will be needed to build a mammal body appears to be made quite early in development. Some of the best hints about the size-determination process come from work with early embryos. One of the earliest multicelled stages is called the *blastula* and consists of a ball of cells that forms from the earliest divisions of the fertilized egg. The ability to separate and manipulate cells from this structure has played a crucial role in genetic engineering. If these cells are removed and maintained in suspension in a culture dish bathed by an appropriate growth medium, they can continue to divide over and over again without beginning the first steps toward differentiation that, in a normal embryo, would eventually lead to a nondividing final form. In this undifferentiated state, they can be placed into another blastula where they will be recruited to differentiate along with its own cells, and there they can come to assume any of the possible cell types of the developing host body. But one feature of their development appears to be invariant: whether placed in an embryonic body or an adult body, they appear to develop on a schedule that reflects the rate at which they would normally develop, irrespective of whether placed in a fetal, adult, or different species context. At the start of the differentiation process, in whatever context this begins, the cells activate an internal developmental clock that determines the number of cell divisions that are allowed before becoming committed to final cell fates and ceasing mitosis.

The regulation of the relationship between cell differentiation and mitosis probably involves an interaction between gene products in the cytoplasm of the cell and the genes within its nucleus. In experiments where the nucleus of a cell destined to become a frog skin cell is reimplanted into a zygote that has had its own nucleus removed, the genes within the skin cell nucleus are able to start the whole embryonic process all over again and produce an entire frog. Presumably, the actions of the genes produce cumulative changes in the contents of the cytoplasm that exert a feedback effect on gene activity. A similar rate-concentration effect also may account for the difference in the setting of the developmental clock in species with different adult body sizes. Slower production or accumulation of these signaling molecules will allow more mitotic cycles to occur between successive transitions to more restricted cell fates. The starting of this clocklike interaction appears to affect numbers of cell divisions primarily by determining when crucial differentiating genes, like homeotic genes, get turned on. Most vertebrate embryos regulate differentiation irrespective of cell proliferation once the process has started. This is demonstrated by removing a significant fraction of the as yet undifferentiated cells from a vertebrate blastula and letting it continue to develop. The result will be a dwarfed but otherwise normal body that develops at a normal rate.

Another clue to this process is provided by species of frogs and salamanders that have enormous amounts of DNA in their genomes (the excess is apparently redundant, noncoding DNA). These species have a few other curious traits in common. They are all very slow-developing dwarves compared to other related species with normal genomes, and they all have very low metabolic rates and very slow cell-proliferation rates. It seems that by forcing the gene transcription and replication processes to sort through reams of superfluous genetic information, the ticking of the developmental clock is slowed, along with metabolism and cell-division rates. Normally, a slowed developmental clock would allow more divisions between sequential stages of development, producing a larger organism. The large genome species are the exceptions that prove the rule. In them, fewer cell divisions take place between the greatly prolonged developmental transitions.

For brain development, the autonomy of the developmental clock means that the number of cell divisions completed by the time the cells of the brain have taken on their final fates is decided before there is any sign of a brain or other body regions in the embryo. From that point on, the size of the brain depends on what fraction of the developing embryo is selected to become neural tissue by the expression of homeotic genes. The relative independence of homeotic parcellation and proliferation processes can be demonstrated by modifying differentiation at the middle stages in development. The induction of neural tube formation[11] can be duplicated in a single embryo if another inducer region from another embryo is implanted into the developing embryo. As a result, a second neural tube will be induced to form, and if allowed to mature, it will produce Siamese twins with linked bodies but separate nervous systems. An important feature of Siamese twinning, for questions of brain size, is that although the mass of the whole body is lower than for two separate individuals, the two brains are generally not so correspondingly reduced in size. A similar independence is also demonstrated by cross-species transplantation (or xenotransplantation) experiments. In one insightful experiment, whole segments of the embryonic neural tube from Japanese quail embryos were transplanted into embryonic chick brains.[12] Japanese quails are much smaller than chickens, and as adults they have much smaller brains. When major parts of the quail brain are substituted for corresponding parts of a chicken brain shortly after neural tube formation, the resulting chimeric birds grow up with quail brain structures that are diminutive compared to the host brain structures. This tendency to grow to appropriate size even when in an inappropriate developmental context applies to other body structures and other combinations of animals as well. In our laboratory, we have transplanted dissociated cells from embryonic pig brains into adult rat brains. Even in this altered context, they still develop on a pig schedule to become pig-size neurons (which are slightly larger, with longer axons and dendrites, than rat neurons; see Figures 6.8 and 6.9).[13]

These unnatural manipulations demonstrate that early setting of the developmental clock results in a kind of preestablished growth harmony among all subsequent organ systems. The setting of the clock thus appears to differ from species to species, though not from cell to cell within an individual. This explains why the growth of the brains and bodies of different species tends to be so similarly orchestrated and produces almost mathematical regularity in allometries. Embryos that are initially carved up in the same way but whose cells have different settings of the clock follow a common growth plan that is merely extrapolated to produce systematically different end points.

A number of evolutionary theorists in the first half of the twentieth century, noticing that in certain ways human adults resembled ape fetuses (large brains/small faces for our bodies; hairlessness), suggested that this might reflect a sort of arresting or retardation of human development resulting in the retention of fetal-like traits. But the superficial resemblance

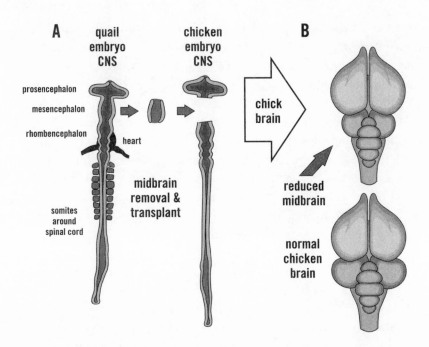

Figure 6.8 *Results of transplanting a segment of the neural tube from the central nervous system of a developing quail embryo to that of a developing chicken embryo. Since quails are smaller than chickens, by the time the birds hatch the inserted quail region has grown to quail proportions but not chick proportions, indicating that the developmental clock is intrinsic to the cells.*

of human proportions to fetal-ape proportions is no more than a resemblance, because there is no global sense in which humans appear to have slowed development or differentiation of cells. However, there is a more subtle sense in which developmental timing mechanisms may have been reset in human evolution. The rate of human brain maturation is appropriate for the size brain we have, but it is prolonged compared to other primates of our same body size. In this regard, our brain's developmental clock has been extended, though the effect is not equivalently mapped to cell-proliferation and differentiation processes throughout the body.

Where does all this leave us in our analysis of the disproportionate growth of the dorsal forebrain of human embryos? At the present time, we have taken this reductionistic analysis down as far as the clues, and our current knowledge of the relevant processes, allow. But now that we know the "signature" of such a segmental change in cell production, we can begin to look for other examples involving other species, other brain structures, and other genes, and to use these as models to investigate the sort of molecular mech-

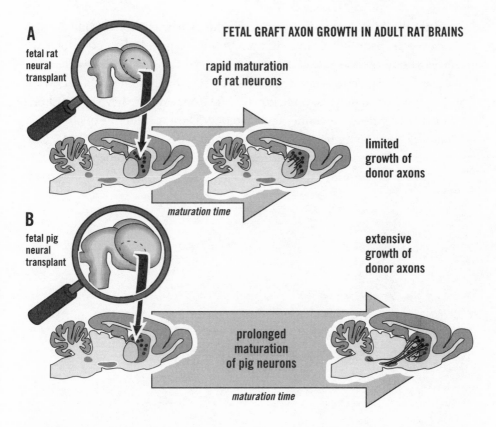

A

fetal rat
neural
transplant

rapid maturation
of rat neurons

limited
growth of
donor axons

maturation time

B

fetal pig
neural
transplant

extensive
growth of
donor axons

prolonged
maturation
of pig neurons

maturation time

Figure 6.9 *Correlations between size and the setting of the developmental clock for brain structures are demonstrated by fetal neural grafting experiments.*

A. *Transplantation of neural cells from a rat embryo brain into an adult rat brain does not result in full regrowth of long axonal connections because the adult brain inhibits axon extension and thus slows growth so that when the cells reach maturity (after a few weeks at most), the axonal connections have only grown a fraction of a millimeter.*

B. *Xenotransplantation experiments demonstrate that the prolongation of the developmental clock of larger brains is even reflected in the development of isolated neural cells, irrespective of context. Fetal cells from a larger species (pig) transplanted into an adult brain of a smaller one (rat) actually grow further to reach distant targets (bottom right) than do rat donor cell axons (top right). This appears to be the result of the much longer time (more than 4 months) during which pig axons remain immature and able to grow.*

anisms that must have been recruited in our own evolution. Despite the lack of a mechanism fully to explain human brain segment disproportions, we have a considerable base of developmental and comparative data at our disposal from which to reconstruct how these changes affected brain evolution, brain structure, and brain function. The critical clues have already been provided by developmental neurobiology. These can help explain why it matters how the human difference in brain size was achieved. So we turn to these other neurodevelopmental processes in the next chapter.

CHAPTER SEVEN

A Darwinian Electrician

We all agree that your theory is crazy. The question which divides us is whether it is crazy enough.

—Niels Bohr

Musical Chairs

Brain evolution should be impossible! Darwinian explanations require that incremental stages in the evolution of any adaptation always must be ends in and of themselves. Each must be useful. Classic theories of brain evolution have long been haunted by a conundrum posed by this requirement. If brains evolved piecemeal, structure by structure, then how could a new structure be added or modified in any significant way and have any hope of being useful, given that it must be linked up in a highly systematic fashion with hundreds of thousands or millions of other neurons in dozens of other brain regions? This would seem to require the simultaneous occurrence of matching mutations in a large number of independent structures. So many parts of the brain are connected so intricately to so many others that it should be astronomically unlikely that adding new brain regions or even modifying old ones could produce a result that worked well together, much less provide a functional advantage. Even changes in the periphery are difficult to reconcile, since for them to be functional a correlated change must occur in the way the brain handles the input it provides. A mismatch between brain and peripheral organs would be less than useless. In far less complicated mechanisms, like computers or TV sets, slight changes are far more likely to degrade the function of the whole than improve it.

Luckily, brains aren't designed the way we design machines. Neither incredibly fortuitous and intricate mutations, nor immense amounts of evolutionary time, nor intermediate functions that bridge the gap turn out to be necessary to address this riddle. The reason is that evolution builds brains using evolution itself as a design tool. As it matures, a brain literally adapts to its body.

It would not be too much of a simplification to say that the size and shape of my hand and the types of cells that compose it were mostly determined by processes that took place in my hand during its development. Morphogenesis of most body parts is a result of local cell-cell interactions in which signaling molecules from one cell affect neighboring cells. So that when a genetic accident produces a mutated hand, it is fair to assume that the damaged genes expressed in hand cells produced this effect. Cells are small, so most cellular communications act over short distances. But this model of development is inadequate to explain brain development. It is not necessarily the case that a modification in the size or shape or function of part of the brain is determined by the actions of cells within that region. Indeed, the processes that determine where functions will come to be located may depend more critically on what is going on in a number of other very distinct parts of the brain during its development. This is because the developmental assignment of neural functions to different regions of the brain is in many respects *systemically* determined. In a very real sense, the brain as a whole participates in designing its parts. The implications of this unusual developmental logic are only beginning to be appreciated for brain evolution.

Unlike other cells of the body, neurons can be in direct contact with many cells that are located quite far apart from one another, by virtue of their long output (axons) and input (dendrite) branches. Because neurons are specialized for cell-cell communication over long distances, they can utilize an additional level of structural information over and above the regional segregation of tissues and cell lineages to help organize their functions. As it matures, a neuron will send out a long axonal process with a specialized tip (a growth cone) that selectively extends the axon to contact cells in other regions guided by molecular signals expressed along the way. Eventually, growing axons will make contact with target neurons, often located in distant parts of the brain. Within target regions, axons establish functional connections (synapses) with target neurons. Such connections are the basis for the specialized form of cellular communication—neurotransmission—that underlies the brain's information-processing functions.

In the early stages of making synaptic connections, these synaptic links between nerve cells are also a force for structural differentiation. Because axonal extension allows populations of cells located distant from one another in the brain directly to interact and influence one another, it superimposes a nonlocal developmental logic on top of the local regional differentiation that preceded it. This complex interplay of local and distant cell interactions is capable of producing far more cellular heterogeneity, and therefore far more potential for functional differentiation, than is possible in any other organ system. Neurons born in distant regions and following very different developmental trajectories can directly and specifically communicate with each other, and so affect each other's differentiation. This introduces a whole new level of cell- and tissue-differentiation possibilities that makes brain development particularly complicated and counterintuitive. It makes it possible for the nervous system as a whole to participate actively in its own construction. It also offers an important source of variations and adaptations that can play a role in brain evolution.

Classic neuroanatomical theories assumed that the differentiation of each brain structure was an independent trait, which would mean that different parts of the brain could be subject to independent evolutionary influences. But this turns out to be unlikely. One of the most important insights to come from neurobiology is that nongrowth processes play major roles in determining the size, organization, and function of brain regions. Many of these processes are actively self-destructive. For example, cell death—sometimes spontaneous, and sometimes driven by competition between cells for local resources—turns out to be a very important mechanism for the developmental sculpting of parts of the nervous system, serving to match the proportions of one to another (see Figure 7.1). The logic of this process is essentially a Darwinian logic: overproduction of random variants followed by selective support of some and elimination of most. It is similar to building a door by first building a wall and then later removing the portion of it that will serve as the doorway. Such a strategy, while appearing somewhat wasteful of material, is highly efficient in its use of informa-

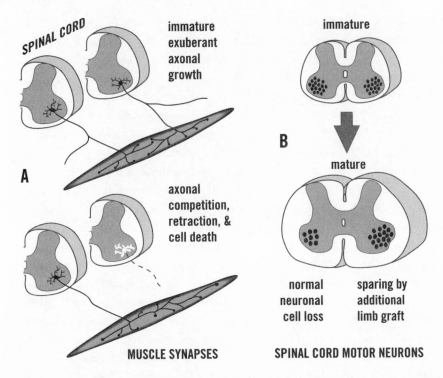

Figure 7.1 *Darwinian processes in development of the peripheral nervous system.*

A. *During fetal development, axons growing out from spinal cord motor neurons compete for access to synaptic targets on muscle fibers. Initial axonal growth is essentially nonspecific (though constrained by such factors as distance and substrates) and overlaps in its projections. Competition ensues in which only one neuron will occupy synapses on one muscle fiber; those that end up losing all their connections die out. The result is that the neuronal population is matched to the muscle cell population.*

B. *Motor neuron cell bodies in the ventral horn of the spinal cord are initially produced in greater numbers than will survive to adulthood. But if a supernumerary limb is grafted onto the developing embryo (this can be done with chick and frog embryos), fewer cells are eliminated.*

tion. It circumvents the difficulties of planning ahead and allows development to proceed with a minimum of design or regulatory mechanisms.

Since distant neuronal structures must engage in complex cooperative interactions in mature brains, the functional matching of cell populations and connections is particularly important. Programmed neuronal death plays a first role in this process by matching different but interconnected cell populations to one another. Some of the first evidence for the role of selective cell death in the nervous system came from studies of peripheral nerves and their external connections. Normally, a significant fraction of the initial population of motor neurons is eliminated from the spinal cord. In addition, animals from which muscles or whole limbs have been experimentally removed early in embryogenesis lose an even larger proportion of the motor neurons of the spinal cord and brain stem that would have projected to these peripheral structures. But in graft experiments where additional tissue is added to a developing embryo (e.g., an extra limb), fewer neurons than normal are lost from the areas connected to these grafts. Extra neurons were not initially produced, but neurons that would normally die were spared by the enlarged target (see Figure 7.1). Nature prefers to overproduce and trim to match, rather than carefully monitor and coordinate the development of innumerable separate cell populations.

This logic also describes the development of connectivity. Fetal axons have only rather general target attraction and avoidance information available to them, and so they don't "know" with any precision where to grow or on which cells they should terminate (see Figure 7.2). Recent studies have begun to demonstrate that a variety of guidance mechanisms assist a growing axon in locating a target region elsewhere in the brain. Among these are guide filaments extended from non-neuronal cells, regional differences in cell-surface adhesion, spatial patterns of attraction and repulsion molecules, mechanical properties of tissues, and specific growth factors released by cells within target regions which help support axons that have arrived at the correct destination. These mechanisms are sufficient to bias axon growth toward selected general target regions, but they are insufficiently precise to specify any but the most global target distinction. More specific connectivity is instead specified post hoc, so to speak, as many connections become selectively culled in response to functional processes.

One reason for this lack of specificity is that the amount of information necessary to specify even a few percent of neural connections between cells would demand incredible amounts of genetic information. Further evidence that there must be an outside source of brain-wiring information is provided by the relative constancy of genome size across vast differences in brain size and correspondingly astronomical differences in connections. Although the human brain probably possesses hundreds or even thousands of times the number of neurons that are in some of the smallest vertebrate brains and millions of times more connections, it does not appear that this has correlated with a significant increase in genome size.

But there is another constraint on the developmental process that has dictated this design strategy. The genetic signaling mechanisms (regulatory

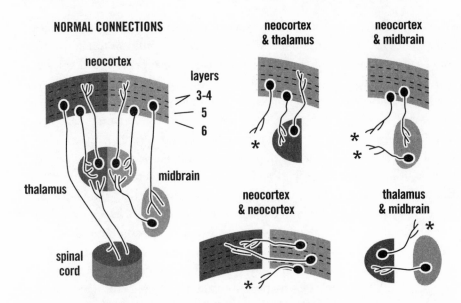

Figure 7.2 *Four kinds of evidence that suggest that different regions of developing cerebral cortex all begin without specific input and output affinities, though cortical layers do show connectional specification, and that spatial position and competitive interactions between axons determine local function and connection patterns that distinguish functional areas.*

Left. Normal connection patterns between cortex and other subcortical structures.

Right. In tissue culture (in vitro) *experiments, where explanted cerebral cortex and thalamus (the major input nucleus) slices are allowed to grow axons into one another, there are no preferred associations between specific thalamic nuclei and cortical regions. There are, however, other more selective patterns between cortical layers. For example, layers 3 and 5 can connect to any cortical structure (lower left); layer 5 can also project to midbrain tissue, but not to thalamus (top); and layer 6 neurons do not project to cortex, only to thalamus (top). Relationship between thalamus and midbrain is predicted from the other combinations and natural patterns, but has not been demonstrated by experiment. These patterns suggest many predicted patterns that are untested. The figure summarizes studies from Molnár and Blakemore (1991) and Yamomoto, et al. (1989; 1992; 1995), among others who have corroborated these results. Asterisks indicate axons that fail to grow into the co-cultured tissue.*

genes) that initially partition the neural tube into major regions operate by cell contact and the diffusion of large signaling molecules, and there is an upper limit to the size of the embryo within which such molecular diffusion processes can operate effectively. Since essentially identical molecular processes are involved in establishing the initial brain divisions in animals as different in brain size as mice and humans, the same partitioning of neural segmental divisions must be extrapolated over a vast range of sizes. Larger brains will be less able to rely on genetic mechanisms to determine structural differences and more prone to proportional perturbations due to extrapolated growth processes. The added design information has to come from somewhere else in larger brains, but little of the extra information seems to come directly from genes. In the same sense that Darwinian processes have created new design information for building organisms during the course of the evolution of life, Darwinian-like processes in brain development are responsible for creating the new information required to adapt large brains to themselves and to their bodies.

Striking evidence for the generality of the information involved in specifying connections in the brain comes from neural transplantation experiments that cross species boundaries. It seems reasonable to assume that the determination of circuitry in different species' brains should rely on different guidance signals for specifying connectional development. For example, it is often assumed that human brains are different because human neurons receive different genetic instructions about where to grow and where not to grow, and which cells to connect with and which not to connect with. Taking a bit of brain tissue from one species and placing it in another species' brain should therefore mess up the neural "switchboard," so to speak, by flooding it with misrouted connections. The chimeric brains that result from such interspecies transplantation experiments should be dysfunctional in the extreme, and the greater the species difference, the greater should be the disruption of function.

Surprisingly, this turns out not to be the case. In xenografting experiments (see Figure 7.3) in which fetal neural cells or even whole sections of embryonic brain tissue are transplanted from one species into another, the donor neurons make connections that are not only appropriate but functionally integrated with the host. In one series of experiments, we transplanted cells from selected regions of fetal pigs' brains into various regions of adult rats' brains. When we analyzed them later, we discovered that the growing axons of pig cells correctly interpreted the signals provided by their rat host's brain to grow to regions of the brain that would have been normal targets for corresponding rat neurons.[1] They apparently utilized the very same generic connection information that rat axons use during normal rat brain development. So, at this level of target specification, it appears that the signals for guiding the formation of neural connections in rat and pig brains are interchangeable. The implications for brain evolution are challenging. Rat brains are not pig brains, nor are pig brains merely bigger. It

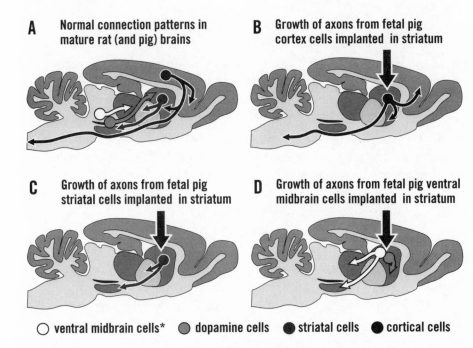

A Normal connection patterns in mature rat (and pig) brains

B Growth of axons from fetal pig cortex cells implanted in striatum

C Growth of axons from fetal pig striatal cells implanted in striatum

D Growth of axons from fetal pig ventral midbrain cells implanted in striatum

○ ventral midbrain cells*   ◐ dopamine cells   ● striatal cells   ● cortical cells

Figure 7.3 *Experiments implanting embryonic pig cells into rat brains (see also Chapter 6, Figure 6.9) demonstrate that initial axon guidance clues are not highly species-specific, but instead are shared by species as different as pigs and rats. Even though the donor neurons are from a very different species, and are sometimes implanted in places in the brain where such neurons are not normally found, their axons are still able to use the host brain's signals to guide their growth to appropriate targets (as in A). Neural cells from the embryonic cortex (B), striatum (C), and ventral midbrain (D) are shown transplanted into the striatum in separate experiments. It even seems probable that many of these guidance signals are shared with nonmammal species. This evolutionary conservatism suggests that species differences in brain design may not be determined by specific "wiring instructions," but by other less direct mechanisms. The asterisk indicates that the ventral midbrain contains both dopaminergic (light gray) and nondopaminergic (white) neurons, which are transplanted together but whose axons grow to very different targets.*

is of course possible that there are unique target signals in each, but the similarities suggest that the majority of connectional differences between pig and rat brains mostly emerge without cell-by-cell instruction. When a pig neuron grows up in a rat brain environment, it integrates with other neurons according to rat rules. We must look elsewhere for the source of their differences—to a sort of micro-ecology of axon growth.

So, how do precisely organized neural circuits develop from information

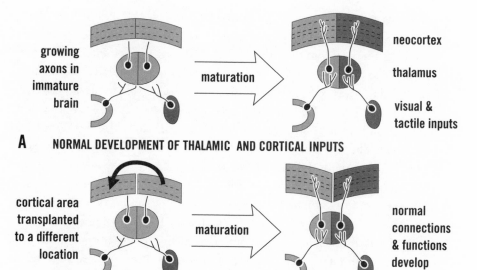

A   NORMAL DEVELOPMENT OF THALAMIC AND CORTICAL INPUTS

growing axons in immature brain

maturation

neocortex

thalamus

visual & tactile inputs

cortical area transplanted to a different location

maturation

normal connections & functions develop

B   DEVELOPMENT AFTER HETEROTOPIC CORTICAL TRANSPLANT

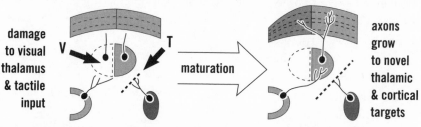

damage to visual thalamus & tactile input

maturation

axons grow to novel thalamic & cortical targets

C   DEVELOPMENT AFTER INPUT INTERRUPTION AND NEARBY TARGET DAMAGE

Figure 7.4 *Manipulations in developing animals that involve altering normal axonal competition relationships demonstrate how this can alter the wiring pattern of the brain.*

A. Normal pattern of connections.

B. Transplantation of the frontal cortical (motor) region to the occipital (visual) region, prior to the development of axonal connections, causes the transplanted sector to develop connections appropriate to the visual, not the somatosensory area (Stanfield and O'Leary, 1985).

C. Removing inputs destined for specific thalamic targets can allow other inputs to take over the vacated region and induce this part of the thalamus to transmit different sensory information to the cortex changing its function (Sur and Benson, 1988; Frost and Metin, 1985). This indicates that the projections to thalamic nuclei also lack specific restrictions as to which specific nuclei they can enter.

that is this generalized and vague, and how do species brain differences arise if not by specific instructions about which connections to change? Neurons overcome the problem of underdetermined target specificity by the same sort of logic that is used to match cell populations: selective elimination. They tend to overproduce branches of their growing axons, and these sample a large number of potential targets during the early stages of development, though only a fraction of these connections are retained into adulthood. The remainder are eliminated in a competition between axons from different neurons over the same synaptic targets (Figure 7.4). This Darwinian-like process is responsible for much of the fine-tuning of neural connection patterns that accounts for the adaptive precision of brain functions.[2] Like Darwinian evolution, the adaptive structure of neural circuitry emerges out of the selective promotion and elimination of specific variant patterns. By initially overproducing connections that have been spread to a wide variety of targets, and then selecting from among these on the basis of their different functional characteristics, highly predictable and functionally adaptive patterns of connectivity can be generated with minimal prespecification of the details. This design logic also provides a means by which adaptive structural differences in neural circuitry can evolve in different species with a minimal number of correlated genetic changes. All that is required are changes that bias either the initial growth and variety of axonal connections or changes that bias the selective processes that cull some connections in favor of others.

Biases influencing axonal selection can arise from both functional and quantitative factors. The central rule underlying this selection process is believed to be the degree of temporal correlation of firing patterns of input axons and output neurons. A neuron generally receives hundreds or thousands of inputs from other neurons. No one of these inputs is sufficient to cause the receiving neuron to initiate an output signal; only the near-synchronous firing of many inputs will succeed in activating the recipient to fire. By a simple cellular mechanism (initially hypothesized by the psychologist Donald Hebb),[3] axons that regularly release their neurotransmitters in synchrony with the firing of the recipient cell (which indicates synchrony with a large fraction of the other input axons) will tend to have their links to that cell strengthened, perhaps by the release of some growth factors. Those that tend to fire out of synchrony will, conversely, tend to lose support, and eventually may be eliminated. Though initially proposed as a mechanism for learning, this mechanism can account for more than just the strengthening or weakening of connectional influences. In the context of the developing brain, where the numbers of connections are significantly

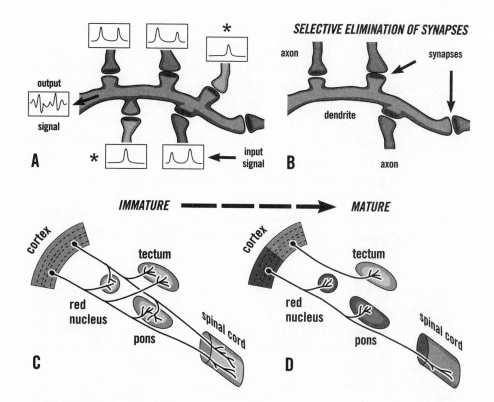

Figure 7.5 *Top: Presumed basis for synaptic competition between axons is the correlation of their firing patterns with other axons synapsing on the same cell. Axons firing relatively more out of sync (*) with the majority (A) will tend to be eliminated (B), probably due to failure to receive growth factors from the receiving cell.*
Bottom: Development of cerebral cortical projections to the brain stem and spinal cord begins with most areas of cortex projecting nonspecifically to most potential targets (C). During subsequent development, collateral projections that do not both receive and connect to other systems with similar signalling characteristics are eliminated, leaving connections with functionally segregated topographically organized connection patterns (D).

in excess of what will be maintained to maturity, it determines which connections will "win" in a biological variant of the children's game "musical chairs," where the numbers of viable targets decrease over time (Figure 7.5A and B).

The best illustrations of this competitive determination of connectivity and functional parcellation in brain regions come from studies of cortical development in mammals. Let's begin by considering the functional and architectural specificity of different sectors of the cortical surface. The adult mammal cerebral cortex is divided into a number of distinct regions, defined in terms of cell structure, functional specificity, and connections to other brain structures. In the rhesus monkey brain, for example, the distinct visual cortical areas alone may number in the dozens. The developmental processes that are responsible for this "parcellation" of cortex involve some genetic specification of regionality, but surprisingly little.

Another fetal transplant paradigm in which fetal cortex from one region is moved to another region has been employed by Dennis O'Leary and his colleagues at the University of California in San Diego in order to probe for factors controlling the establishment of long-distance connections of cortical neurons. These studies have provided some of the most dramatic evidence of nonspecificity and selection processes in brain development. When immature tissue is transplanted from one site to another in the rat cortex—at a stage before any connections have been established—the connections don't follow the move, but connect as though nothing had happened, apparently uninfluenced by the site of origin of the transplanted tissue (see Figure 7.4B). Related studies by many researchers have demonstrated the mechanism that underlies this flexibility.[4] In the early stages of cortical maturation, axons from every region of the cerebral cortex grow to all the potential targets for a cortical neuron. Subsequently, cortical neurons in different regions selectively lose their connections with some target regions but maintain their connections with others, and complementary patterns of retained and lost connections develop in different cortical regions (Figure 7.5C and D). By the later stages of development, strict regional differences in connectivity remain so that cortical efferents from one area project to targets that all have similar modality of function (i.e., all-motor or all-visual structures).

But what biases the competition so that one region of cortex ends up specializing in visual and one in motor outputs? The complementary bias comes from input projections that nearly all arrive by way of the thalamus. But thalamic inputs to cortex also exhibit nonspecificity. This has been shown by growing brain tissues together in a dish (explantation as opposed to transplantation). When randomly chosen chunks of embryonic fetal thalamus and cerebral cortex are grown near to one another in tissue culture, different sectors of cortex and thalamus are equally likely to grow linking connections (see Figure 7.2). This is not totally nonspecific growth because thalamic connections will not grow to most other brain structures in similar conditions. Such growth appears more affected by physical proximity and the physical constraints inadvertently created by neighboring axons than by any spatial signal specificity (see Figure 7.5).

In normally developing brains, however, not every thalamic structure is likely to get an even chance to innervate every cortical structure. Differences in timing of axonal growth, molecular gradients, and proximity may each offer subtle growth biases. The topology (i.e., relative spatial position of points) of connection maps from brain region to brain region tends to be maintained with only subtle variation from individual to individual within a species (and even to a large extent across species); also simply because projections tend to have difficulty crossing through one another in different directions, and tend in most systems to separate into parallel bundles. Nevertheless, this spatial organization is only approximate in the early stages, and many divergent connections are made; but as development proceeds, most of these spatially aberrant projections are culled by competition. Within specific projection areas there is even a further degree of shaping up of connectional specificity. Projections that initially ramify locally over a many-millimeter-wide area eventually become confined to very narrow columns of recipient neurons to create precise maps that exhibit sensory and motor topography. Thus, beginning from crude spatial and temporal gradients which contribute initial biases to global afferent connection patterns, competitive processes at progressively more localized levels amplify these into what eventually become fine-grained, point-to-point maps. Intrinsic factors clearly play a role and may introduce additional functional biases, but these are expressed in the context of many levels of previous biasing and shaping processes that have roughed out what is connected to what.

Cells in different areas of the brain are not their own masters, and have not been given their connection orders beforehand. They have some crude directional information about the general class of structures that make appropriate targets, but apparently little information about exactly where they should end up in a target structure or group of potential target structures.

In a very literal sense, then, each developing brain region adapts to the body it finds itself in. There is a sort of ecology of interactions determined by the other brain regions to which it is linked that selects for appropriate brain organization. This process provides the answer to the problem of correlated adaptations in different parts of such a complex system as brain and body. There need be no "preestablished harmony" of brain mutations to match body mutations, because the developing brain can develop a corresponding organization "on line," during development. Paradoxically, the initially crude determination of targets is what allows the topographic precision of connections within the brain, because it allows for progressively more detailed fine-tuning. After adapting to invariant features of the rest of the brain and body, neural connections can further take advantage of the wealth of invariant patterning that is intrinsic to the stimuli the organism encounters in its environment as well, by virtue of the micro-changes in synaptic distribution and strength linking individual neurons that constitute most of learning in mature organisms. Learning, then, is only the late-stage expression of a fine-tuning process that progresses from patterns involving the whole brain to those involving its smallest cellular branches.

Thus, contrary to a century of speculative accounts of brain evolution, phylogenetic differences in the sizes and functions of particular cortical or nuclear regions cannot generally be attributed to the addition of cells to that area or to changes in gene expression in that area. This sort of evolutionary phrenology is inconsistent with the processes that underlie differences in brain organization. The relative sizes of the different cortical areas, the specific connections that they have with other brain structures, and even attributes of local cellular architecture are not locally determined. If a cortical region appears to have changed size or function in the course of evolution it is likely because of a systemic change affecting a number of brain regions whose connections happen to converge on it. This indirect and distributed determination of brain structure and connectivity radically changes how we must think about the evolution of the nervous system.

Evolution is thus provided with a power tool for adaptive flexibility. The brain does not have to be redesigned every time that the body is restructured. The eyes can converge or diverge, the olfactory apparatus can shrink or expand, limbs can be reduced to vestigial proportions or radically restructured for different forms of locomotion, or tactile receptors can be concentrated into the tips of sensitive digits or tails over the course of phylogeny, and *the very same neural developmental mechanisms can produce a brain that is appropriate.* This explains why pig or human fetal neurons transplanted to the brain of a rat host grow to appropriate targets and produce appropriate functional consequences—for a rat. Each species does not need its own revised axonal growth instructions. The developmental information is highly conserved precisely because it can be general, relying on Darwinian-like developmental processes to produce the detailed adaptations of neural networks to one another. This introduces an evolutionary logic that runs counter to many of the most basic assumptions of classic theories of brain evolution. We do not need to invoke all sorts of specific and highly improbable mutations of brain structure design in order to account for changes in the relationships between the parts of brains. But we are also not able to call on genetic micro-management of neural wiring to explain brain and cognitive differences between species either.

Displacement

Because the construction of neural circuitry has a crude and initially un-specified character, subtle differences in such biasing influences as developmental timing, neuronal numbers, and correlated activity patterns can be a source of species differences in brain function. If the determination of which connections persist and which are retracted depends on the cor-related activity of other axons projecting to the same target region, then the relative quantities of projections that arrive in any target region are a par-ticularly important selective bias affecting which connections will be elim-inated or not. Among competing structures, the structure that sends the greatest number of axons to a particular target will tend to drive the activ-ity patterns of cells in that target more effectively, and this will give con-nections that are from a larger source population a "voting" advantage in determining which connections will remain. This has very important im-plications for understanding the patterns and processes of brain evolution, because it means that modifications of the relative proportions of periph-eral and central nervous system structures can significantly alter connec-tion patterns. So, although genetic tinkering may not go on in any significant degree at the connection-by-connection level, genetic biasing at the level of whole populations of cells can result in reliable shifts in connection pat-terns.

This suggests that regional brain and peripheral nervous system size ef-fects play a major role in mammalian brain evolution.[5] I call this evolutionary mechanism *displacement*. In general terms, relative increases in certain neu-ronal populations will tend to translate into the more effective recruitment of afferent and efferent connections in the competition for axons and synapses. So, a genetic variation that increases or decreases the relative sizes of competing source populations of growing axons will tend to displace or divert connections from the smaller to favor persistence of connections from the larger (see Figure 7.6). Differences in the relative sizes of alternative target structures will have a complementary effect. The relative enlarge-ment of one target or another will tend to attract connections away from the smaller since the competitive elimination is more fierce within the smaller structure than in the larger structure.

The notion of axonal displacement helps explain a number of cases of

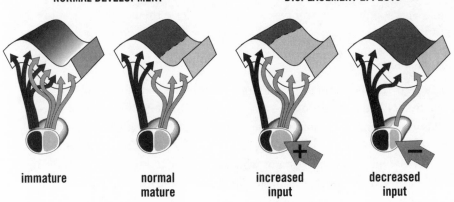

NORMAL DEVELOPMENT　　　　　**DISPLACEMENT EFFECTS**

immature　　　normal mature　　　increased input　　　decreased input

Figure 7.6 *Examples of some general mechanisms of cell and axonal displacement effects influenced by relative size changes in different brain or peripheral neural structures.*

Left: Normal culling of projections from a source to a target (idealized thalamocortical connections are depicted) results in a parcellation of the target structure into functionally discrete regions (two shades of gray).

Right: Changes in either target or source cell populations can shift the balance of competition and produce displacement effects, with cascading consequences in other brain structures. Relatively increasing (+) or decreasing (-) the numbers of inputs from different peripheral sources (e.g., due to normal cell proliferation differences in different species or due to early damage) effectively reduces one of the competing input sources and modifies the parcellation at subsequent levels as well. Enlargement or reduction of the target with respect to its sources of inputs can also bias the axonal competition in complementary ways. Change in the relative population of neurons in a source or target can even induce projections to invade "phylogenetically new" territories.

species-specific brain reorganization described in the literature of both developmental and comparative neuroanatomy. Some of the most labile components of the nervous system, in terms of size, are the peripheral sensory and movement systems. Once the sizes and spatial geometries of these structures have been determined in each species independent of the brain, they contribute stable biases whose influences can propagate through many levels of central nervous system organization. Thus, changes in the total numbers or proportions of sensory receptors (size of retina, increased receptor density in specialized tactile surfaces, etc.) should have significant organizing consequences in the developing brain. This is exemplified in a series of studies with newborn ferrets in which the ascending projections from tactile centers of the spinal cord were cut so that few reached their destination in the thalamus, the major relay station between most inputs and the cerebral cortex, and also in which the major thalamic targets for visual projections from the retina were damaged (see Figure 7.4C). These combined insults caused the undamaged visual projections—which lacked a normal target—to reroute to the region of the thalamus that otherwise would have received the now-severed tactile projections. As in the cortex, different parts of the thalamus were equally willing to accept almost any of the class of thalamic inputs, irrespective of the type of information they carried. Since the thalamus still sent connections to the cortex, the information that arrived in a region that otherwise would have handled touch was visual information—and, remarkably, this tactile area of cortex became responsive to visual stimuli. Similar studies involving newborn rats, ferrets, and hamsters have shown that the rerouted information can even induce structural changes in the cellular organization of cortex that are appropriate for the change in modality of inputs.

Dennis O'Leary has completed this loop of inferences by showing that experimentally produced changes in inputs that modify the pattern of cortical maps can also contribute to the patterning of the cortical output connections as well. Cortical regions that get rerouted visual inputs and lack normal somatic and motor information eventually lose all but those output connections that contact to visual system targets of the tectum. (Compare Figure 7.4C to Figure 7.5 C and D.) Input information is essential for sculpting the output connections. Neither the inputs nor the outputs are predetermined intrinsically.

In an analogous manner the potential for developmental displacement harbors a potentially important source of variability that can be unmasked during the course of evolution in different species. For example, transient connections that are eliminated during the development of an ancestral species might not be eliminated under differently biased conditions that could arise in a later lineage. An interesting example of such a possibility has been discussed by O'Leary.[6] He notes that fibers projecting in a bundle called the fornix, which originate from the hippocampus (a limbic cortical structure of the cerebral cortex) and terminate in targets mostly in the diencephalon (including the mammillary bodies of the posterior hypothalamus and the anterior nuclei of the thalamus), also project beyond these normal targets to reach central midbrain regions during a transient period in the early development of the rat brain. They are culled as the rat matures. But in some brains, this connection appears to be re-

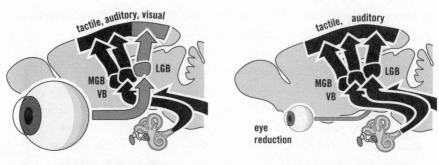

TYPICAL RODENT PATTERN   BLIND MOLE RAT PATTERN

Figure 7.7 *The blind mole rat* (Spalax) *provides an example of neural displacement effects in evolution of the brain. This species' nearly vestigial eyes provide so few visual projections to the thalamic visual targets that during development they are outcompeted for this target by auditory inputs. As a result, a normally visual nucleus (the lateral geniculate nucleus, LGN) receives auditory projections that would usually be confined to an adjacent nucleus (the medial geniculate body, MGB), and passes auditory signals to cortex to take over territories usually involved in visual sensory analysis. The figure summarizes findings from Heil, et al. (1991), and Doron and Wollberg (1994).*

tained into adulthood. Elephant brains (and some human brains), for example, have a clear postmammillary fornix that persists in this midbrain region in adults. It is entirely possible that the size differences in these species is what contributes to a different competitive milieu that favors retention of this connection (and conversely biases against it in small brain species).

More generally, sensorimotor specializations of the brain may be achieved in different species simply by the enlargement or reduction of peripheral structures. In cave-dwelling and subterranean species, we see a trend toward a reduction of eyes. This is the natural counterpart to lab experiments involving interrupting visual inputs in early development. Displacement is clearly evident in these cases. For example, in a species of "blind" mole rat (*Spalax*), the cerebral cortex is almost entirely lacking visual areas, while somatic, motor, and auditory areas seem to have expanded to take over the territory that vision would have occupied in a nonblind rodent counterpart (see Figure 7.7).[7] This displacement of functional regions clearly corresponds to the nearly complete reduction of the eyes and the massive expanse of muscles of the head and neck that are important for digging. Analysis of connections in the brain of this species demonstrates that there has been axonal displacement at the level of the visual thalamus and below. Although the lateral geniculate nucleus (typically the primary retinal receptive nucleus) of the thalamus projects appropriately to the posterior dorsal cortex (the location where visual cortex normally would reside), it receives projections from the inferior colliculus, the major midbrain auditory analyzer that normally would project to the neighboring medial geniculate nucleus of the thalamus. Thus, as in the experimentally manipulated perinatal brains, radically altering the proportions of peripheral projections from different systems can produce axonal displacement effects that cascade throughout the developing brain, spontaneously matching the central organization to the peripheral specializations.

But not all quantitative shifts of inputs and outputs are caused by evolutionary modifications of peripheral sensorimotor organs. Since body plans for both the smallest and the largest mammals must be scaled up from similar embryonic beginnings, significant allometric shifts in body and brain proportions may result from this extrapolation process. These, too, should contribute displacement effects, but they will tend to correlate with scale. One of the first patterns of cortical "evolution" recognized by neuroanatomists at the beginning of the twentieth century was the progressive decrease in the relative proportion of highly specialized sensory cortical areas (koniocortex) and the relative increase in more generalized cortex (the so-called association areas) in the presumed evolutionary "ascent" from small insectivores to large primates. This has long been interpreted to mean that there had been a trend toward more elaborate "higher-order" cognitive functions in brains with a higher proportion of association cortex. But this pseudo-evolutionary trend is highly correlated with size. The relative proportions of koniocortex and association cortex can be predicted from brain size alone.

In the last chapter, I suggested that scaling patterns of different brain regions might in part reflect internal information-handling problems (informational allometry and analogous to the increase in middle-level management with business size). This might be achieved by displacement effects. For example, larger animals do not have proportionately larger eyes. Their eyes and the number of cells on the retina are negatively allometric with body size and even with brain size, so as animals get larger, these retinal projections probably become a smaller and smaller fraction of the projections to the thalamus during development. If the allometry of brain and spinal cord is indicative, this also seems to be true of somatic inputs and skeletal motor outputs, since the brain is larger with respect to the spinal cord in larger species. As result, direct peripheral projections probably diminish in proportion to intrinsic projections to the thalamus. The decreasing proportion of "projection" cortex and the increasing proportion of "association" cortex with increasing brain size might thus reflect a cascade of size-correlated displacements.

The displacement process is likely to be most informative in the case of the human. As we have seen, the most robust differences distinguishing human brains from other primate brains have to do with size. In the mystery of the human brain, the displacement process provides the crucial link between global changes in brain growth patterns and changes in functional organization. Our relatively larger brain and its comparatively prolonged and out-of-sync growth suggest that displacement may have played a crucial role in restructuring the relationships within it, ultimately resulting in some very different functional relationships from those in other primates and other mammals generally. Major shifts in proportions between the enlarged human brain and the peripheral nervous system structures of a relatively unenlarged body should have produced a cascade of displacement effects within the brain. Major shifts in the initial proportions of major brain divisions should also have added unique internal displacement effects, none of which have clear precedents in other primates. Unlike speculations about unique human functions based upon hypothetical special mutations modifying the local wiring of this or that brain structure, or adding this or that new region—all of which appear to be ruled out by the Darwinian nature of brain development—there are quite specific rules of inference we can apply to the analysis of quantitative changes and their effect on neural circuitry. Knowing something about the general patterns of early target determination, and the types of biasing influences that translate these into final architecture, we are in a much better position to extrapolate from other, better-understood mammal brains to human brains.

Now we can see why it matters *how* the human brain became enlarged with respect to the body. And why the level of encephalization exhibited in adulthood does not necessarily have consistent neural or cognitive correlates. What distinguishes human encephalization from that of other primates and from chihuahuas is not so much the extent of encephalization, but at what point during development the size differences appeared, and which structures accounted for the growth disproportions. The major displacement processes that shape the proportions and patterning of neural circuitry are completed around or shortly after birth. At this point in development, growth of most of the body is still at an early stage. Proportional variations that develop later in life will thus have a minimal impact on brain structure. The chihuahua, which develops along a brain/body growth trajectory very similar to that of larger dogs *in utero* and only significantly deviates from this pattern after birth, does not develop its neural architecture in the context of significantly modified central versus peripheral proportions. In a chihuahua, while in the womb and just around birth, the brain is adapting to a typical-size fetal dog body.

But the situation for primates as compared to chihuahuas and to most other mammals is different. Even if primate encephalization is a consequence of reduced body growth and not accelerated brain growth, as is suggested by developmental growth curves, the shift in these proportions is present throughout fetal development. Immature primate brains do indeed adapt to bodies that are proportionately smaller than in most other mammals. The deviation of primate brain organization from more typical mammalian patterns, then, depends on exactly how these proportional differences are distributed in the fetal body. Is the primate head, including the brain, eyes, and ears (but not face, mouth, or nasal region), spared from this postcranial reduction, or just the brain itself? If, for example, primate eyes do not participate in this postcranial reduction, then we would predict that visual projections will exert greater success at recruiting sensory brain regions when compared to tactile and motor systems. I suspect that this is the case, given the comparatively large percentage of primate cortex that is visual; however, neither fetal body segment data nor adequate brain region data are currently available to test this possibility. In whatever way this shift in proportional growth partitions primate bodies, it will initiate a series of competitive biases that ramify through the primate brain during development, and cause it to differ systematically in connection patterns and regional proportions from those of other mammals. Even if we are not willing to conclude that primates are more intelligent than other mammals as a result, we can confidently predict that their brains will operate with very differently distributed sensory, motor, and cognitive resources.

Human brains, too, will have a unique shift of cognitive resources, distinct both from other primates, whose encephalization is the result of embryonic reduction of body size, and from chihuahuas, whose encephalization is the result of dwarfism. The disproportionate expansion of human dorsal forebrain structures from a point in early fetal development should have produced a unique signature of connectional and functional shifts as well. Some, due to the global shift in brain / body proportions, will further extrapolate the primate pattern of brain structure deviations. Others, due to the segmental differences between brain regions, will have no nonhuman counterpart. Both are relevant to the evolution of language.

An Alien Brain Transplant Experiment

While attending a conference in Santa Fe, New Mexico, I came across an advertisement for a lecture by a New Age guru. The lecture promised revelations that would unite both the biblical and evolutionary versions of the Creation story and explain many mysteries from human nature to the Egyptian Pyramids. Though I was unable to attend, I learned that the lecturer presented evidence that humans are an artificial species, created by a race of aliens/gods as a result of an advanced genetic experiment. Presumably, this experiment involved a modification of the brain, perhaps the genetic analogue of a brain transplant. Though I am not in any way suggesting that we seriously consider an alien experiment scenario for human origins,[8] there is at least one sense in which the thought experiment of a hominid brain transplant leads to interesting consequences. The altered growth process that produces a human brain can perhaps best be described as though a human child is growing a brain from a much larger primate species. The human pattern of brain growth is appropriate for a gigantic ape, while the pattern of body growth is appropriate for a large chimp. So, imagine that alien scientists literally transplanted the brain from the embryo of a giant ape into the head of a chimpanzee embryo. The species known as *Gigantopithecus*—an extinct eight-foot ape that left fossils in Asia and Europe perhaps as recently as a few hundred thousand years ago—would be a good choice for the donor species. When we try to imagine how this fanciful experiment would affect the host's brain development, some interesting predictions result.

The large adult size of the transplanted fetal brain compared to the adult size of its host body would radically alter the normal competitive "balance" between connection systems originating peripherally and centrally. Projections from peripheral organs like the host eye and the tactile sensory inputs from the host body would recruit target populations of neurons within the donor brain that were appropriate for the number of inputs they supply. But since the chimp body would only be a fraction as large as would normally carry a brain this size, the space within this brain that these inputs would recruit would be significantly reduced compared to what would happen in a normal *Gigantopithecus* body. Similarly, as in embryos where limbs are removed prior to motor innervation, there would probably be greater cell loss in motor-output systems and less recruitment of central brain areas for motor functions than in a typical *Gigantopithecus* brain. In general, the disproportions of body and brain would ramify through the developing brain as each stage in competition for connections was influenced by prior biases. The adult brain that would result would be quite different from either the donor or host species' brains. Many structures and functional divisions of the transplanted brain would be smaller than expected for a *Gigantopithecus* brain, but others would be larger, if they inherited neural space from those that were more constrained by peripheral connections.

So let's return to reality. Though a chimpanzee-*Gigantopithecus* chimera is a product of science fiction, it offers a remarkably close analogue for understanding the special problems of human brain development, and a good predictor of human brain structure proportions. This science fiction scenario leads us to expect some curious discrepancies in our brains compared to more typical primate patterns. In general, we should expect that brain structures more directly tied to or dependent upon peripheral systems should be the most constrained in size to match them, and that those most synaptically removed from the periphery should be the least constrained in the competition for neural space.

For over a century neuroanatomists have been collecting and comparing data on the sizes of brain structures in human and nonhuman brains. However, without an adequate understanding of how growth processes and developmental competition effects interact to produce these quantitative results, interpretations of these data often did not consider the ways that relative size differences at all levels may be linked. The result has been that brain evolution was treated as a sort of mosaic affair, in which changes in different parts were considered independently. If one structure appeared enlarged, it was thought to have become more important or a more powerful information processor (following a gross or net functional logic). The *Gigantopithecus* brain transplant analogy suggests, however, that local structural changes can be developmental consequences of large-scale changes in relative size of the brain and body, and so may not be independent adaptations, even if all may influence neural processes.

Consider the visual cortex as an example of this linkage between global and local size effects. The Columbia University paleoneurologist Ralph Holloway was one of the first researchers to notice that an unexpectedly small proportion of the human cerebral cortex was taken up by visual projections.[9] Though not small in absolute terms, in comparison to other primate visual cortices it was smaller than would be predicted for a primate brain as big as the human brain. Thus, to extend the analogy, when corrected for our large brain size, our primary visual cortex appears to occupy less surface than it would in a super *Gigantopithecus'* brain. Holloway offered this as evidence that the human brain had not just evolved increased size, but had become differently organized as well. But does this suggest that in some way vision became less important, or that other nonvisual areas became more important than this one and took up more space in our evolutionary past?[10]

The imaginary example of the alien brain transplant experiment provides another way to look at these proportions. Consider the interrelationships between the eyes and the visual structures to which they project. Both the lateral geniculate nucleus, which receives direct inputs from the retina, and area 17 of the cerebral cortex, which receives lateral geniculate inputs relayed from the retina, are significantly smaller than would be predicted in a primate brain that reached human proportions. This is because an ape of the immense proportions that would normally carry such a large brain would also have had much larger eyes than we have; the number of visual projections to that brain would have been far greater than for the human retina. In the *Gigantopithecus*-chimp chimera, the proportions of these visual structures would be matched to a relatively reduced input, and would be more typical of chimp proportions than *Gigantopithecus* proportions. A similar argument can be made for the human visual analyzers. Our visual cortex is not small for our brain size because of some reduced importance of vision or because of the independent addition of other nonvisual areas to the cortex. The human brain does not have a reduced visual cortex, but the appropriate amount of visual cortex for its retina (see Figure 7.8).

A similar pattern characterizes motor and tactile systems as well. The primary motor cortex is also closely linked to the periphery, but by efferent (output) rather than afferent (input) connections. The boundaries of this cortical region are clearly demarcated in microscope sections by the presence of unusually large output neurons located in layer five of the six-layered cortex. These are termed *Betz cells*. The limiting factor that determines how many Betz cells are in the motor cortex, and thus its total size, is probably the number of long direct axonal connections that can be established with primary motor neurons in the ventral horn of the spinal cord. The number of spinal motor neurons thus determines the number of cortical motor neurons that will take on the morphology characteristic of these giant long-distance projection neurons. The number of spinal motor neurons in turn is determined by their competition for muscle fibers. This peripheral competition between motor axons is

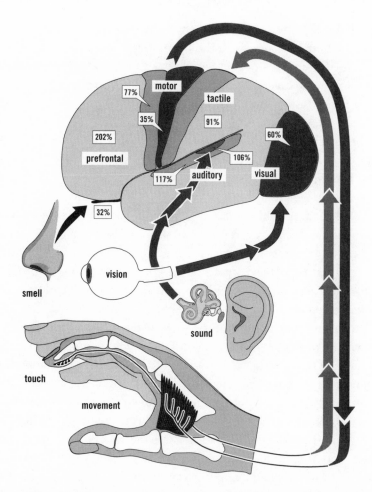

Figure 7.8 *Quantitative deviation of human cerebral cortical regional areas as a function of the sizes predicted for a "typical" ape brain of human brain size (this is analogous to studying the brain of a 1,000 lb. ape: the size that a primate would be if it had a human size brain but fit the typical brain-body trend for monkeys and apes). Many of the data are incomplete and insufficient for statistical tests, but for prefrontal and occipital cortices the values are sufficiently deviant and the data are sufficiently complete to know that these structures are significantly different in size than expected. The figure points out a global pattern to these proportional changes. Structures with relatively direct peripheral links (from eyes, nose, touch senses, muscles) tend to be constrained in size by the size of these sources or targets (which are proportional to the body). Prefrontal cortex lacks direct peripheral connections, as do auditory and parietal areas for which data are not available. This pattern is also observed for relay nuclei between these areas and the periphery. Thus, the human cerebral cortex reflects significant displacement effects, due to the shift in brain/body proportions in our evolutionary history.*

probably even better understood than that which occurs in the central nervous system. As we saw in the beginning of this chapter, it results in both a retraction of supernumerary (outcompeted) projections and the death of some spinal neurons in the process of matching neuron numbers to muscle fiber numbers during early development (see Figure 7.1 and earlier discussion). Studies of other peripheral sensory systems that have been rewired from birth suggest that this effect will be passed on up the line, and that the size of the cortical motor region (as identified by the presence of Betz cells) will, in turn, be indirectly determined by the production of fetal muscle fibers.[11] Cells in primary tactile, motor, and visual regions of the cortex are just one synapse removed from the corresponding peripheral representations, and the corresponding thalamic nuclei of these two sensory areas are only one synapse away, so it is not surprising that these structures are among the most appropriate in size to their peripheral connections and the most reduced with respect to typical predictions for brain size (Figure 7.8).

So, a bigger brain in the same-size body will also be a very different brain, but it will be predictably different, and the developmental causes and consequences of this should not be difficult to trace to changes in body structure and proportions. The relative sizes of the brain's functional divisions are determined in a systemic competition for space driven ultimately by peripheral constraints and shifts in segmental proportions determined in early embryogenesis. Though this does not rule out functional and adaptational consequences of individually enlarging or reducing brain structures, it forces us to understand such size variations in terms of systemwide effects:[12] they are not isolated adaptations. With peripherally specialized input and output systems recruiting less synaptic "space" than would be expected in such a large ape brain, some other systems must stand to benefit instead. Those nuclei and cortical areas that receive little or no input from peripheral nervous structures should stand to inherit extra space because they are relatively insulated from peripheral constraints.

Among cortical areas it appears as though the prefrontal cortex may have inherited additional territory in the human brain, perhaps from reduced motor areas nearby. According to separate extrapolations derived from independent data sources (see Figure 7.9), I have estimated that human prefrontal cortex is roughly twice the size that would be predicted in an ape brain as big as ours.[13] This is probably the most divergently enlarged of any large brain region (the olfactory bulbs are probably the most divergently reduced). The magnitude of this expansion of the prefrontal lobes has been so extensive that it has been recognized by researchers since the late nine-

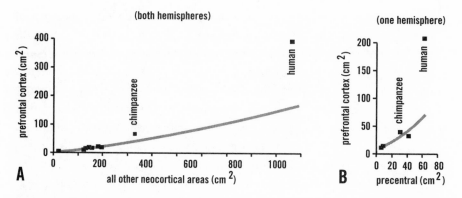

Figure 7.9 *Human prefrontal cortex allometry assessed using two independent data sets and corrected for part/whole artifacts.*

A. *Scaling of prefrontal cortex surface with respect to the remaining surface of the cerebral cortex shows that, for the trend extrapolated from other monkeys and apes, the human proportions are anomalous (data from Brodmann, 1912).*

B. *Human disproportion of prefrontal cortex with respect to the adjacent premotor cortex (data from Blinkov and Glezer, 1968) compared to the trend extrapolated from the other apes and a baboon.*

teenth century, based on gross comparisons of human and nonhuman brains and skulls. To get some idea of how this compares to systems that are constrained by peripheral representations, consider that it is roughly six times larger (in absolute size) than the chimpanzee prefrontal cortex, though chimpanzee and human bodies are nearly comparable. Prefrontal cortex receives only very indirect relayed information from peripheral systems, mostly via connections from other cortical areas. Its thalamic inputs come from nuclei that receive limbic and dorsal midbrain information conveying arousal and orienting signals. All these inputs are from brain structures that have been part of the embryologically expanded brain regions, so it is not surprising that the prefrontal cortex should be among the least constrained in size. Though this relative size increase is an indirect consequence of these many converging competitive processes during development, it nevertheless contributes one of the more extreme shifts in net human brain function, compared to other species.

So in all these ways, the *Gigantopithecus*-chimp brain transplant analogy provides a reasonable model for a variety of human brain structure proportions. However, neither the whole brain nor even the whole forebrain has enlarged in humans, only the dorsal portions of the forebrain (as described in the last chapter). Thus, the effects on connectional organization are somewhat more complex than are envisioned by this imaginary extraterrestrial experiment. The divergence in size of the human dorsal versus the ventral forebrain has shifted connectional patterns in ways that are radically different from other species' brains. These turn out to be critical to understanding two of the most central features of the human language adaptation: the ability to speak and the ability to learn symbolic associations. Specifically, I will argue that the ability to produce skilled vocalizations can be traced to changes in motor projections to the midbrain and brain stem, while the ability to overcome the symbol-learning problem can be traced to the expansion of the prefrontal cortical region, and the preeminence of its projections in competition for synapses throughout the brain. The functional consequences of these displacement effects will be discussed in some detail in the next two chapters. But before we can consider these issues, we need to consider briefly some general problems of interpreting the relationship between size and brain function that we encountered in the previous chapter.

Beyond Phrenology

This developmental approach to brain morphology gives a new meaning to a classic theory concerning the relation of brain structure to function: the concept of "proper mass." As we saw in the last chapter, the concept of "proper mass" is usually invoked along with the assumption that larger brain structures are more powerful analyzers or more spacious storage devices than smaller ones. So, the presence of a comparatively large auditory region might have suggested that its possessor had a greater capacity for sound analysis and a heavier demand for sound analysis in its niche. This "bigger-is-more-powerful" perspective has thus been applied at the regional as well as whole brain level of analysis. Indeed, the regional interpretation is probably older. It was the basis for one of the first theories of localized brain functions—phrenology—proposed at the turn of the nineteenth century by Franz Josef Gall. Gall and his student Spurzheim reasoned that the shape of brains and the effects of localized brain damage suggested that brain functions were organized in local centers. He further hypothesized that there should be individual differences in brain structure that corresponded to differences in talents, propensities, and other personality traits, and that these should be reflected in the relative sizes of the corresponding centers for these functions. Greedy people should have a large aquisitiveness center, musicians should have a relatively large musical center, and so on. This also led to the prediction that head shape should reflect underlying enlargements or reductions of the various centers, allowing phrenologists to collect evidence for their theories by studying patterns of cranial geometry. Though Gall's choice of functions to assign to brain centers seems curious from a modern perspective, the logic of his underlying theory is alive and well.

These phrenologically derived notions are to some extent in conflict with developmental data which suggest that the sizes of individual brain structures are not determined in isolation. Both peripheral-central matching processes and region-by-region competitive effects play a role, along with initial more global cell-production processes. A larger retina or a more densely innervated tactile receptor surface will require a larger network in order simply to break even, in information-processing terms, and their projecting axons will compete in proportion to their numbers with other projections for the synaptic space that corresponds to that information-processing need. This rule of thumb can be extended to other brain regions that do not directly link to the periphery, though the analysis becomes complicated by the cascade of converging parcellation influences from many different sources.

The logic of the developmental processes that determine brain structure sizes may offer some initial clues about the resulting functional consequences. Since a Darwinian-like process in brain development determines the relative sizes of functional brain regions and their patterns of connection, there is reason to suspect that a Darwinian-like functional consequence should result. The brain's wiring is determined by virtue of the interaction of information conveyed by its connections, so the way that information is analyzed ultimately becomes reflected in the way brain regions are "designed" by this activity. This is a competitive selection process, a sort of rapid local evolutionary process on a microscopic scale. It makes sense, then, to predict that the principal effect of changes in relative scale within the brain will also have a competitive-selective consequence on function. The relative enlargement of one structure with respect to another may give the larger some sort of competitive advantage in the battle for influence over target synaptic activity. In other words, if one brain structure is relatively enlarged compared to another, this should translate into both displacement of connections during development and displacement of computational influence in adulthood with respect to other competing inputs from other brain structures. More inputs equals more votes influencing the computational outcome.

Parcellation is a zero-sum game. When one structure become partially enlarged, another is reduced. Applying this analogy to cognitive functions would suggest that shifts in relative proportions will be translated into functional trade-offs. The class of neural computations supported by an enlarged region will tend to have more influence over final global outputs than that supported by a reduced region. This argument has analogies to some of the effects of brain damage. When a structure is damaged, not only is there a loss of function but there is also inevitably a gain of function—though not an improvement—in the form of "released" behaviors, which appear to be disinhibited by the removal of a competing influence. A zero-sum size effect is also suggested by studies of the effects of quantitative differences in connections. For example, differences in the number and extent of fibers projecting between two major structures within the hippocampus[14] (a limbic structure that has been strongly implicated in the consolidation of certain kinds of memory) correspond with inverse patterns of learning improvements and impairments on complementary tasks (such as maze learning versus passive avoidance learning). Differences in relative hippocampal size with respect to the rest of the brain have also been correlated with food-caching behaviors in birds and their ability to recall a large number of observed hiding places in tests. If this too reflects a trade-off, one might predict that species with relatively smaller hippocampi would outperform their food-storing counterparts in some other complementary mnemonic tasks, though I do not believe this has been studied.

But this zero-sum logic which applies to parcellation processes in development also applies to growth processes, which are also involved and are of special importance for the expansion of the human brain. Differential growth alters the context of functional competition and parcellation so that certain classes of connections—those with input and output connections to other enlarged regions—gain an unfair competitive advantage. For them, the competition is less severe and the game is thrown in their favor. Thus, the developmental interpretation that led us to assume a functional interpretation of size relationships cannot strictly apply to these systems. In developmental terms, these changes are addition effects that are prior to and independent of the competitive parcellation processes. By analogy, however, we could argue that larger recruitment fields reflect some "virtual" competitive parcellation. It is *as though* they were being recruited by some massive input projection, or were supported by a massive output target from outside, though in fact this is not the case. From an adaptational perspective, the source of the competitive bias is irrelevant. The final connectional proportions are what will determine function.

Recasting the problem of net versus gross function of different brain regions in this way provides a new tool for thinking about the functional consequences of proportionately enlarged human brain structures, not in terms of their being more "powerful" computers, but rather in terms of shifting the balance of computational influences in the whole brain: size difference as a source of cognitive bias. It is as though the enlarged structure's share of the workload of information handling has been increased, and its ability to recruit and dominate the computations going on in other structures has also increased. The relatively enlarged regions of the human brain should thus tend to shift the balance of information processing towards the sorts of sensory, motor, and mnemonic processes that characterize these structures in other species.

In summary, the Darwinian nature of neural connectional development suggests a new interpretation of the functional consequences of neural allometry—that functions will be modified in response to proportional changes by a kind of functional displacement of some computational tendencies by others, not just increase or decrease of localized functional capacities. Considering the displacement of connections that distinguishes species of different sizes and relative proportions, and the displacement of competing mental computation processes that results, forces us both to extend and modify the classic notions of proper mass as applied to brain structure, by extrapolating the analysis to circuitry relationships.

The displacement theory offers a powerful predictive tool for interpreting the significance of quantitative differences in brain structure. In particular, displacement effects can help answer the troubling question raised by the difference in development of chihuahuas, primates, and humans. When and how such disproportions arose and how they are distributed in the developing brains and bodies of these different species have everything to do with how they become wired up in the end. Chihuahuas' brains and bodies are growing along typical dog trends during the phase when axonal competition divides up the brain, so their brains should not diverge significantly in design from other dog brains, but primates depart from the brain/body trend of other mammals right from the beginning of embryogenesis, so their brains should depart in some interesting ways, as well. The initial changes in proportions within the human brain also are expressed early in development and so inevitably produce a novel pattern of internal structural proportions and correspondingly shifted distribution of functional relationships. The details of these displacement processes, though not yet directly analyzable in human brains, should be at least broadly predictable from well-known general processes of brain development. These differences should be the best clues to the functional demands that selected for them. The resultant shifted pattern of regional parcellation in human brains can be interpreted as though enlarged systems were deluged with some massive new set of peripheral inputs. This input is not supplied from the periphery but internally, as a result of shifts in the early production of neurons. It is an evolutionary response to a sort of virtual input, with increased processing demands. This suggests that the difference between human and nonhuman brains may be far more complex and multifaceted than simply an increase in extra neurons over and above the average primate or mammal trend. Working backwards from these brain changes to shifts in function offers the best hope of developing a model of just what sorts of virtual inputs were responsible for these changes in the first place.

The Talking Brain

Silence is deep as Eternity, speech is shallow as Time.

—Thomas Carlyle

Hoover's Brain

One evening in the mid-1980s my wife and I were returning from an evening cruise around Boston Harbor and decided to take a waterfront stroll. We were passing in front of the Boston Aquarium when a gravelly voice yelled out, "Hey! Hey! Get outa there!" Thinking we had mistakenly wandered somewhere we were not allowed, we stopped and looked around for a security guard or some other official, but saw no one, and no warning signs. Again the voice boomed, "Hey! Hey you!" It now sounded a bit thick-tongued. Perhaps it was the half-conscious territorial threat of a wine-drinking vagrant sitting in a dark corner. The words were repeated again, and again we searched for the origin, calling back with no reply besides the repeated commands. As we tracked the voice we found ourselves approaching a large, glass-fenced pool in front of the aquarium where four harbor seals were lounging on display. No guard, no drunk, no prankster; just harbor seals. Incredulous, I traced the source of the command to a large seal reclining vertically in the water, with his head extended back and up, his mouth slightly open, rotating slowly. A seal was talking, not to me, but to the air, and incidentally to anyone within earshot who cared to listen. Perhaps it was a practical joke and someone was hiding nearby. I stood there for a long time trying to figure out the trick, expecting to discover a slip that would show this seal to be a ventriloquist's dummy, an unwitting accomplice to someone hiding in the shadows, laughing at the gullible passers-by. But there seemed to be no voice when he was underwater or when he closed his mouth, and no place for a speaker to hide. Still, just to make sure, the next morning I placed a call to the aquarium.

"Oh, yes," I was told. "That's Hoover, our talking seal. He's become our star attraction." I was dumbfounded. How come this wasn't national news? Were scientists studying him? Was the ability to train seals to talk common knowledge, except to scientists? When I was put in touch with the staff in charge of the animals, I learned Hoover's brief life story. It seems he was discovered as an orphaned and sickly pup by a fisherman in Maine, who took him in and nursed him back to health. But harbor seals grow quickly and require incredible amounts of food. In no time, he was getting into everything and eating his benefactor out of house and home. As a result, he was named for the famous vacuum cleaner, and the Boston Aquarium was asked to become his foster home soon after.

Hoover did not talk from infancy, not until many years later, in fact, as he approached puberty. During the interim he became a sort of mascot of the aquarium staff, continuing to get into things and live the life of a pet. He was also very sick more than once, and it was thought he might even have suffered a bout with encephalitis. He was weakly, and did not seem to do well with other harbor seals, especially males. But most peculiar were his efforts to vocalize. What started out as rather odd but unclear vocalizations as a growing pup developed into a few reasonably clear phrases that he used in no particular order and to no apparent purpose, perhaps just out of the boredom of captivity. These included "Hoova!" "Hey!" "Hey hey hey hey!" "Hey you!" "Get outa there!" and a sort of gurgling or laughing sound like "Yadda yadda yadda" that no one could seem to translate. He also had the standard seal vocalizations, which he used socially with other seals. Opinions were mixed on where and how he learned these phrases. Some were convinced that he learned them from the staff, or was taught by them as he began to vocalize in ways that sounded speechlike; but the story that seemed to ring true was that he sounded just like the old fisherman who originally took him in, years before. I thought from the beginning that he had sort of a down-east, old-salt accent. And something else fit with this story.

It had been known for more than a decade that many songbirds learn a dialect version of their species' song when they are nestlings, but only start singing when they approach puberty. Their first singing, called *subsong*, is a rather crude version that they soon hone to reflect their parents' song, irrespective of the songs currently being sung around them. Apparently, they have a sort of auditory template, remembered from the earlier stage, that they try to match (a mechanism verified by isolating and deafening birds at different ages and seeing how this affects their song). Though we will never know for sure, the image of Hoover guzzling the food in the cupboard and the old fisherman yelling, "Hey! Hey! Hoover! Hey you! Get outta there!" has a persuasive feel, or twisted irony.

Staff members made some attempts to study his speech, and some effort to train him to produce it more predictably, for a fish reward. I too recognized that Hoover was worth paying attention to. It took little convincing to find a Harvard undergraduate who was interested in studying this curious fellow with me. The student, T. H. Culhane, and I began to videotape Hoover's speech and other behaviors for later analysis. Over the course of an academic year we followed his speech and behavior patterns. When did he speak? What else he was doing when he spoke? Did he speak to other seals, or to people, or to no one? And how did he respond to training efforts? There were some interesting patterns. For example, he did not seem to speak more during the mating season, though adult male seals are most vocal at that time. If anything, his speech decreased and his normal seal vocalizations increased at that time, and he did not lack for the normal repertoire of harbor seal barks, grunts, and cries. His response to training was not very promising, either. Not only did he not learn any new words or phrases; our data suggested that training might be causing him to increase the proportion of short phrases and decrease the more complex ones, probably Hoover's way to increase the rate of fish thrown to him. In other words, his speech did not appear to be made up of modified seal vocalizations. It was something independent.

What if Hoover *had* been taught to speak? Seals (usually sea lions) are often trained to perform rather complicated tricks, even easily trained to bark on command. The aquarium staff had many expert trainers and they presented some marvelous seal and dolphin acts, though Hoover wasn't involved. Had these trainers found a way to train a new trick? I don't think so. Hoover was one of a kind. Efforts to train other young seals to mimic speech have not, to my knowledge, produced any new talking seals. Nor does just being around people during development seem to get other seals to speak. It turns out that there are good reasons to suspect it would not. And this is what makes Hoover's speech so curious.

Hoover's speech, like our own, begs a question: why are other mammals so poor at vocal learning? It is a curious fact that, except for some dolphins and whales, other mammals have quite limited vocal abilities. Mammals are poor at learning to produce any new vocalizations that are not minimal variants on their innate ones, and even their natural vocalizations tend to be limited in articulatory variety and complexity. They are no match for most birds in this regard. The vocal repertoires of birds appear far more flexible and are far more dependent on learning to develop normally. Many species of birds learn distinctive local song dialects as nestlings, and some have the ability to learn complex arbitrary sound patterns. Humans, some species of dolphins, and some species of whales seem able to learn more than the vocal calls they are born with, but even these other vocal species seem to learn and use their new vocalizations in quite restricted ways. More important, it is no accident that the other highly vocal mammals are all cetaceans (the order of mammals including all the dolphins and whales), because they don't produce sounds the ways that other mammals do. Their sound production appears to depend on air passing through specialized sinuses leading to their blowholes, which are modified nasal openings on the top of the head, and not on vibrations of the vocal folds of the larynx. The details of this unique mechanism are still poorly understood, but it almost certainly involves structures and muscles that have no similar function in terrestrial mammals or even seals. But Hoover's throaty speech was not exactly produced like ours, either. He would lean his head back, prop his mouth open, and with only a slight movement of his tongue and lips, like a novice ventriloquist, he would say his piece. No special muscles or vocal structures were involved, probably fewer than we normally use to produce similar sounds. This made his speech sound a little slurred, but everyone who heard it recognized it as speech, and it took little imagination to understand him.

The single mockingbird that wakes me each morning sounds more like a carefully directed choir of birds, with one bird species after another singing its short phrase. This is probably what mockingbird singing was selected for during their evolutionary past: providing the illusion of a territory overfilled with birds. Sometimes I am even convinced that I hear a cat's meow or the approximation of a car door squeak. The mocking bird's ability and propensity to mimic the sounds it hears around it are unusual, even for birds, but many other unrelated bird species have independently evolved this trick, which is an extreme variation on the more widespread ability to modulate singing behavior dependent on the variants of song sung by one's local population. But although many bird species are chosen as pets because of their ability to copy human speech, no mammals are, or could be. Such abilities are not just rare, they are nearly nonexistent. Why? It's not because mockingbirds are particularly bright, or that dogs, cats, and monkeys are less bright. Monkeys and apes are also well known for their mimickry, except when it come to vocalization. Theirs is not a failure of general cognitive abilities, but the lack of a special skill for producing diverse vocal sounds and of a predisposition to match them to some model.

So, why can't most mammals learn to sing or talk? Why can birds? And why can we? What crossing of connections or shift of emphasis in neural function or unprecedented developmental experience changed this one seal's brain to give him an isolated fragment of human uniqueness. What treasure was locked up in Hoover's brain?

The story of Hoover has an unfortunate end—for Hoover and for science. A little less than a year after I first heard his gravelly imperatives, Hoover died. It was during the annual fall molt, when a harbor seal's summer fur is shed and he can become prone to skin infections and other more serious complications. By the time I was told of Hoover's death, his body had been autopsied. And as parts of that talkative old fellow were sent out to determine the cause of death, I lost track of the one clue to his anomalous capabilities, his brain. The autopsy report noted that Hoover had died of an infection. In addition, the veterinarian who examined his body noted in passing that there was an unusual degree of calcification in his cranium, probably associated with early encephalitis or other brain damage. To my knowledge, no further analysis of Hoover's brain has been done. For the pathologist and the aquarium staff, the case was closed. Hoover died of natural causes. No one's negligence was to blame.

For me the case remains open and unsolved. This "believe it or not" oddity has never been explained. Had congenital or infantile brain damage contributed to his curious foray into the world of speech? Was Hoover's speech the result of just a few critical short circuits in his brain? Though Hoover's brain will not provide the answer, it's reasonable to suspect that by some misfortune it was modified in a way that parallels our own. This prompts us to ask whether there might be some common aspect of brain organization that is shared in common by all articulate species? If so, it might provide a clue to what happened both to Hoover and to *Homo sapiens*.

Visceral Sounds

It turns out that communicative behaviors in a wide range of vertebrate species, including members of every class from fish to birds to mammals, are funneled into a final common pathway that consists of structures in the central midbrain and brain stem. The midbrain is the transition zone between the brain stem and the forebrain. It is the first point, ascending from the spinal cord, where the long-distance senses, vision and hearing, are integrated with information about touch and movement. As a result, we find that structures of the midbrain effectively provide the first tier of complex behavioral control. Together, the behavioral programs built into midbrain structures comprise what might be called "higher-order" or "second-level" reflexes. These include a variety of automatic tracking and orienting responses of the senses, as well as an integration of head and body movements to support these automatic processes.

The midbrain contains sensorimotor-orienting systems involved in automatically shifting gaze and attention, and turning the head in the direction of interesting stimuli. It also contains descending motor pathways and a number of intrinsic motor-control systems. One of the most crucial but still poorly understood motor systems is distributed within the reticular formation (an amorphous collection of nuclei, located about midway between the dorsal and ventral midbrain). The reticular formation surrounds a centrally located canal through which cerebrospinal fluid flows between forebrain and hindbrain, and it derives its name from the crisscrossing fibers of the sensory and motor systems that enter and leave at this level. But these are not just fibers passing through. The reticular formation monitors inputs from most sensory nuclei and sends outputs to most motor nuclei of the midbrain and brain stem. Most neuroanatomists believe it to be the locus of arousal and a gateway to the majority of mid-level innate motor programs, including everything from chewing and swallowing to threatening or cowering.

At the core of the midbrain, surrounding the fluid-filled central canal that connects the forebrain ventricles (fluid-filled chambers) with the fourth ventricle and spinal canal, lies a cell-dense structure called the central gray (or periaqueductal gray) area of the midbrain (see Figure 8.1). The central gray area appears to be endowed with numerous hormone receptors, and passing nearby are ascending pain-projection systems and other ascending

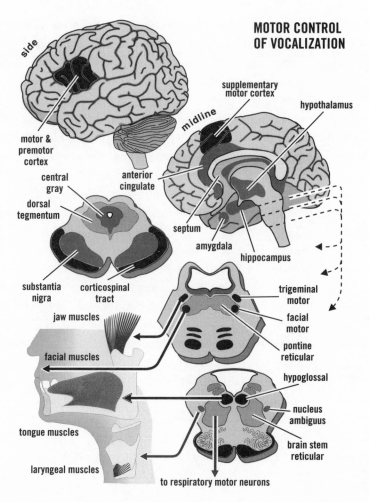

MOTOR CONTROL OF VOCALIZATION

Figure 8.1 *Output control systems for the major muscles of the vocal tract used to vocalize in humans and primates. The trigeminal and facial motor nuclei are typical skeletal motor output nuclei for controlling face and jaw muscles, respectively. The hypoglossal nucleus is a special motor region that controls the tongue muscles. The nucleus ambiguus controls laryngeal muscles. The reticular regions contain circuits that are substrates for many automatic movement patterns, and reticular regions around the nucleus ambiguus and further down the brain stem are premotor regions for breathing. The muscles of the tongue and face are more directly accessible to cortical motor and premotor control, whereas the muscles of the larynx and respiratory system are indirectly influenced by limbic structures such as the hypothalamus, amygdala, and cingulate cortex. Many of the limbic structures were once called the rhinencephalon ("nose brain") because the olfactory input is most direct. Stimulation of most limbic structures (not the olfactory bulbs) can elicit stereotypic vocalizations from mammal species, but motor cortex stimulation cannot.*

pathways carrying autonomic nervous system feedback on their way to the forebrain. Descending inputs arrive from areas of the forebrain limbic areas (hippocampus, amygdala, and cingulate cortex) and hypothalamus, which are specifically associated with emotional state and autonomic arousal. Among the output targets of the central gray area are both respiratory premotor regions of the brain stem reticular region and motor neurons that control the larynx (nucleus ambiguus). The central gray area is also a source of outputs controlling other visceral "behaviors" via the autonomic peripheral nervous system. Together with the surrounding dorsal tegmentum, the central gray area is thus part of a core system for coordinating relatively automatic vegetative and arousal functions of the organism, including ingestion and respiration.

Not surprisingly, behaviors that communicate arousal and emotional states in most vertebrates depend on this core midbrain center. The majority of communicative signals that animals produce, whether vocal or nonvocal, are associated with sensory/motor systems of the head. As the mobile "turret" on which all the long-distance sense organs are mounted, and also often the carrier of some of the most formidable "social" weapons (e.g., teeth, antlers), the head is a crucial source of information about an animal's direction of attention, motor readiness, and emotional state. Such important indicators of one's present and future propensities are crucial bits of information to make available to members of one's social group, in order to help mediate collective behaviors. Situated at the top of the spinal cord, the midbrain and brain stem are the locus for input-output systems of the head, and so it is not surprising that they should play crucial communicative functions.

Vocalization, in particular, has a special association with these midbrain systems because it involves the interaction of the oral and respiratory tracts. To organize vocalization requires the coordinated activation of clusters of motor neurons that control the muscles for breathing, the tension of the larynx, and the movements of the oral and facial muscles. The motor neurons controlling all of these are located in the upper brain stem. Even premotor neurons controlling the diaphragm and intercostal muscles for breathing can be found in this region, though the primary neurons for these muscles are located in the upper spinal cord. The premotor neurons for respiration include a distributed set of "pacemaker" neurons in the lower brain stem (the medulla oblongata) that control the breathing cycle. These receive inputs relaying information about oxygen and carbon dioxide levels in the blood, and are responsible for automatically adjusting the rate and volume of breathing during different states of activity and metabolic demand. Like most systems of the brain that control visceral functions, the pacemaker neurons comprise a network that is spontaneously active and autonomous. This enables breathing to be completely reflexive and automatic. The pacemaker system in turn controls motor neurons in the spinal cord which directly control the diaphragm and intercostal muscles that inflate the lungs.

But the oral tract has to serve two partially conflicting functions: both breathing and ingestion of food. Since swallowing and breathing are exclusive functions, the larynx plays a pivotal role as the gateway to the respiratory tract. When a bolus of food is pushed back by the tongue into the pharynx, an automatic swallowing reflex becomes engaged. With the initiation of this automatic movement pattern, breathing is momentarily inhibited and the larynx is briefly closed. Though well protected by the physical position of the epiglottis, which forces food behind and around the opening in most mammals, the laryngeal constriction prevents air movement from inadvertently diverting the passing food into the airway. Many infant mammals and some adult mammals can continue to breath, albeit in a slightly altered pattern, while swallowing liquids (e.g., during suckling), but this capability is unavailable to adult humans because of a much lower position of the larynx in the throat. As adults, we are much more dependent on laryngeal closure to protect the air passage from food and liquid than are other mammals (a curious difference we will return to in Chapter 11).

There are other reasons for adjusting the flow of air through the larynx. By constricting the air passage to a greater or lesser extent during different modes of breathing, pressure within the lungs can be adjusted. This can affect oxygenation during exertion, as well as modulate rigidity of the trunk in synchrony with locomotor activities that also need to make use of trunk muscles.

These problems of coordinating swallowing and breathing, and adjusting subglottal air pressure, have dictated an ancient linkage between the motor control of such processes. The muscle systems involved in eating and breathing are largely automatic, like many other visceral muscles, such as those that control digestion. Breathing, for example, needs to continue whether we are conscious or not, and so it defaults to an autonomous reflexive behavior. But these prepatterned behaviors can sometimes be intentionally inhibited and modified in response to changing circumstances, and there are times when it may become critical to stop breathing for a second or two. For the most part, however, the links among swallowing/breathing behaviors are not modifiable. The reason for all this should be obvious. If we needed consciously to inhibit inspiration of air while swallowing, sooner or later we would make an error and choke.

Not infrequently, we hear of cases where even the automatic system fails. This is probably far more common in human beings than in any other species because of the peculiarly long distance from mouth to larynx, and so has made the Heimlich maneuver an important addition to an otherwise automatic protection. Nevertheless, even animals that are less susceptible to choking because of a higher position of the larynx tend to alternate breathing and swallowing, or significantly modify their breathing during swallowing.

The functional organization of the breathing-swallowing-chewing systems is also reflected in the anatomical organization of the nuclei that control these functions in the brain stem. The groups of neurons that control the muscles for these behaviors are lined up in side-by-side columns, or nuclei, within the brain stem. At one end is the skeletal motor column (controlling voluntary muscles of the mouth and face), at the other end the visceral motor column (controlling automatic muscle systems for swallowing, breathing, and heart rate); and between them is the branchio-motor column, which connects to muscle systems that have both automatic and voluntary functions (i.e., controlling the tongue). The branchio-motor nuclei get their name from their position near the branchial arches that form where the embryonic head meets the neck. In fish embryos, these arches develop into gill arches.

The visceral motor systems were probably coopted for communication during terrestrial vertebrate evolution because changes in respiration patterns provide some of the most useful indices of arousal state. Accentuating the symptoms of respiration by making a hissing sound—produced as air passes through the narrow constriction of the throat and mouth—may have provided some of the earliest forms of vocal communication. The subsequent differentiation and specialization of sound-breath patterning has likely built upon this otherwise primitive index of arousal. Because of this ancient pattern, vocal communication has inherited many of the organizational features of a partly automatic, partly controllable motor system. These include the automatic, reflexlike production of sounds when in an appropriate arousal state; stereotypic vocalizations; and a minimal role for learning.

Vocalization, like swallowing, requires linked synchronous activity of the oral, vocal, and respiratory muscles, and so depends on a more central site that connects to all. This convergence center appears to be the central gray area of the brain. Electrical stimulation of the central gray area in cat, squirrel monkey, and macaque monkey brains appears to be capable of producing the full range of vocal calls for these species. The corresponding structure has been shown to elicit communicative gestures, postures, and sound production in essentially every vertebrate species that has been studied, including fish, frogs, reptiles, birds, and mammals. Stimulation of other linked brain stem sites (e.g., individual motor nuclei involved in vocalization) only produces isolated movements of the corresponding muscles. The central gray area is the critical central link among the surrounding reticular nuclei and brain stem motor nuclei comprising a distributed network for innate sound production.[1] It seems clear that call-production "programs" are actually embodied within this network of structures linked with the central gray area. Damage to components of this system can alter the form and sound of calls,[2] and transplantation of the entire midbrain from quail to chick embryos can "transfer" some of the quail's calls to the chick.[3]

Vocalizations do not occur in a behavioral vacuum. They are inevitably one component in a more complex visual-auditory display that also includes postural and gestural information (e.g., facial expressions). Although the networks that embody the innate motor programs underlying species-typical vocal behaviors are located in midbrain and brain stem circuits, they can be activated by arousal from a number of higher structures, including the hypothalamic and limbic system structures located deep within the forebrain. The limbic system contains the brain regions mostly responsible for emotional experience, motivation, and attentional processes, as well as for their associated autonomic and hormonal responses. Rather than an isolated behavior, innate vocalization is one of the outward manifestations of an integrated state of emotional and behavioral arousal. Other output systems that tend to be simultaneously activated may include postural, autonomic, and hormonal changes. Decades of electrophysiological and lesion studies of these structures in monkey brains have mapped out the forebrain sites that contribute directly and indirectly to call production. In general, the further these sites are removed from the midbrain, in connectional terms, the more integrative and less direct their effects on vocalization. This is often reflected in considerable delay between electrical stimulation of a site and the production of a vocalization. In some regions, vocalizations are only produced after the electrical stimulation is turned off, producing a kind of rebound effect. These forebrain systems provide multiple levels of control over vocalizations.

The motor programs for most innate mammalian vocalizations are highly modular. They are relatively invariant from birth; learning plays little or no role in their form and often can do little to influence when a vocalization is produced or inhibited. The links between specific vocalizations and distinct states of emotional arousal are also highly automatic and invariant. Appropriate calls are produced when arousal rises above some threshold in certain fairly stereotypic contexts. Different arousal states, such as fear or sexual arousal, are correlated with very different activity patterns within the structures of the limbic system and the hypothalamus. In some sense, the circuits that are active at a given time provide a signature of a particular emotional state. Not surprisingly, different limbic circuits project their outputs to the midbrain vocalization system through independent pathways, which are divided along lines that correspond to alternative arousal states. When directly stimulated, these pathways can also elicit vocalizations. Since each distinctive arousal state has a characteristic activity signature, its pattern of outputs can serve as a sort of code sent to the midbrain that specifies which vocal program to run. This linkage has other consequences as well. Because of this relatively invariant association, vocal calls are quite literally symptoms of specific emotional and arousal states.

The reflexlike links between perceiving and producing calls, and the emotional states associated with them, are made evident by the "infectiousness" of some of our own species' innate calls, specifically laughter and crying. Newborn babies lying in a maternity ward demonstrate this innate primitive linkage as they are induced to cry by hearing the cries of other babies, but many of us are also familiar with the personal experience of being induced to tears by the crying of others over the death of someone we do not know, or made to laugh by the irritating but effective "laugh track" that underlines and "sells" the jokes on television's situation comedies. Like the motor systems underlying innate vocalizations, the call analysis process also works on many levels at once. Before even arriving at forebrain auditory centers, auditory inputs are relayed in midbrain analyzers. Here, an initial categorization may occur, sufficient to activate initial orienting responses to evolutionarily important classes of stimuli. The emotional response elicited by innate species calls is also probably mediated by cortical analysis, since the cortex is the major source for sensory input into limbic structures. Links between midbrain structures may, however, provide the substrate for an independent tendency to call in response to a call, that, in effect, short-circuits higher-order analysis.

Why Don't Mammals Sing Like Birds?

The key to the vocal difference between birds and mammals can be traced to some interesting differences in their anatomy. Though human speech depends critically on structures of the cerebral cortex and rapid, skilled movements of the oral and vocal muscles, this is not true of vocalizations in other mammals. The human, cetacean, and bird "solutions" to skilled vocalization are quite different, but some interesting parallels offer important clues. What these exceptions to the rule share in common is what they avoid: leaving vocal communication under the control of the visceral motor systems of the brain.

In order to learn and produce skilled movements, many regions of the brain must become involved that are unnecessary for innate gestures and vocalizations. The visceral muscle systems that are most directly involved in mammal vocalizations are not well suited to this task, precisely because these systems must be capable of operating autonomously, according to a few set motor programs. Such programs are the epitome of modular brain functions that are closed to any interference from other systems. Once they are activated, they tend to run their stereotypic course irrespective of our awareness or monitoring of the process. In addition, such modular programs tend to be quite specifically localized to circuits interconnecting midbrain and brain stem nuclei. Skilled behaviors, in contrast, are associated with the skeletal muscle systems, such as the limb muscles. Since locomotor needs are so varied and unpredictable in most species, this system has to be capable of considerable plasticity and modifiability. Ongoing behaviors must be capable of being monitored and must be open to interruption and modification. Just the opposite of autonomous and modular. Nevertheless, once a behavior program has been learned, it can be "offloaded" to other motor systems that allow it to be treated as an unanalyzed modular program. Consequently, highly skilled behaviors can be as rapid and autonomous as any innate behaviors. Not surprisingly, the systems of the brain that support these plastic motor abilities are quite widely distributed. The cerebral cortex and cerebellum of mammals are among the most critical structures for consciously elicited movement, as well as for the development and modification of skilled behaviors.

Damage to the primary motor cortex in a mammal brain (motor cortex lies along a vertical strip running down either side of the cortex just ahead of the halfway point from back to front) can produce a complete loss of movement on the opposite side of the body that corresponds to the position of the damaged region in an inverted map of muscles from legs and feet at the top, to head and mouth at the bottom of the motor strip. Damage to the cortical area for oral and facial muscles on one side produces a partial paralysis that can affect the use of the mouth and tongue during eating, and bilateral damage can completely paralyze these muscles. The other parts of the brain that are involved in skilled behavior, such as the basal ganglia and cerebellum, play a complementary role to the cortex, and appear to be particularly important for fine-tuning and automating skilled movement patterns. It is probably accurate to view the motor cortex as helping to compose and test a behavioral subroutine that will eventually be run with little cortical intervention by these subcortical systems. In birds, although there is nothing quite like a motor cortex (in fact, no cerebral cortex as in mammals), there are forebrain nuclei that play a similar role in orchestrating intentionally monitored movements and programming deeper brain systems to run the programs for well-learned behaviors.

The superior vocal abilities of birds are probably a consequence of a peculiar quirk of anatomy having to do with the adaptation to flight. The evolution of flying provided intense selection pressure for weight reduction. One aspect of this was an elimination of the heavy teeth and bones of the mouth and jaw, which were replaced by a much lighter bill and a reduced tongue. In this process, the larynx was also simplified to accommodate special breathing problems imposed by flight. In birds, unlike most mammals, the larynx has become vestigial and is not the major regulator of air flow to and from the lungs. Birds control air flow with a paired muscular structure called a syrinx, which is located somewhat deeper in the chest cavity near the divergence of the bronchial tubes projecting to the two lungs. The syrinx functions like the larynx in mammals by selectively pinching off the flow of air; but being situated lower, it provides the possibility of independent control of the air flow through each bronchial tube. Tensing the syringeal muscles is also what produces the tones of birdsong. Some of the complexity in such song stems from the interaction of sounds produced by the two syringeal pathways. But why should this have also conferred more flexible vocal abilities and the possibility of vocal learning?

This difference between birds and mammals is related to the difference between the tongue and laryngeal muscle control in mammals. The larynx, as we saw, is controlled by the visceral motor system, which mostly produces stereotypic preprogrammed movements. In contrast, the tongue is controlled by systems intermediate between visceral and skeletal motor systems, and is capable of both stereotypic movements and some more deliberate movements. The muscles of the tongue run both radially and longitudinally. Radially oriented tongue muscles control its elongation by contracting at right angles to one another and squeezing the larynx in on itself, like the way that squeezing a balloon elongates it. Longitudinally oriented muscles of the tongue extend into it from attachments behind the tongue and aim the extended tongue to either side or up and down. One right and left pair of these connect at their posterior ends to the paired styloid processes, little spikes of bone extending from the frontal base of the skull, and another pair connect to the hyoid bone—a small, horseshoe-shaped bone that encircles the top of the larynx and bottom of the epiglottis. The tiny hyoid bone is not directly attached to other bones, but "floats" freely in the throat, suspended by long, thin ligaments and muscles that attach it to the base of the skull and the tip of the jaw. Contraction of the hyoid-tongue (hyoglossal) muscles retracts the tongue with respect to the larynx, and contraction of the hyoid-jaw (geniohyoid) muscles raises the larynx and epiglottis, at the same time compressing the tongue against the roof of the mouth and the back of the throat to push food along during swallowing. Contracting the muscles that link the hyoid to the jaw also aids in extending the tongue, by pulling it forward in the mouth.

All of these tongue muscles are controlled by fibers coursing through the hypoglossal nerve (named for its position "beneath the tongue"), which carries impulses from motor neuron cell bodies in the middle of the brain stem. These motor neurons form a cluster of brain stem columns collectively known as the hypoglossal nucleus (after the nerve that grows from it). A few years ago, a colleague (Alan Sokoloff) and I put some of the finishing touches on a half century of anatomical investigations of this most complicated muscle of the body by showing that in monkeys, each class of intrinsic and extrinsic tongue muscles is innervated by its own columnar subnucleus of the hypoglossal nucleus.[4] This point-to-point organization is what makes possible the remarkable articulatory control that is necessary for food preparation and for speech, at least in humans.

As far as we can tell, the human and monkey hypoglossal nuclei contain entirely corresponding columnar divisions, and these have parallels in many other vertebrates as well. One part of this motor map of tongue muscles stands out as particularly relevant to the bird-mammal comparison. Among the muscles involved in tongue movement, the muscle between the hyoid and the jaw (the geniohyoid muscle) is the exception. It is the only muscle that controls tongue movement without actually attaching within the tongue body. The motor neurons that control it occupy a column of the hypoglossal nucleus that is correspondingly separated from the rest, at the ventralmost tip of the nucleus. Since mammals lack a syrinx, the comparison is difficult, but it appears that the syringeal muscles are not simply laryngeal muscles that have been shifted to a lower position, and their innervation reflects a sort of mix of mammalian larynx-and-tongue control systems. Different syringeal muscles are controlled by neurons in both the nucleus ambiguous (larynx) and the hypoglossal nucleus (tongue). Curiously, the ventral hypoglossal subnucleus that controls the geniohyoid muscle in mammals appears in the same relative position as the subnucleus that is the major con-

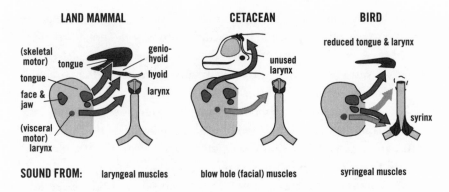

LAND MAMMAL　　　　**CETACEAN**　　　　**BIRD**

SOUND FROM:　laryngeal muscles　　blow hole (facial) muscles　　syringeal muscles

Figure 8.2 *Comparison of structures that control the musculature of sound production in terrestrial mammals, dolphins and whales, and birds.*

Left: In terrestrial mammals there is a clear segregation of the skeletal motor nuclei that control the movement of the tongue and the visceral motor nuclei that control the respiratory tract, including the laryngeal muscles. This limits coordination, control over automatic vocalization patterns, and vocal skill learning.

Middle: Dolphins and whales may get around this limitation because they probably do not make much use of their larynx in sound production and instead appear to use muscles associated with sinuses and the blowhole. These are derived from skeletal muscles of the face and thus have links to skilled volitional motor systems.

Right: In birds, though their brains are organized quite differently, there is considerable participation by forebrain auditory, motor, and association regions.

Different muscles of the tongue in mammals are controlled by subnuclei in the hypoglossal nucleus. One of these divisions controls an extrinsic tongue muscle in mammals that links the small horseshoe-shaped hyoid bone with the the inner tip of the jaw and can cause the whole body of the tongue to protrude. It is controlled by a subnucleus of the hypoglossal nucleus that is situated just ventral to the rest of the nucleus (compare bird and land mammals). In birds, in contrast, vocalization is produced by contraction of a paired muscular structure at the branch point of the paired bronchial tubes. By a curious twist of evolution, this structure is controlled by the ventral subnucleus of the hypoglossal nucleus (designated XIIts), which probably corresponds to the part that controls the geniohyoid muscle in mammals. Thus, a function subserved by visceral control in mammals is provided with skilled motor control in birds.

troller of muscles of the syrinx in birds (see Figure 8.2). Perhaps with the reduction of the skull, the loss of the jaw and teeth, and the modifications of the upper respiratory tract in birds, the corresponding nerves and muscles were recruited to control syringeal tension rather than tongue protrusion. By whatever means this occurred, this subdivision of the tongue-control nucleus appears in birds to have taken on the problem of breath control and sound production, thus handing over a visceral motor function to a skeletal motor control system.

This shift to a more deliberate and monitored control of air flow is probably an adaptation to the special demands of flight. For birds, flying and breathing are intimately linked. With massive flight muscles strapped to the sternum and alternately pulling against the rib cage, expanding and collapsing the rib cage in order to breathe could become complicated if the two functions were not coordinated. Unfortunately, this cannot be accomplished by simply linking breathing and wing flapping together in some stereotypic motor program. Flight requires the ability continually to adjust wing movements in the face of the unpredictable demands of changing wind conditions and obstacles. An ability to adjust related breathing and wing movements thus required involving the skeletal motor systems also in the control of air flow. Since the control of air flow is the basis for sound production, this too has come under more deliberate control as a side effect of the evolution of flight.

In addition to our own species, and perhaps Hoover, there is one group of mammals that exhibits some significant degree of vocal flexibility and learning ability: the cetaceans (dolphins and whales). In many ways, they are the exceptions that demonstrate the rule of visceral versus skeletal motor control. Though much of cetacean sound production remains poorly understood, it is generally believed that the many sounds they are capable of emitting probably are not produced by the larynx. Instead, dolphins and whales appear to generate squeaks, clicks, and whistles within an elaborate system of sinuses that are located in the front of the skull (behind the large, sound-focusing "melon" in dolphins) and that feed into the blowholes on the top of the head. This is probably controlled by passing air through constrictions between sinuses tightened by the contraction of underlying muscles. These blowhole muscles correspond to face muscles in other mammals, and are almost certainly controlled by the skeletal motor nuclei of the brain stem (probably the facial motor nucleus). Deliberate control of air flow to and from the blowholes is of course particularly important in these aquatic mammals. Like birds' adaptation for breathing in flight, but recruiting an entirely different muscle system, this special mammalian adaptation for adjusting air (and perhaps water) flow seems to have incidentally opened the door for more controllable sound production as well.

So, how might this anatomical logic shed light on our own vocal facility? Humans have not entirely shifted phonation to skeletal muscle systems, as have birds and cetaceans. Like other mammals, we produce most vocal sounds by constricting our laryngeal muscles. And yet a large fraction of sound variation in speech is only minimally explained by changes in laryngeal tension. Most speech sounds involve both muscle systems: a modification or interruption of laryngeally produced sounds by movements of the skeletal muscles that control the jaw, lips, and tongue. Some speech sounds—such as fricatives (*sss* or *fff*), plosives (*p*), and clicks (sounds not used in European languages but used by, for example, the !Kung San Bushmen of Africa and produced by variously snapping the tongue off the palate, cheek, or teeth under a slight negative pressure)—are produced entirely by oral muscles. So there are some interesting anatomical parallels between cetaceans, birds, and humans in the flexibility of sound production. It demonstrates the general rule: only when the skeletal muscle control system enters into the process is there any significant capacity for flexibility, learning, and the intentional control of sound production.

But a shift of most sound manipulation to oral muscles is not the whole story. Ours is a dual or hybrid vocalization process. The articulation of the tongue in the mouth must be precisely orchestrated with the production and modification of sounds produced in the larynx. Not surprisingly, the linguistic roles of variation of fundamental frequency and gating of sound production are mostly related to sound continuance. Though many languages (Chinese is one) use a few patterns of tone shifts as phonemes, only a very small number of such distinctions (e.g., rising, falling, stable) ever play a major role in speech. Most tonal variation plays a paralinguistic role in speech prosody—the tonal, amplitude, and rhythmic variations that convey attentional and arousal information in speech—and most of this occurs subconsciously and automatically with the corresponding shifts in affect. In this regard, laryngeal activity functions much as it does in the vocalization of most mammals: as a symptom of limbic arousal.

But singing demonstrates the extent of our capability systematically to control specific tone production by an independent mechanism. Such a precise control of relative vocal frequency and timing exemplifies an unprecedented capacity for laryngeal control that is only minimally incorporated into any language. Moreover, the crucial role of laryngeal sound production in distinguishing phonemes (e.g., by differences in voice-onset time) to partition words and mark phrases must be coordinated to millisecond accuracy with the movements of the lips and tongue, and these carefully choreographed combinations must be learned so that they can be deployed at a moment's notice in a vast array of contexts. Clearly, then, more than just a shift in emphasis to oral muscles underlies our speech abilities.

There must also be a difference in our capacity to control laryngeal movements that reflects differences not just at the periphery but within the brain.

Humans have a degree of voluntary motor control over the sound produced in the larynx that surpasses any other vocal species. Indeed, this degree of voluntary control is otherwise found only in motor systems controlled by cerebral cortical and cerebellar motor pathways projecting ultimately to skeletal muscles. This suggests a difference in the neural control of the human larynx that is the neural equivalent of the shift from visceral to skeletal muscles that underlies vocal skill in birds and cetaceans. Though direct evidence of such a change in the human brain's motor output to the larynx is not yet available, the circumstantial evidence is abundant. One way or another, the human larynx must be controlled from higher brain systems involved in skeletal muscle control, not just visceral control. Additional evidence for such a shift is supplied by the partial decoupling of sound production from emotional arousal states in language, as it is also in birds and cetaceans.

This separation is particularly evident in bird species that are capable of mimicking other animals' sounds or being taught speech. Though it may be necessary to be sufficiently aroused to vocalize at all, a particular vocalization is only arbitrarily linked with a particular emotion. A similar dissociation of specific vocal sounds from specific emotions is also characteristic of humpback whale songs, which change annually (see Figure 2.1 in Chapter 2), as well as of many of the vocalizations that dolphins produce. This dissociation from specific affective states is an essential requirement for learning novel vocalizations. It is my suspicion that Hoover's speech was also dissociated from any specific emotional states, except perhaps just being bored. His speech did not replace any typical harbor seal vocalization. He seemed to produce speech less often in threatening social contexts or during the mating season, when he used typical seal vocalizations instead. And his most vocal periods seemed to be at times when he was aimlessly "pacing" back and forth (in his case swimming repeatedly back and forth) in his enclosure and periodically punctuating this boredom by reclining back in the water and speaking.

But the motor/emotion dichotomy is not absolute. Not only do we exhibit a mix of arousal-independent learned vocalizations (speech and song) and highly stereotypic vocalizations that are innately linked to certain emotions (laughter and sobbing), but our visceral-emotional and skeletal-muscle-skill-learning systems often compete for access over vocalization, as well as often complementing each other in the same vocal behavior. It is as though we haven't so much shifted control from visceral to voluntary means but superimposed one upon the other.

One of the more intriguing examples of the interactions between intentional motor systems and automatic calling tendencies has been provided in an observation by Jane Goodall.[5] Chimpanzees often produce food calls when they come upon a new food source. This stereotypic call attracts hungry neighbors to the location, often kin who are foraging nearby. Goodall recounts one occasion where she observed a chimp trying to suppress an excited food call by covering his mouth with his hand. The chimp had found a cache of bananas she had left to attract the animals to an observation area, and as she suggests, apparently did not want to have any competition for such a desirable food. Though muffling the call as best he could with his hand, he could not, apparently, directly inhibit the calling behavior itself.

People are often faced with circumstances of this sort. For example, there are many times when humorous events threaten to make us laugh, but for reasons of politeness we feel compelled to stifle the tendency. To suppress an irresistible laugh, we resort to such tricks as gritting our teeth, clenching our jaws and lips, putting our hands over our mouths, or simply turning so as not to face someone who might take offense. Even for humans, the essentially automatic and unconscious nature of many stereotypic calls causes them to erupt without warning, often before there is time indirectly to interfere with their expression. Some degree of self-consciousness may be a prerequisite for such control. This curious conflict between simultaneously produced intentional and unintentional behaviors offers a unique insight into the nature of language. The superimposition of intentional cortical motor behaviors over autonomous subcortical vocal behaviors is, in a way, an externalized model of a neural relationship that is internalized in the production of human speech. It graphically portrays the functional bridge that links the vocal communication of primates to the speech of humans. The evolution of speech effectively occurred at this neurological interface. Understanding this relationship is a first step in the process of deconstructing language into its evolutionary antecedents.

Though the cerebral cortex plays little role in the production of vocalizations in most mammals, some cortical systems appear to regulate call production by superimposing or relaxing inhibitory control over these spontaneous tendencies. The only cortical areas involved in call production in primates are located in the medial frontal cortex (anterior cingulate cortex).[6] These areas are intermediate between limbic cortex and neocortex, and are otherwise involved in sustaining attention and initiating intentional actions. Extensive bilateral damage to this region in human brains produces at least a temporary immobility and mutism, though not actual paralysis or language loss. Damage to nearby medial and ventral frontal cortical areas may also produce disturbances of affect and emotional expression in both monkeys and humans. Though most mammal calls are not learned, these cortical regions appear to be critical for learning under which circumstances one needs to facilitate or inhibit the tendency to call.

The cerebral cortex in most mammals includes areas that control movements of the oral and vocal muscles, but these cortical areas are probably most involved in controlling movements of the mouth, tongue, and lips during grooming, food preparation, and eating. When these areas are damaged, or their output nerves are severed, paralysis of the facial and oral muscles occurs. Bilateral destruction of these areas in a monkey's brain makes it impossible to eat, and yet, though paralyzed and severely incapacitated in other uses of the oral pathway, these animals can still produce calls whose form is not radically altered from normal. Despite the fact that cortical motor damage does not disrupt call production in the monkeys that have been studied, motor cortical areas may nevertheless play an indirect role. Projections to oral and vocal motor nuclei in the brain stem may offer a route for direct intentional inhibition of calls. The direct superimposition of other competing motor output signals can help to inhibit or block call production under circumstances where arousal cannot be suppressed, but where such calls might result in unfortunate consequences. Consider, for example, the tendency to produce distress calls when separated from a group, or fright calls when startled. An animal that suspects that a predator is nearby may need to be able to suppress these tendencies in order to avoid giving its location away.

Again the comparison to bird brains is instructive. Forebrain structures are intimately involved in both birdsong production and learning (although bird forebrains are organized quite differently from mammal forebrains, correspondences become more obvious in midbrain and brain stem). Many of the structures that contribute to singing can be compared to limbic structures in mammal brains and contribute to arousal thresholds for initiating singing, but a number of other structures have been shown to be essential for song learning and even to the organization of the song structure. These include the forebrain auditory nuclei, and motor nuclei as well as structures that play integrative roles between them. When the motor nucleus of the bird archistriatum (RA) is damaged, for example, singing is disrupted; but if more anterior auditory and association structures are damaged, song learning and song structure may be disrupted while singing remains possible. This difference from most mammals is analogous to the difference between humans and other primates. Only in humans does damage to or stimulation of the cortical auditory and motor cortex produce an analogous disturbance of the structure of vocalizations.

One reason why calls are largely unaffected by cortical motor damage is that they do not generally involve complex articulatory movements of the mouth and tongue. Instead, vocalization is produced along with specific stereotypic positions and shapes of the lips and tongue, sometimes with the addition of simple repetitive stereotypic movements of the mouth. Segmentation of calls is mostly marked by the onset and offset of air flow and the associated frequency shifts (e.g., on and off sound production, or rising and falling tonality). These features are also well exemplified by the two most distinctive human innate calls, laughter and sobbing. Both involve rhythmic sound production and very stereotypic breathing patterns that segregate the sound into repeating units (laughs and sobs). Both are also associated with relatively stable mouth-tongue positions and no articulation of the lips, or of the tongue against other oral structures. The fixed stereotypic positioning of the mouth and tongue during these human calls often makes it impossible to speak at the same time.

Speech, in contrast, involves extensive and rapid articulation of the tongue with lips, tongue, and palate, along with simultaneous changes in lip shape, and opening and closing of the jaw. Though phonation is often interrupted by articulatory movements that momentarily cut off expiration, and tonality may be varied slightly as we speak to mark distinctive features or to provide emotional or attentional information, vocalization during speech is fairly continuous and invariant compared to the stereotypic on and off patterns demonstrated in calls. Breaks in sound production may be necessary to take a breath or may signal a break between thoughts or sentences, but individually vocalized words or syllables separated by silences are not normal, and may occur as a symptom of some types of brain damage (see the discussion of Broca's aphasia in Chapter 10).

In neurological terms, calls are characterized by foreground visceral motor programs on a background of relatively stable oral facial postures. The skeletal motor system is subordinated to the visceral motor system. Speech inverts this relationship. When humans speak, the skeletal motor components of the oral tract take the foreground against a comparatively more stable vocal-respiratory background. Most of the information that distinguishes individual speech units is encoded in rapid articulatory movements. These are superimposed on a background of constant expiratory pressure and more slowly varying tonality. Thus, although both calls and speech require the coordinated action of the subcortical motor systems that control respiration and vocalization as well as the cortical skeletal motor systems that control the jaw, lips, and tongue, the patterns are inverted. Humans are perhaps at an intermediate stage between birds, who have usurped vocal control from visceral motor systems, and the chimp with his hand over his mouth trying to stifle a call produced with visceral systems using a limb controlled by voluntary cortical systems. Cortical motor movements during speech are superimposed over a relatively stabilized visceral motor activity. In this case, oral muscles rather than hand muscles are involved in modifying the vocal output, and so the control is more complete and the modifications more facile and subtle. The fact that chimps (and indeed the vast majority of mammals) cannot exercise a more direct cortical motor control over either their phonation or their oral movements at this time demonstrates that even at this low level in the neural control of sound production, human brains must be unusual.

A Leveraged Takeover

In all mammals, the cortex includes neurons (called "pyramidal cells" because of their shape) that send output axons to deeper structures of the brain. In the motor cortical region, these output neurons are particularly large, and send their axons into the brain stem and spinal cord. There they contact the dendrites of premotor interneurons (one step removed from final output neurons) and motor neurons responsible for muscle activation. The brain stem of a typical mammal also receives direct cortical inputs into regions of the reticular motor system. These are widely distributed premotor structures, some of which send their outputs to the motor neurons that directly control the muscles of the face and jaw (the facial and trigeminal motor nuclei, respectively). In the nonprimate mammals we and others have studied, relatively few if any cortical projections contact the output motor neurons directly. The separate reticular regions relay signals from the motor cortex to the tongue motor nucleus (the hypoglossal nucleus), and to the laryngeal motor nucleus (the nucleus ambiguus).

Compared to most other mammal species, however, the projections from the monkey cerebral cortex are more extensive (Figure 8.3). The cortical inputs to the spinal cord in monkeys include the intermediate premotor layers that are innervated in most mammals, but also extend all the way to include the ventral-most layers of the spinal cord, where they likely contact motor neurons directly. In the brain stem, there is a similar pattern. Not only do axons from the motor cortex connect to reticular regions, they also

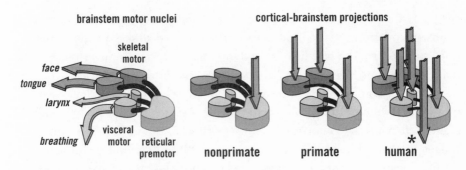

brainstem motor nuclei **cortical-brainstem projections**

skeletal
motor

face

tongue

larynx

breathing

visceral
motor reticular
premotor

nonprimate **primate** **human**

*

Figure 8.3 *Two evolutionary shifts producing increasing cortical control over motor output from brain stem articulatory and vocal systems. These shifts were produced by an increase in the proportions of the cerebral cortex in comparison to these brain stem structures.*

Left: Basic relationships between brain stem nuclei (neuronal cell clusters) controlling face-jaw-tongue and larynx-respiration muscles.

Right: This shows the progression from a typical terrestrial mammal plan (far left of the three) with minimal direct cortical input to these systems (mostly indirect via the reticular regions of the brain stem) to the monkey-ape condition (center image), where increased forebrain size gives cortical projections a favorable competitive bias, to the possible human condition (far right), with extensive recruitment of even visceral motor systems due to the massive cortical projection to these regions. The asterisk indicates increased projection to spinal respiratory motor neurons as well.

connect directly with the facial, trigeminal, and hypoglossal motor nuclei. Not contacted by cortical inputs are the visceral motor nuclei. The nucleus ambiguus, for example, still only receives indirect projections via the reticular regions, as it does in other mammals. The consequence of direct cortical motor neuron connections in monkeys is probably their increased voluntary control over the movements of hands and fingers. The importance of this for primate locomotion and foraging is obvious. Shifts to more voluntary control of the facial and mouth muscles could also be important for preparing food and communicating by gestures.

How has this takeover of motor nuclei by cortical inputs been achieved in primate brain evolution? Primate forebrain expansion provides a sort of vertical rendition of the displacement process responsible for the connectional changes in the brain of the blind mole rat, *Spalax*. As a result of the reduction in proportions of the postcranial body compared to the head and brain there has been a change in relative proportions between the forebrain as compared to the brain stem and spinal cord, and this embryological shift in neural proportions is a recipe for displacement. With so many more descending axons vying for space in the primate motor system, the more numerous cortical axons displace the less numerous local connections, since these displaced connections arose from systems that were scaled for a smaller body. In nonprimate brains, the initially overexuberant and somewhat nonspecific cortical projections to these brain stem motor nuclei are outcompeted by local projections and pruned back during development to leave only those projecting to premotor regions of the brain stem and spinal cord. In primate brains, on the other hand, the initial cortical projections are so numerous that they outcompete the local connections and persist in far greater numbers in many additional motor nuclei.

This same logic can be extrapolated to the human case. The primate/human difference is characterized by an additional increase in cortical/brain stem disproportion. Though of different embryological origin, this additional human shift in embryonic neural numbers should provide an analogous displacement, compared to the typical primate pattern. Because of their greatly expanded numbers, human cortical axons should enjoy an even greater ability to recruit brain stem and spinal cord targets during development. But where else are there targets for human cortical axons to invade? They will almost certainly increase in proportions in face and tongue muscle nuclei, with the consequence that the voluntary control of these systems will be greater than in other primates. In addition, however, the more extensive human cortical projections have probably also invaded nuclei in the brain stem and neurons in the spinal cord that even primates do not have voluntary control over: nuclei-controlling visceral muscle systems. Two of these, in particular, are relevant to speech: the motor neurons that control the larynx (the nucleus ambiguus), and those that control breathing (in the brain stem and upper spinal cord).

Neurologists have long assumed that humans have direct cortical control over the larynx because of the tendency of cortical motor damage to produce muteness. Uwe Jürgens, Detlev Ploog, and their colleagues at the Max Planck Institute have provided additional circumstantial evidence for a primate/human dichotomy in this regard by demonstrating that monkeys lack direct cortico-ambiguus connections, and by showing that cortical motor damage does not eliminate a monkey's ability to vocalize.[7] They surmise that such a difference might explain the behavioral difference in response to brain damage, and might also explain the well-known difficulty in getting monkeys and apes to vocalize from learned associations, much less form articulate words. But gaining volitional control over laryngeal functions is only part of the story. In order to produce complicated skilled vocal sounds, the laryngeal muscles need to be coordinated with breathing, and with muscle movements of the tongue, lips, and jaw. The human ability to speak is probably a consequence of all these systems being brought under common cortical control.

One of the more interesting sources of evidence for a shift in motor control over vocalization and breathing is suggested by studies of laughter. Laughter is not only highly stereotypic in form and similar in sound structure in all normal people, it is also quite different in some important respects from corresponding vocalizations in other primates. Though chimpanzees also produce a laughterlike vocalization—often as a result of tickling or related tactile play—which is made up of similar short, rapidly alternating inhalations and exhalations, chimps vocalize on both the outbreath and the inbreath, whereas humans vocalize almost exclusively on the outbreath (though there are individual differences in this regard).[8] This may not seem to be particularly noteworthy, but consider the curious fact that although many other primate vocalizations also involve this in-out vocalization pattern, as though the larynx is just set at a fixed tension irrespective of breathing, no typical human vocalizations, whether calls or speech, have this character. Why not? We tend to relax the larynx on inhalation, even when we are not thinking about it. It is indeed quite difficult and almost painful to speak while inhaling.

For speech, this asymmetry has obvious communicative value. A large number of units of sound distinction (phonemes and syllables) and the sound units of meaning (words and morphemes, like word stems and prefixes) can be produced during a single exhalation by modifying the resonances of laryngeal sound production by complicated sequences of upper vocal tract movements. These are far less effectively produced during inhalation (try reading this sentence aloud while inhaling). Because of this, during speech we rapidly inhale between phrases in order to set up for each subsequent prolonged, controlled, sound-producing exhalation. This rapid inhalation is facilitated by completely opening the laryngeal folds. Rapid alternation between orally modified vocalization and nonvocalized inhalation increases the information transmission rates of spoken information by effectively shifting the focus of information transmission from breath units to articulatory units. It appears, then, in the interest of speech, that the relatively inflexible link between laryngeal movements and breathing has been broken and subordinated in a rather thorough way to skilled articulation. Human laughter and sobbing are like primate calls in that they tend to be based around alternation of the presence and absence of vocal sounds, superimposed on relatively more stable mouth postures. They are bottom-up, viscerally driven patterns. In speech, this relationship is reversed. Relatively slower tonal changes and exhalation patterns become precisely timed and subordinated to match the rapidly fluctuating articulatory movements of the mouth and tongue. Speech is thus comparatively top-down in its control. The skeletal motor systems driving mouth and tongue movements dictate breathing and laryngeal patterns.

The predominance of cortical projections to these visceral motor systems also explains another, often ignored oddity of human vocalization: the unprecedented babbling of human infants. Within a few months of birth, human infants begin spontaneously and incessantly experimenting with sound production, sampling most of the range of possible phonemes that speech will later employ. No other mammal species' babies produce even a tiny fraction of the sort of unstereotypic vocal play that human babies produce. And human infants don't need to get particularly excited or upset to babble; in fact, they only babble while emotionally calm. When they get upset, their babbling is interrupted by a more stereotypic call: crying. This propensity is a clear indication that human babbling is activated differently from other innate vocalizations. It is the first sign that human vocal motor output is at least in part under the control of the cortical motor system. Even the timing of the first appearance and maturation of babbling and vocal mimickry corresponds to the maturation of the cortical motor output pathways. Although newborns do engage in some vocalizations other than crying, the kind of vocal manipulation that characterizes babbling is not produced until a few months after birth. At this time the cortical motor tracts are just beginning to become encased in a sheath of cells filled with myelin (a fatty substance which acts like a sort of neural insulation, promoting fidelity of the signal and more rapid signal transmission). By the time a child produces its first words and its first steps (around one year of age), the projections that carry voluntary movement information to the brain stem and spinal cord have nearly reached adult levels of myelin.

Since the displacement of noncortical inputs by cortical inputs in brain stem motor nuclei is an indirect consequence of the shift in brain and body proportions in our species, information about relative brain proportions can fortuitously provide an index of the degree of cortical vocal control in our fossil ancestors. Though fossil speech sounds weren't left behind in the fossil record, nor even fossil vocal tracts, we do have numerous measurements of the brain and body sizes for each major hominid species. These data suggest that human vocal skills first exceeded the capabilities of any living nonhuman primate at least 2 million years ago, in the hominid fossil species *Homo habilis*, since this species marks the first significant upward shift in relative brain size (see Chapter 11 for discussion). Since the trend toward larger brains continued from that point until about 200,000 years ago, we can predict with some confidence that vocal abilities were enhanced continuously over this entire extended period of hominid evolution. These data suggest that it is unlikely that speech suddenly burst on the scene at some point in our evolution. The ability to manipulate vocal sounds appears to have been in a process of continual development for over 1 million years.

However, there is an additional source of evidence for a shift in cortical control over respiration that is provided by hominid fossils. One correlate of our greater cortical control over breathing appears to be an enlargement of the thoracic (chest) spinal cord as compared to other primates. This is the region of the spinal cord that contains motor neurons that control the intercostal (between-rib) muscles and other trunk muscles involved in breathing. This enlargement may reflect a comparative increase of the numbers of motor neurons in this segment of the cord (because of reduced cell death?) and possibly an increase in descending connections terminating in this area. In their recent book on human evolution, the Johns Hopkins paleontologists Pat Shipman and Alan Walker suggest that the evolution of speech may be contingent on this enlargement. They looked for this feature in the relatively complete vertebral column from a fossil *Homo erectus* boy, but could not find a corresponding thoracic enlargement. They conclude from this that speech had not yet evolved at this stage in human evolution. Given the neurodevelopmental information suggesting that cortical takeover of breathing control is a quantitative allometric effect—and not likely the result of a pair of all-or-none mutations that added new neurons to this part of the cord and retargeted cortical axons to match—I suspect that an intermediate interpretation is more likely. With a relative brain size intermediate between modern apes and modern humans, this *erectus* boy likely also had an intermediate level of cortical control over respiration, supported in part by an increase in cortical projections to thoracic motor neurons as well as to other higher respiratory centers.

Did the use of speech have to wait until these modifications reached a modern level, or did the demands of vocal communication contribute some of the selection pressure that led to these modifications? These are questions that we will return to later (in Chapter 11), but in general, we can conclude that the modern human level of control over vocalization did not evolve overnight, and that upper vocal tract articulatory capabilities were probably always in advance of laryngeal capabilities, due to the target bias in favor of skeletal motor over visceral motor nuclei. This means that if this *erectus* boy and his contemporaries did communicate using something like a language, it may well have been one that was more reliant on orally produced sound variations than on laryngeally produced ones. In other words, it likely employed fewer vowels and rapid tonal variations, relied more heavily on consonants and oral clicks, may have been limited to short phrasing, and so probably required more nonverbal support as well.

So, what about Hoover's speech? If most other species of mammals are unable to talk because the connections from the cortical motor centers to the vocal nuclei in the brain stem are pruned away during early development, isn't it possible that perinatal damage to the limbic-midbrain projections in Hoover's brain turned the tables in favor of cortical motor projections to vocal systems? If more typical projections to vocal control systems had been damaged early enough, the usually transient cortical projections to these same nuclei might have been able to persist into adulthood. Brain damage could indeed have provided Hoover with more direct motor control over his tongue and larynx. In any case, it is probably not coincidental that Hoover was an aquatic mammal. Unlike terrestrial mammals, for whom breathing can be left to run on autopilot, aquatic mammals need to have direct control over when and when not to breathe. Overriding automatic respiration via cortical control is almost certainly a prerequisite from an early age. How this is accomplished is not known, but we now can make some confirmed guesses. Though we will never know Hoover's secret, he may have taken a road toward speech that happened to converge with ours at these critical neural intersections of breathing, sound, and oral movement control.

Symbol Minds

If our brains were simple, we would be too simple to understand them.

—Mario Puzo

Front-Heavy

A gradual superimposition of cortical motor control on visceral motor systems for vocalization has made possible our modern speaking abilities, but the ability to produce articulate sounds is not in itself sufficient to make language possible. An articulate verbal ability is not sufficient to lift parrots across the symbolic threshold, nor has merely shifting to manual signing enabled apes to become articulate members of human society, despite the hopes of some researchers. Even Kanzi—who more than any other nonhuman has demonstrated the extent to which language comprehension abilities are possible without any explicit language adaptations, and how providing a computer speech system can enhance access to symbolic communication—will probably not progress beyond the level of symbolic sophistication exhibited by a three-year-old human child (though he still has many years in which to prove me too pessimistic). This is because the more fundamental constraint affecting language evolution derives from learning and memory difficulties, not limited motor abilities. The evolution of vocal abilities might more accurately be seen as a consequence rather than the cause of the evolution of language. The changes in brain organization that have provided humans with the support necessary to surmount the symbolic-learning threshold are more critical. Symbol acquisition abilities provide the pacemaker for language evolution from which other adaptations for language must derive their usefulness. A greatly enhanced vocal articulatory ability would be of minor adaptive value if not coupled with symbolic abilities.

Human brains are not just large ape brains, they are ape brains with some rather significant alterations of proportions and relationships between the parts. We ought to expect a significant part of this shift in proportions to reflect adaptations to the unusual cognitive demands imposed by symbol learning. In other words, if the human brain is a more language-friendly ape brain, then the special demands of the language adaptation are likely to be reflected in the ways that human brain structure diverges from nonhuman brain structure. Given the completely unprecedented and indeed countervailing demands imposed by symbolic learning, as contrasted to all other forms of learning, it seems inevitable that some equally unprecedented neuroanatomical changes must lie behind the human symbolic facility. The most extreme deviations in brain structure between human and other primate brains may thus offer clues to the neural computations that most distinguish human from nonhuman minds.

As we saw in Chapter 6, these radical structural deviations in the human forebrain have resulted in a disproportionately enlarged human prefrontal cortex and a shift in connectivity favoring prefrontal connections in other systems. Let's examine this most dramatic brain renovation for clues to the kinds of cognitive demands that drove human brain evolution (Figure 9.1). What functional correspondence could link this remarkable neural disproportion to our unusual mode of communication? Does this most modified structure reflect the most intense and unusual demands imposed by language? If so, it is not the production of everyday speech and the comprehension of others' speech that is the main problem. These processes are only minimally affected by damage to prefrontal cortex; such damage seldom produces permanent difficulty producing speech, comprehending speech, or analyzing grammar. Disruption of these basic language abilities— aphasias—typically results from damage to areas more closely associated

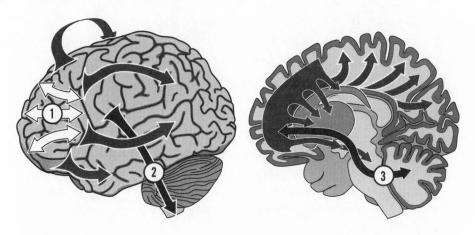

Figure 9.1 *Idealized depictions of the predominance of prefrontal influence over other brain structures. White arrows in the left image represent the relative expansion of prefrontal cortical areas (1). Gray arrows in both images represent expanded prefrontal projections to other cortical areas, basal ganglia, and thalamus. Black arrows signify projections that have expanded so as to invade novel targets: including (2) increased prefrontal projections into midbrain and brainstem vocalization circuits (discussed in the previous chapter) with respect to other limbic and midbrain sources; and (3) expanded interconnectivity of prefrontal regions and lateral cerebellar cortex, probably including connections with ventral prefrontal cortex (which has auditory-oral cortical links). For the first time in evolution, this would bring cerebellar systems into the loop of higher-order control of vocal-auditory processes (discussed later).*

with motor and auditory analysis: Broca's and Wernicke's areas, respectively (these will be the subject of the next chapter). Prefrontal areas seem instead to be recruited during such tasks as planning complex behaviors. But the range of specific cognitive processes that are altered by prefrontal damage cannot be described in such simple terms, and searching for a common underlying pattern in prefrontal syndromes leads to an indirect and far more interesting connection to language and symbolic cognition.

Just as prefrontal expansion can only be understood as a function of dynamic systemic interactions between many of the brain's structures during development, its structural and functional consequences also require an understanding of the systemic consequences of changes in size. This prefrontal enlargement as compared to the majority of brain structures in the cortex and elsewhere is a consequence of the developmental competitive advantage that its afferents have over other types of cortical afferents. These afferents come from such thalamic nuclei as the medial dorsal nucleus (which receives a majority of inputs from the midbrain tectum and dorsal tegmentum) and the anterior nuclei (which receive their inputs from limbic cortex), and from a very wide range of other cortical areas, including all sensory and motor modalities.

Of particular interest are the widespread cortical outputs that project back from the prefrontal cortex to every modality of cortex, including limbic cortex and cortical projections onto the basal ganglia and into the midbrain, particularly to the tectum and dorsal tegmentum. Given the very large size of the prefrontal cortex compared to the size of its targets, it could be expected to recruit a far greater proportion of synapses in these structures during development than do other structures that send competing afferents to these targets. Compared with more typical primate and mammal brains, then, we should expect that in the human brain, prefrontal synapses will be in greater proportions within its targets. Consequently, prefrontal information processing will likely play a more dominating role in nearly every facet of sensory, motor, and arousal processes. Irrespective of whether this structure has more "capacity" in some information-processing sense because of its size, it simply "has more votes" in whatever is going on in those regions of the brain to which it projects. In general terms, human information processing should be biased by an excessive reliance on and guidance by the kinds of manipulations that prefrontal circuits impose upon the information they process. We humans should therefore exhibit a "cognitive style" that sets us apart from other species—a pattern of organizing perceptions, actions, and learning that is peculiarly "front-heavy," so to speak. But how can this be described in neuropsychological terms?

Although, during development, the prefrontal region is probably carved out as a single projection field, in the mature brain the prefrontal cortex is not a single homogeneous structure with a single function. There is a danger of extrapolating from studies based on one prefrontal area to claims about the whole prefrontal region. Different prefrontal regions receive diversely different cortical inputs and outputs which provide general hints concerning their functional differences. Many regions receive inputs from specific sensory or motor modalities, and others receive converging inputs from more than one modality. No prefrontal area, however, receives direct input from primary sensory or motor cortices. One reason the prefrontal region remains to some extent mysterious is that its map structure is difficult to discern. Unlike the cortical topography of most sensory areas, positions within prefrontal regions do not seem to correlate with the peripheral topography of any sensory receptor surface. Nor is there a clear map of motor topography.

One hint concerning the sort of mapping of functions within the prefrontal regions, however, comes from studies of visual attention and the subcortical structures that underlie it. Patricia Goldman-Rakic and her colleagues at Yale University have demonstrated that one portion of the prefrontal cortex in monkeys (the dorsal lateral prefrontal region, or *principalis region,* named for its location surrounding the principal sulcus) maps the direction of attention-driven eye movements with respect to the center of gaze.[1] As a result, damage to some sector of this region can selectively block the ability to learn to produce or inhibit directed eye movements in a particular direction, or in response to cues in a particular direction. Not surprisingly, this subregion of the prefrontal cortex is located close to a region known as the frontal (motor) eye field, which directs eye movement. The eye-movement/attentional features of this region of the prefrontal cortex are also not surprising when considered in the context of its input/output association with the deep layers of the superior colliculus. And it has extensive connections with temporal and parietal visual areas.[2]

Just above and below this prefrontal visual attention zone are regions that are reciprocally connected to the auditory and multimodal auditory/somatic cortical areas of the temporal lobe.[3] They likely also send projections to tectal regions where auditory information is initially processed in the deep layers of the superior colliculus and the inferior colliculus. One way to understand the frontal cortical auditory areas is that they "map" auditory-orienting processes in ways analogous to the way principalis cortex "maps" visual orienting. In addition to the sensory-associated prefrontal subdivisions on its lateral surface, there are also orbital and medial regions of the prefrontal cortex. These are less easy to find map correlations for. These have predominantly limbic and adjacent prefrontal cortical connections, and show output pathways that include structures more associated with visceral and arousal functions than sensory motor functions.

The function of the lateral divisions of prefrontal cortex must, in part, be understood in terms of attentional mechanisms, both with respect to collicular systems and with respect to cortical systems to which they project outputs, whereas the function of orbital and medial divisions of prefrontal cortex must in contrast be more involved with arousal, visceral, and autonomic functions. These two systems are not only structurally interconnected but likely also to be functionally interdependent as well. Arousal, orienting, and attending are all part of the same process of shifting motivation to regulate adaptive responses to changing conditions. The lateral divisions may provide a substrate for intentionally overriding collicular-orienting reflexes, using orienting information as cues for working memory about alternative stimuli, or to select among many sensory configurations for further sensory analysis. The orbital and medial divisions may provide correlated shifts in arousal and autonomic readiness both to support shifts in attention and to inhibit the tendency for new stimuli to command attentional arousal.

So, what does the prefrontal cortex do? This is no simple question. In fact, it remains one of the more debated questions in neuropsychology.[4] The reason it is difficult is that the explanation cannot be tied directly to any sensory or motor function. When prefrontal areas are damaged, there are no specific sensory or motor problems. Surgeons who performed prefrontal lobotomies used to point out that it didn't reduce their patients' IQs either. The consequences of prefrontal damage only show up in certain rather specific sorts of learning contexts; but these can be extensive and ultimately debilitating. Understanding these associations is also difficult because there is not one type of prefrontal deficit for the simple reason that there is not one homogeneous prefrontal area. Because different prefrontal areas are connected to different cortical and subcortical structures, when they are damaged, they produce slightly different types of impairments. Not only are there numerous competing theories that attempt to explain individual types of prefrontal impairments, there is no account of the family resemblances that link the many different deficits associated with different prefrontal subareas.

Let's begin this global overview of prefrontal functions, not by treating the prefrontal cortex as homogeneous, but rather by searching among prefrontal areas, connections, and deficits for common themes and family resemblances. I am encouraged that there are some common threads because of the global similarities in connecting architecture that link these areas with the rest of the brain. Like the numerous subareas of the visual cortical system, I suspect that the different prefrontal regions share a common computational problem, but have broken it up into dissociable subtasks in large brains, perhaps separated according to differences in modality.

Let me begin by surveying a variety of interesting examples of tasks that are affected by damage to the prefrontal cortex in monkeys (see Figure 9.2). It is appropriate to start with the classical prefrontal task identified by Jacobsen many decades ago.[5] The highly schematized figure depicts the delayed response or delayed alternation task. In this task, a food object is placed in a covered container as the monkey watches; then the monkey is distracted for a few seconds, often by pulling down a blind; and finally he is allowed to retrieve the food object by uncovering it. This is no problem. But on a succeeding trial, the hidden food object is placed in the alternate

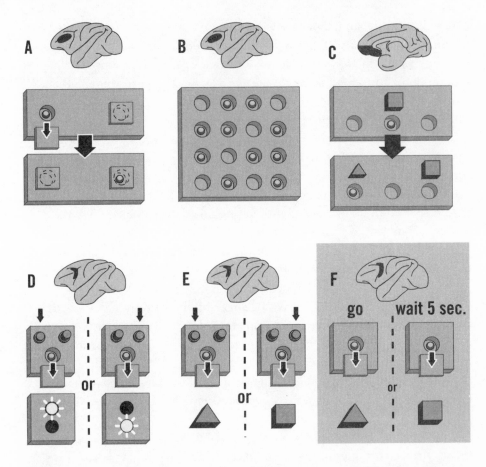

Figure 9.2 *Diagrammatic depiction of six different cognitive deficits shown in monkeys with frontal lobe damage to different subregions (indicated by blackened regions). A: Delayed response (delayed alternation) task associated with dorsal lateral prefrontal damage (Jacobsen, 1936). B: Self-ordered sampling task associated with dorsal lateral prefrontal damage (simplified from Passingham, 1985) C: Delayed non-match to sample task associated with ventral medial prefrontal damage (Mishkin and Manning, 1978). D and E: Conditional association tasks (spatial versus nonspatial cues, respectively) associated with periarcuate prefrontal damage (Petrides, 1982; 1985). F: Go/no-go task associated with periarcuate premotor damage but not prefrontal damage (Petrides, 1986).*

container, again in full view. Now, however, after the delay period, rather than looking in the new hiding place, the prefrontally damaged monkey again tends to look in the place where he found food before, not where he saw it being hidden.[6]

Some have explained this as a problem with short-term memory. The monkey might be unable to use information from a past trial to influence his choice in a future trial. A simple memory problem, however, would tend to produce random performance. In general, the animal's perseveration indicates that he does remember the previous successful trial, all too well, it would seem. Apparently he either can't inhibit the tendency to return to where he got rewarded the last time, or can't subordinate this previously stored information to the new problem. Historically, interpreters of prefrontal deficits have split evenly over whether they define this response in terms of memory or response inhibition. But before taking sides, let's consider a few additional examples.

Another, more sophisticated version of the same task has been investigated by Richard Passingham.[7] It offers some insight into how this task might have real-world adaptive consequences. As in the simple delayed-response experiment, food is placed in food wells while the monkey watches (although the observation is not a necessary factor); but in this case food is hidden in all or many of a large number of wells. No delay is necessary. The monkey must simply sample through the wells to retrieve all the food objects. Monkeys with efficient sampling strategies won't tend to sample the same wells twice. Once food has been located in one place and taken, there is no reason to go back and check it again. Prefrontally damaged monkeys, however, fail to sample efficiently. They perseverate, return more often to previously sampled wells, and fail to sample others at all. Again, it is not clear whether to consider this as forgetfulness or as a failure to inhibit a repetition of past responses. The practical significance of such an ability is clear, however. This is precisely the sort of problem that might be faced by an animal foraging in many different places. Once all the food has been eaten in one place, it makes no sense to go back looking for more, even if the food tasted particularly good there.

Turning now to monkeys with medial frontal damage, we find a slightly different kind of deficit. Although such monkeys succeed at the tasks described above, they fail at tasks where the shift in food location is cued by a shift in stimulus as well. The food object is hidden, and the location is cued by some stimulus. After the monkey succeeds at this task, the food object is rehidden and the hiding place is marked by a new stimulus, with the previous stimulus now marking no food. Thus the monkey must learn that the food will always be hidden where the new stimulus of the two is placed. Medial prefrontal damage appears preferentially to affect this kind of learning, but not delayed response, delayed alternation, or sampling tasks that are sensitive to dorsal lateral prefrontal damage.

Compare this to tasks that are sensitive to posterior prefrontal regions.

Such tasks have a multipart form, but common to them all is a dependency between two classes of cues or cues and behavior options. One cue stimulus indicates that the food is hidden in the well with the lighted light, whereas the alternative cue indicates that the food is hidden in the well with the unlit light. The pattern is "If X then Z; if Y, then not Z." It is a conditional relationship in which one stimulus indicates the relationship between another stimulus and the position of the food. Other variants that involve a similar dependency relationship, but differ in terms of the kind of response (e.g., choice of buttons to push instead of positions to check), also are sensitive to damage in this region. Unlike the previous task, there is no spatial difference in food position; but like the last task, there is a conditional relationship in which the stimulus indicates which of two alternatives is associated with the food. Depending on what stimulus is presented, the monkey has to reverse his expectation about the association between the reward and some behavioral option.

These tasks are all different and yet they share a number of common features. Although all involve some apparent inability to inhibit responding, this is probably not itself the locus of the deficit. Prefrontally damaged animals do not show a problem with simple go/no-go tasks that require withholding a response. In this sort of task, the presentation of one of two different stimuli indicates whether an immediate response or the same response just delayed a few seconds will produce a reward. Animals who have difficulty suppressing a response will be unable to learn such tasks because they will fail at the no-go task. This difficulty is, however, demonstrated by animals with premotor lesions, which additionally produce problems with motor sequences and skilled movements.

Another whole class of learning abilities undermined by prefrontal damage are those involving transfer of information from one learning task to another—often referred to as transferring a "learning set." The analogy to the prefrontal tasks we've just been considering is that the transfer requires using information from previous trials, but divorcing it from specific stimuli. Probably the most difficult learning-set problems involve transfer of an inverse pattern of associations from one task to another. Only some of the larger brained primates succeed at such doubly demanding tasks (see Rumbaugh, et al., 1996). This suggests some interesting species differences that may be correlates of both the size and relative proportions of prefrontal cortex.

One might argue that all of these tasks involve holding information in mind while not acting on it—a function some have called working memory. Arguing against a simple mnemonic interpretation of these errors is the fact that the deficit pattern always involves perseverative failures, that is, errors that result from repeating the previously successful performance. We could, for example, interpret the failures as resulting from information in short-term memory that inappropriately dominates the tendency to respond. Although a trace from previous trials must be retained in order to succeed at these tasks, what must be "held in mind" in these tasks is not just prior information, but information about the *applicability* of that information in a different context. One of the most salient common features of these tasks sensitive to prefrontal lesions is that they all in some way or other involve shifting between alternatives or opposites, alternating place from trial to trial, shifting from one stimulus to a new one, or from one pairwise association to another depending on the presence of different cues. Tasks sensitive to prefrontal damage thus involve short-term memory, attention, suppression of responses, and context sensitivity. But they all have one other important feature in common: each involves a kind of negation relationship between stimuli or stimulus-behavior relationships. They all have to do with using information about something you've just done or seen *against itself,* so to speak, to inhibit the tendency to follow up that correlation and instead shift attention and direct action to alternative associations. Precisely because one association works in one context or trial, it is specifically excluded in the next trial or under different stimulus conditions. An implicit "not" must be generated to learn these tasks, not just an inhibition.

Similar deficits are well known in human patients,[8] though associations between specific tasks and different prefrontal subareas are less well worked out in humans. For example, human prefrontal patients often fail at card-sorting tasks that require them to change the sorting criteria. They also tend to have trouble generating word lists. In trying to generate lists of words according to some criterion or instruction, they barely get past the first few names of things before getting stuck or repeating already named items. These tasks are formally similar to conditional association and sampling tasks, respectively. To generate a word list, it is necessary to maintain attention on the selection criterion and on previously generated words in order not to produce them again, but rather to elicit additional alternatives appropriate to these same criteria.

Prefrontally damaged patients also often have difficulty learning mazes based on success/failure feedback, making plans, spontaneously organizing behavior sequences, and performing tasks that require taking another's perspective (*allocentric* as opposed to *egocentric*). The last are analogous to using a mirror. In order to think in allocentric terms, one is required to perform a systematic mental reversal of response tendencies, and so egocentric information must continuously be used as the frame of reference, but responses based on it must be specifically inhibited and inverted. In general, tasks that require convergence on only a single solution are minimally impacted by prefrontal damage, but those that require generating or sampling a variety of alternatives are impaired. This capacity has been called *divergent thinking* by J. Guilford,[9] and it may explain why prefrontal damage does not appear to have a major effect on many aspects of paper and pencil IQ tests. Prefrontally damaged patients exhibit a tendency to be controlled by immediate correlative relationships between stimuli and reinforcers, and this disturbs their ability to entertain higher-order associative relationships. In summary, these human counterparts to different frontal lobe defects in monkeys also involve difficulties in using information negatively. Impairment of the neural computations provided by these cortical areas makes it difficult to subordinate one set of associations to another, especially when the subordinate associations are more immediate and salient.

Making Symbols

These general insights about the contributions of prefrontal cortex in primate and human problem solving may not answer the riddle of prefrontal function in general, but they offer some useful clues to the significance of the disproportionately large size of this structure in human brains. What might an anomalously large prefrontal region have to do with that other human anomaly, language? Or to pose the question more specifically: Is there something about language that requires a predisposition for working with difficult conditional associative relationships, maintaining items in working memory under highly distractive conditions, or using negative information to shift associative strategies from concrete stimulus driven links to abstract associations?

Phrased in these terms, the parallels with the cognitive processes required for symbol acquisition become obvious. The contributions of prefrontal areas to learning all involve, in one way or another, the analysis of higher-order associative relationships. More specifically, judging from the effects of damage to prefrontal regions, they are necessary for learning associative relationships where one associative learning process must be subordinated to another. These are the most critical learning problems faced during symbol acquisition, and less effective regulation of competing learning tendencies by the prefrontal cortex is almost certainly the primary roadblock to symbol learning in nonhuman species. Conversely, the expansion of this structure in human brains may reflect the advantage of exaggerating the predisposition to employ this learning strategy and giving it added strength over competing learning tendencies in order more efficiently to cross the threshold from indexical to symbolic associations. Though experiments with chimpanzees have shown how it is possible to supplement the special attentional and mnemonic demands of symbol learning from the outside, the constant demand for this analytic process throughout human evolution has apparently selected for a pattern of brain reorganization that shouldered more and more of this burden internally.

The prefrontal cortex helps us inhibit the tendency to act on simple correlative stimulus relationships and guides our sampling of alternative higher-order sequential or hierarchic associations. Its role in language and symbol learning in particular is not, however, merely to increase something we might call prefrontal intelligence. Rather I suspect the importance of the size change can be thought of in displacement terms, in patterns of cognition as in patterns of brain development. Prefrontal computations outcompete other cognitive computations and tend to dominate learning in us as in no other species. In simple terms, we have become predisposed to use this one cognitive tool whenever an opportunity presents itself, because an inordinate amount of control of the other processes in the brain has become vested in our prefrontal cortex. The way the parietal cortex handles tactile and movement information, the way the auditory cortex handles sound information, the way the visual cortex handles visual information are all now much more constrained by prefrontal activity than in other species.

We should not, however, make the mistake of thinking that prefrontal cortex is the place in the brain where symbols are processed. It is not. Massive damage to the prefrontal cortex does *not* eliminate one's ability to understand word or sentence meaning. The symbolic associations that underlie the web of word meanings are probably much more dependent on the mnemonic support of sensory-based "images." This is supported by the high incidence of semantic disturbances after posterior cortex damage, especially damage to areas surrounding the posterior temporal cortex. It is also supported by our intuitions of mental imagery generated as we read stories, or of sound motor "images" as we search our memories for the right words to say or the correct names to match with familiar faces. It would be misleading, however, to suggest that these images are all there is to symbols, any more than the words on this page suffice in themselves to convey their meanings. They are merely neurological tokens. Like buoys indicating an otherwise invisible best course, they mark a specific associative path, by following which we reconstruct the implicit symbolic reference. The symbolic reference emerges from a pattern of virtual links between such tokens, which constitute a sort of parallel realm of associations to those that link these tokens to real sensorimotor experiences and possibilities. Thus, it does not make sense to think of the symbols as located anywhere within the brain, because they are relationships between tokens, not the tokens themselves; and even though specific neural connections may underlie these relationships, the symbolic function is not even constituted by a specific association but by the virtual set of associations that are partially sampled in any one instance. Widely distributed neural systems must contribute in a coordinated fashion to create and interpret symbolic relationships. The prefrontal cortex is only one of these.

The critical role of the prefrontal cortex is primarily in the *construction* of the distributed mnemonic architecture that supports symbolic reference, not in the storage or retrieval of symbols. This is not just a process confined to language learning. The construction of novel symbolic relationships fills everyday cognition. A considerable amount of everyday problem solving involves symbolic analysis or efforts to figure out some obscure symbolic association. As soon as language processing leaves the realm of relatively habitual phrases and uses, it, too, often involves some level of novel symbol construction. Almost certainly the confusing and idiosyncratic ways that information is weaved into the sentences and paragraphs of this book impose significant demands for symbolic analysis on its readers. It is in these circumstances that the language impairment of patients with damaged prefrontal cortex shows up. One of the most important uses of language is for inferential processes, for taking one piece of information and extrapolating it to consequences not obvious from the information given. This is essentially using symbols to elicit or construct new symbols.

The prefrontal cortex is not totally uninvolved in ongoing language processing, and it is particularly important in language processing, where corresponding sequential, hierarchic, and subordinate association analyses are required. The recursive embedding of sentence clauses within one another, which provides language with its expressive economy, also takes advantage of our remarkable facility for manipulating conditional association hierarchies. Such syntactic constructions are specifically cued by sentence morphology (using position, word, or inflectional markers) and so probably require only minimal prefrontal support to sort them out. Nevertheless, there may be occasions where the syntactic constructions are particularly tortuous, or where semantic cues directly interfere with the syntactic analysis (e.g., "the cat the mouse killed"), which require the special contributions of prefrontal functions. Moreover, as word association tasks reveal, prefrontal control is likely critical for guiding word choices and shifts in logic during discourse. It is often noted that despite their control of the mechanics of language production, prefrontally damaged patients are characterized by a disturbed "flow" of ideas and word choices, not to mention a sort of "concreteness" in their interpretations of sentence meaning.

The more complicated the combinatorial relationships, or the more easily confused the correlated relationships, the more prefrontal systems are taxed. This is clearly demonstrated by imaging studies of the metabolic correlates of different cognitive tasks in human subjects.[10] Complex sorting problems and difficult word association tasks have been shown particularly to activate prefrontal metabolism. There is also indirect evidence that the difficulty of a task determines how much of prefrontal cortex gets recruited to that task. Electrical stimulation studies of awake neurosurgery patients have shown that patients with lower verbal IQs tend to have larger regions of prefrontal and parietal cortex susceptible to disruption of language tasks.[11] In addition, when a difficult word association task (such as the example of producing appropriate verbs for rapidly presented nouns) is practiced so that the task format and the words presented are no longer novel, the level and extent of metabolically activated ventral frontal and cingulate cortex is significantly reduced (see discussions of PET scan studies in the next chapter). Numerous tasks involving other cortical areas demonstrate a similar pattern of recruitment of additional cortical space to respond to novelty and difficulty, and a reduction of activity and extent of areas involved after practice. This might offer another correlate of cortical size–related effects. Having a larger prefrontal cortex is analogous to having a large voting contingency in reserve to guarantee a majority even when the competition is stiff. A larger prefrontal region in general might increase the territories that can be easily recruited for diverse tasks that tax specific prefrontal computations.

The most critical effect of a disturbance of prefrontal functions to language should be most evident in the processes of symbol construction during initial language learning. Prefrontal impairment early in life should thus be far more devastating than later in life, because it would make the critical shift away from rote, stimulus-bound learning of words and phrases far more difficult. A brain-damaged child, limited in his or her ability to pull attention away from the surface correlations between stimuli, would require much more extensive external support in order build up each minimal group of symbol-symbol relationships, and so might find language acquisition particularly rough going, for precisely the same reasons that other species do.

Direct evidence that prefrontal damage in young children impairs language learning comes from a survey study by Bates, Thal, and colleagues (1994). They found particular impairments in development of both vocabulary and grammar after prefrontal damage in the critical period between nineteen and thirty-one months of age. The fact that such damage affects both semantic and syntactic aspects of language is consistent with a general symbolic impairment. Bilateral prefrontal damage at this age produces a permanent deficit, but damage to either side alone can produce language delay.

Language-learning difficulties are often associated with a reliance on rote-learned phrases and overly concrete interpretive responses. Such tendencies are particularly characteristic of Down's syndrome and autism. The language-learning problems associated with these more global cognitive deficits are often compared with an inverse developmental pathology. Williams syndrome (WS) is one of the most enigmatic examples of selectively spared language learning. It is characterized by highly verbal individuals who seem adept at storytelling and recitation of verbal information, but who also exhibit major cognitive deficits in analyzing thematic-level language processes, very poor problem-solving abilities, and very impaired spatial reasoning. Though they may have IQ scores in the range of 50, their vocabulary and speaking skills may test above normal at early ages. Along with their peculiar mix of language abilities and disabilities, they are also intensely social and gregarious, exhibiting an almost hypersocial personality, and they have a slight facial abnormality that is often described as pixielike, with a constant wide grin.

The paradoxically spared language skills of WS children caught the attention of Ursula Bellugi, one of the pioneers in the study of deaf signers. She and her colleagues have mounted an extensive effort to study these individuals, and have compiled both extensive testing data and some of the first evidence concerning their brain abnormalities. WS children are especially interesting because their somewhat precocious verbal skills follow curious patterns. When asked to tell what they know about a familiar object, they can produce a long list of physical characteristics, places you might find one, things they are good for, and so on. But if you then ask whether the same object could be used for a novel purpose that could be simply figured out from its physical characteristics, they are often baffled. Normal children of the same age might not produce nearly as many descriptive characterizations when answering the first question, but would find the second part easy to guess. The WS children's knowledge of words might be likened to that of someone who had memorized entries from a dictionary or an encyclopedia but had never had any experience with the things they know about. They have acquired extensive knowledge of linguistic associations, but only a fraction of the web of additional experiential associations that link words to the world.

The fact that WS children can appear to have well-developed verbal abilities and yet very low IQ scores has been used to bolster arguments about the modularity of language functions. On the surface, WS children demonstrate spared grammar and syntactic ability, but their understanding of language is in the context a somewhat shallow pragmatic understanding, and shows minimal problem-solving ability. For this reason, they have been cited as evidence of innate grammatical knowledge that is independent of other cognitive abilities. This interpretation is also consistent with the minimal consequences of aphasia on intelligence measures that exclude verbal performance tests. But it would be a mistake to think of these two ways that language and general intelligence dissociate under pathological conditions as reflecting an independence of language from intelligence. Intelligence isn't a unitary brain function, and language isn't walled off from other cognitive functions. Nevertheless, in Williams syndrome we see a deficit in which cognitive functions are splintered in a manner that selectively spares certain processes most critical to language development, and in this way serves to reinforce the claim that the type of learning biases that favor symbolic learning may be quite different and even antagonistic to those useful in the majority of other learning contexts.

Recently, anatomical data have suggested some curious abnormalities in the brains of patients with WS that may provide a clue to this paradoxical dissociation. Both postmortem analysis and MRI analysis have revealed brains with a reduction of the entire posterior cerebral cortex, but a sparing of the cerebellum and frontal lobes, and perhaps even an exaggeration of cerebellar size. Currently, there is too little data to be sure that all WS patients show an intact frontal cortex and cerebellum, and the underlying pathology may not always produce such gross abnormalities. Still, this systematic sparing may be the most consistent clue to their curious language symptoms (see Figure 9.3).

This pattern of intact prefrontal cortex paired with a subnormal posterior cortex may offer a hint to the paradoxical abilities and disabilities of people with WS. These proportions represent an interesting twist on the human shift in prefrontal proportions, as compared to other primates. My argument has been that the enlargement of prefrontal cortex with respect to posterior sensory and subcortical regions is responsible for the biased learning that enables human beings to employ symbolic representation strategies.

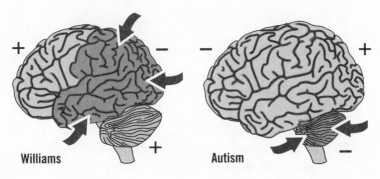

Williams   +   **Autism**

Figure 9.3 *A somewhat idealized comparison of Williams syndrome neurological deficits to autism neurological deficits emphasizing their complementary neurological patterns (as well as deficit patterns). Williams syndrome (which produces relative sparing of superficial linguistic abilities against a background of massive global intellectual impairment) appears to produce sparing of prefrontal cortex and cerebellum but has been associated with significant reduction of more posterior cortical systems (Galaburda, et al., 1994). Some forms of autism (which is associated with relatively severe communication difficulties in language and other modalities) appear often to be associated with apparent depression of prefrontal functions but relative sparing of posterior cortical functions, and have also been correlated with cerebellar abnormalities. Arrows point to areas most likely to exhibit anatomical reduction or abnormality.*

The learning and other cognitive and emotional biases in WS might offer clues to this transition to the extent that it too involves a prefrontal bias, though for quite different reasons. WS is not prefrontal enlargement, only an exaggerated prefrontal predominance compared to normal human brains, due to pathology that spares prefrontal cortex in the context of other widespread developmental damage. At the present time, there is not a sufficient neuropsychological profile of WS to verify this prediction, but it nonetheless might help us to understand the peculiar language abilities of these people. In terms of symbol learning, recall that the contribution of the prefrontal areas is most critical for making the shift away from indexical token-object associations to rely on systematic token-token association patterns to organize reference. This shift is difficult because indexical associations are acquired by constant experience of immediate correlations between events that are highly salient, whereas the token-token patterns are only discovered by comparing their patterns of combination across many interactions with other speakers, and these are obscure because they are distributed across interactions and inconsistently correlated. Prefrontal predominance probably aids this process by biasing both attentional and mnemonic strategies to favor attention shifted from token-object associations to higher-order token-token relationships.

Consider that one consequence of WS is poor learning and problem solving (see Figure 9.3). In terms of referential abilities, this will inevitably translate into poor indexical learning. Physical and pragmatic relationships between physical signs, objects, and their attributes—which for most people and animals are rapidly and effortlessly accumulated in memory, and provide a dense network of indexical references—will be limited in individuals with WS. Although the salience of these associations inhibits the development of symbolic abilities in other species, it is also what "grounds" symbolic reference in the first place. Semantic features are predicated on these potential indexical associations. In these terms, it appears that WS children exhibit a precocious ability to seek out and discover the higher-order combinatorial connections among words. This skill may be exaggerated precisely because of their poorer than normal indexical learning abilities. Discovering an alternative mnemonic strategy to hold together a weak network of indexical associations may lead them to rely more heavily on higher-order combinatorial patterns than do normal children. But lacking indexical support, they are far more trapped by word association logic alone. One of the peculiar features of the vocabulary of WS people is an apparent penchant for unusual words. This can be understood in terms of the insufficient information they have at their disposal to value different lexical items comparatively. Their lack of experiential familiarity should provide far less influence over word salience, and thus the relative frequencies of typical and atypical words should differ less. But in addition, since learning frequency of occurrence, without ancillary indexical support as a guide, is itself based purely on correlation as an index of commonality, even this information will be denied them.

Finally, their exaggerated sociality offers a particularly interesting clue both to the functional effects of prefrontalization and to the attentional biases that support symbol learning. It has long been known that prefrontal damage can often produce a disturbance in social behaviors. One of the first described and most celebrated cases of human prefrontal damage is that of a man called Phineas Gage. While he was working on a railroad crew in Vermont in 1848, compacting blasting powder into a hole with an iron rod (a tamping iron), the powder accidentally exploded, propelling the tamping iron through his head, right through the middle of his prefrontal lobes. Gage lived, but was highly moody and emotionally unstable, and behaved quite inappropriately in public. The belief that prefrontal abnormalities might underlie the inappropriate behaviors of schizophrenics later became a justification for the use of prefrontal lobotomy, and the less damaging prefrontal leucotomy (cutting the fibers to and from the orbital prefrontal regions), as a means of controlling psychiatric patients. More recently, research in which the medial and orbital prefrontal areas were removed from the brains of monkeys demonstrated a parallel loss of appropriate social competence. The animals became both asocial, preferring isolation, and socially inept when forced into social contexts.[12] If WS produces a comparative prefrontalization by virtue of posterior impairment, we might explain exaggerated sociality by an inverse mechanism: a preoccupation with and hypersensitivity to social information and social relationships. This may also aid the process of symbol acquisition, by virtue of exaggerated attention on social stimuli, particularly patterns of word use, at the expense of other information. Thus, speech and word-word relationships would be among the most salient of associative relationships.

A related disorder initially appears as a precocious reading ability. It has recently been dubbed hyperlexia. At an age when most children are just learning to form their first words and sentences, a very few children demonstrate incredible feats of symbol identification. They read words off cereal boxes, quote road signs, identify trade-marks, and know the names of all the letters and numbers. Unfortunately these abilities, which might herald the first signs of a literary or mathematical prodigy, turn out to be isolated abilities of a mind that in fact has a severely retarded ability to interpret these symbols. The comprehension of language is minimal, and their ability to reason or see the connections between the things they read is equally poor. Little is known of the neurological correlates of hyperlexia, but we might predict similar patterns of neurological impairment and sparing, and of learning and attention bias. Other pathologies, such as hydrocephaly, can also produce similar patterns of language impairment,[13] so it is probably not that WS is produced by overexpressing some gene for language but rather that it reflects the biased cognition produced by a consistent pattern of preferential damage and sparing of cortex and cerebellum.

Interestingly, there appears to be good evidence for exactly which genes are responsible for Williams syndrome. Geneticists have identified two genes that are apparently deleted in all cases of the disorder. The first to be identified was the gene for elastin, a connective tissue protein which when absent results in a heart defect that is characteristic of WS. In a collaborative study with a colleague at the University of Massachusetts in Lowell, Thomas Shea, we found that elastin may also serve as a substrate on which axons can grow, so its absence could potentially be a problem during brain development. But recently, a more relevant gene located next to the elastin gene and also deleted in WS has been identified.[14] Its protein product is called LIM1 kinase because of its structural similarity to a homeotic gene product called LIM1. Not coincidentally, the genes for both LIM1 kinase and LIM1 are expressed in the head region of developing embryos. Though much remains to be learned about how the absence of the gene for LIM1 kinase may contribute to WS, there is one feature of the counterpart gene for LIM1 that marks it as highly relevant. Transgenic mouse embryos that have had both copies of the LIM1 gene knocked out fail to develop heads altogether, even though the remainder of the body seems to develop normally. (Also see the discussion of this gene in the context of other homeotic genes in Chapter 6). WS is essentially a human LIM1 kinase knockout, which, analogous to this mouse knockout, seems to produce significant underdevelopment of a large contiguous brain region (it also may be responsible for the abnormal development of the face that produces the pixie appearance characteristic of WS).

Williams syndrome provides a distorted mirror image of the genetic changes that must underlie the human symbolic bias. The underdevelopment of much of the brain but sparing of the frontal cortex and cerebellum, which occurs in a person with Williams syndrome, has given these two structures greater control over all cognitive processes. The effect is to exaggerate the bias to learn symbolic associations, even though the capacity to learn nonsymbolic associations is severely impaired. This accounts for the hyperlexic tendencies of these children: their precocial vocabularies and reading abilities, their preoccupation with unusual words, and even perhaps their heightened sense of sociality (see Chapter 13 for a discussion of other social and emotional consequences of prefrontal bias). But this homeotic gene defect in people with Williams syndrome achieves an exaggerated prefrontal bias at the cost of poorer learning in all other realms, with the result that their understanding of symbols is almost entirely confined to lexical relationships—i.e., word-word relationships—and is lacking in supportive indexical links to objects and events. So Williams syndrome demonstrates that high general intelligence is not as critical for crossing the symbolic threshold as is growing up with a peculiar bias in learning tendencies. But Williams syndrome is more than just an exception that proves a rule about the evolution of language abilities; it is also a clue to the sorts of genetic and neural developmental changes that lie behind that evolution.

An interesting contrast to Williams syndrome is autism.[15] Autism is most notably characterized by complete social withdrawal and a lack of communicative ability and/or desire. Severely autistic children avoid eye contact, are often not comforted but even disturbed by physical touch, seem not to notice the presence of others or find their presence disturbing, and tend to live their adult lives as relative loners. These social patterns have led some researchers to focus on their apparent lack of a "theory of mind," by which they mean an awareness of others as conscious agents and feeling beings. It is often said that autistic children tend to treat other people (and animals) no differently from other objects. This has led to theoretical excesses analogous to language organ claims. Recently, there have been a number of hypotheses about a "primary social cognition deficit" or impairment of a "theory of mind module" in the brain of autistic individuals. These suggestions are at best premature. Not only is the modularity claim superfluous, but a focus on the social impairment alone is incomplete. An impairment in social cognition (however it might be represented in the brain) cannot alone explain the failure to develop normal language, nor a range of other attentional deficits and learning biases also common to autism. It seems more likely that the social deficit is a correlate and possibly a consequence of something more basic. Further clues to this are provided by the other cognitive and behavioral predispositions of autistic children and adults. They are highly likely to develop extensive and very stereotyped ritualized behavior patterns, both socially and in the ways they perform mundane tasks and manipulate familiar objects. As adults, they often run their lives by strict schedules and memorized lists that, if violated, throw them into confusion.

The neurological basis of autism is still unresolved. Unlike WS, which is associated with a specific genetic abnormality, there are probably many potential neurological causes for autism. Numerous studies have failed to identify any consistent structural abnormality distinguishing autistic from normal brains, but where abnormalities are observed, there are some complementary patterns to those seen in WS. Remarkably, despite the symptomatology of higher cognitive impairment involving language and learning, there is both minimal and confusing evidence of structural abnormalities involving the cerebral hemispheres. Many researchers have postulated involvement of the limbic system, considering the emotional features of this disorder, but no solid evidence for a limbic pathology is apparent either. Structural abnormalities seem to predominate instead in the brain stem and cerebellum. Autistic brains are often reported to have smaller cerebellar lobes, especially on the midline, and a smaller brain stem than normal. This has led to all sorts of speculations about the role of the cerebellum in social and emotional development. However, at least one study has found that such deficits are not so much diagnostic of autism as of neural pathologies expressed late in fetal and neonatal development. Since the cerebellum is comparatively one of the latest structures to mature and shows the latest large-scale growth patterns, it may also be most susceptible to other less specific, cellular-based pathologies. The only direct evidence of cortical involvement again seems to focus on the activity of the frontal lobes in autistic adolescents and adults. Like gross anatomical factors in autism, frontal lobe signs are also variable; but in studies that have looked for metabolic differences in cortical regions, a number have shown reduced cerebral blood flow and reduced metabolism of the prefrontal region.[16] Recently, a number of psychologists have shown that autistic patients have a quite specific difficulty with tasks that require imagining what is going on in other peoples' minds. Taking others' perspectives or recognizing that others have different information due to their different perspectives on a task seems to be a problem for autistics. Not coincidentally, this too would be predicted for someone with prefrontal cortex damage.

The significance of the cerebellum's being either specially spared (WS) or specially reduced (autism) in these two congenital syndromes preferentially affecting language ability is curious, because the cerebellum is not generally incorporated into theories of language processing, nor is it implicated in aphasia studies. The cerebellum is usually thought to be involved in the regulation of relatively automatic movements—such as preprogrammed ballistic movements like throwing a ball, jumping over a log while running, or playing a well-practiced scale on the piano. There are good reasons for imagining that it plays some role in the articulation of speech, and evidence that cerebellar damage can affect fluency, but this does not seem to be the factor of central importance in these congenital syndromes. Both syndromes appear to involve higher levels of language organization. Corresponding clues to cerebellar language functions come from imaging studies. Though the cerebellum is to some extent activated during speaking, it is most intensely activated by difficult word association tasks (like the rapid production of an appropriate verb for a noun in the task described earlier). Even when the contribution of the cerebellum to speaking is subtracted in PET (positron emission tomography) studies, its independent activation by the associative task is large. The cerebellar activation may aid this process by providing access to relatively automatic word-sequence subroutines, and possibly supporting the rapid shifts of attention that are required (see Figure 10.4 and the discussion in the next chapter).

A central role for the cerebellum in attentional processes in humans has recently been recognized as a result of studies of patients with cerebellar resections and damage.[17] Cerebellar damage in primates can produce impairments for a number of the same tasks that are also sensitive to prefrontal damage, but more generally seems to impair a variety of perceptual and cognitive tasks where timing issues are important. Rather than some totally different function of the cerebellum, this is probably analogous to the cerebellar role in motor functions. Many features of nonlocomotor activities require similar neural operations to those involved in movements, including automatic prediction and preparation. If the cerebellum augments cortical processing by providing a repository for replaying previously learned sequences of activity and providing clocklike regulation of their performance, it could play an important role in the support of a variety of rapid semi-automatic language processes to the extent that it was linked to the appropriate brain structures.

The cerebellum also turns out to be one of the group of brain structures that has been selectively enlarged in humans as compared to other primates. This implies that it, too, may have displaced projections from other systems that are not expanded. Though the cerebellum receives inputs from the spinal cord conveying somatic information about muscle tension and joint position, and contributes a major spinal motor output via the red nucleus in the midbrain, the link between the cerebellum and the cerebral cortical systems is also extensive. Motor, premotor, and prefrontal projections comprise a major portion of inputs to the cerebellum by way of relays in the pons, the enlargement at the top of the brain stem. In humans, the prefrontal component of this input system is likely to have been particularly enlarged, displacing other inputs from the spinal and cortical motor systems. The cerebellar output nuclei also project back to the cortex by way of relays in the ventral lateral portion of the thalamus, which primarily projects upon motor and premotor areas.

In other primates, there is an extensive array of cortical regions that project to the cerebellum, including most motor and tactile sensory regions. There is also a modest projection from the dorsal parts of the prefrontal cortex, probably from regions receiving tactile and motor information from other cortical areas. But in monkey brains, it does not appear that the ventral parts of the prefrontal cortex send any outputs to the cerebellum. The extensive enlargement of prefrontal cortex in humans, however, has likely shifted these connectional relationships so that a greater number of connections from a larger expanse of the prefrontal cortex project to the cerebellum as compared to other primate species. Areas of prefrontal cortex that do not project to the cerebellum in other primates, such as the ventral regions, may thus take part in the cortical-cerebellar-cortical circuits in humans. This sector of the prefrontal cortex receives relayed auditory inputs, and is closely linked with both premotor vocal-articulatory areas of cortex and midline vocalization systems (see the next chapter for details). Bringing the cerebellar system into the circle of these prefrontal functions would introduce a unique computational aid into the analysis of symbolic relationships, and one that has a novel affinity with symbols encoded as sounds.

So, the cerebellum is probably far more involved with sound analysis in human beings than it is in any other species. This may be very important for the generation of word associations at a rate sufficient for speech. At the rate words are presented in speech, the speaker or listener must be able rapidly to generate associated words and avoid letting earlier associations interfere. The cognitive search process must be as rapid but as shallow as possible. Any slight tendency to perseverate would entirely derail the process. It is not a passive analysis that must be performed, but the rapid, controlled generation of relatively novel responses, though only *relatively* novel because word associations are recalled from innumerable previous associations in which they were elicited together in sentences. A word-generation process analogous to conjugating a verb for a noun in rapid alternation must inevitably be part of all sentence production. Succeeding at this task appears to be greatly facilitated by linkage to the brain's rapid prediction computer, the cerebellum. The cerebellum may provide an independent generator of novel but predictable shifts of associations from one context to another, while prefrontal cortex is providing a selective inhibition of all but the one sample that fits the new criteria. In addition to the cerebellum, the cingulate cortex also seems to be intensely activated in a variety of tasks that require top-down intentional shifting of attention.

This array of special prefrontal links with critical aspects of modern language functions—enabling symbol construction, shifting control of vocalization away from emotional systems, enhancing the tendency for social mimicry, and bringing rapid cerebellar prediction systems to the service of auditory-vocal analysis—almost seems like too much of a coincidence. Why should these changes in prefrontal cortex, derived from a global change in relative proportions, just happen to provide what language needs? In fact, these apparent coincidences only appear coincidental when considered in hindsight, in terms of the way language works today.

Considering these supports for our modern abilities to be prerequisites for language reverses the evolutionary logic behind them. In the co-evolutionary interaction between brain and language evolution, both converged toward the easiest adaptation to the cognitive and sensorimotor problems that were posed. As symbolic difficulties were eased by prefrontal cortex enlargement, the shifts in connectional relationships that resulted also incidentally provided increased auditory and vocal abilities and an increased propensity for vocal mimickry. Those symbolic communication systems that took advantage of such enhanced abilities would consequently have been more successful. And this in turn would have fed back on selection for greater elaboration and utilization of such abilities. The adventitious nature of co-evolutionary processes guarantees that when an opportunity presents itself, it will likely be taken advantage of, but in so doing it will set in motion processes that bias future evolutionary trends in this same direction. It's not that our exaggerated prefrontal cortex was an evolutionary prerequisite for these many supportive language adaptations; rather, these features of languages have evolved to take advantage as well of other incidental prefrontal biases that symbolic evolution inadvertently produced.

The curious logic of the co-evolutionary dynamic that has caused many otherwise unrelated language functions in addition to symbolic learning to converge on the same cluster of neurological changes in humans will be explored more fully in the third and final section of this book. But before we can turn to this topic, we need to complete the circle of logic that links the changes in human brain evolution with the way that other language functions utilize the brain after this symbolic threshold has been crossed in childhood. How does this reanalysis of the core human language adaptation alter the way we need to think about language in the brain?

Locating Language

Everything is absolutely impeccable. There is only one, poor thing, which is mistaken inside.

—Mr. L., an aphasic patient

Forcing Square Pegs into Round Holes

A man of about seventy years of age sits at a table holding a pencil in his hand. It is cradled between his fingers as though he is preparing to write with it. "What is this?" a woman asks, pointing to the pencil from across the table. "Uh, it's a wedge for the rain . . . ?" he replies with a sort of questioning tone, and moves it in a vague writing gesture. "What do you use it for?" the lady continues. " . . . For to rain for the scretch," he answers. "Very good," she responds. "Why to am I here to learn about this?" he asks. "We're here to see if we can help you with your words," she says.

In reality, there will be very little she or others will be able to do. This man has suffered a stroke that damaged a part of his left cerebral hemisphere near the auditory areas of his temporal lobe, and as a result he will spend the remainder of his life not quite getting the questions and not finding the right words to answer. The damaged region of his cerebral cortex subserved a specific set of language operations that cannot be recovered, despite the fact that most of his brain and intellect are otherwise intact. He has not regressed to an earlier stage from which he will need to relearn words and meanings. He has not just lost some word memories. He has lost hold of the cognitive tools to map the sounds of words to what they mean. This loss of language abilities resulted from damage to a brain region that has come to be called Wernicke's area (after the nineteenth-century German physician who discovered it, Carl Wernicke). And the deficit seems to suggest that such language functions might be highly localized to just a few regions of the brain.

Historically, most of our information about how the human brain comprehends and produces language has come from insights provided by studying people with localized brain damage that affects language functions. This information has been filtered through psychological theories about the nature of mind and language, and until recently it has been limited by the intrinsic messiness of studying organic pathologies. Now, however, a stream of information has emerged from new methods and techniques for studying both normal and damaged brains, including ways to get images of metabolism, blood flow, and electrical activity in whole, intact, functioning brains. These new windows into brain function are both extending and challenging traditional ideas about brain-language relationships.

Loss of language abilities due to brain damage (termed *aphasia*, literally, "no speech") does not occur in only one way. When language abilities are lost, they do not regress to childlike forms, as some early researchers surmised. Brain-damaged adults have not simply lost the memory of language or parts of language, and they are not analogous to children arrested at an earlier stage of language acquisition. Language tends to break down along distinct componential lines, in which the functional losses reflect specific processing difficulties, and not some generic diminution of language ability or complexity. The two most intuitive ways of dividing language functions follow linguistic and behavioral categories, respectively. In linguistic terms, the most widely accepted dichotomy of language function is between semantic and syntactic structure. In many theories, these are essentially treated as orthogonal dimensions of language function. Since such aspects of language can be logically separated in language analysis, it seems reasonable to expect that these two aspects of sentences might require different neural substrates to process them. A dissociation of disturbance of the two functions after damage to different locations within the brain would strongly support this. Alternatively, an intuitively useful distinction in language impairments can be made between the hearing and speaking functions of language. These two fundamental modalities by which language enters and leaves the nervous system offer a natural division of functions that must be reflected in the neural organization of language. Both dichotomies have found their way into aphasia theories in various forms.

The most influential categorization of aphasias, however, does not follow either of these logics, but rather an anatomical logic. This is the classic division of aphasic disorders into Broca's and Wernicke's aphasia. Although a number of early nineteenth-century physicians suggested the existence of two major classes of language disorders—a difficulty in remembering how words are produced as opposed to a difficulty in remembering how words sound—the systematic analysis of brain-damage language disorders was prompted by the discovery that different patterns of language dysfunction followed damage to different areas of the brain.

In 1861, a French surgeon named Paul Broca demonstrated that damage to the lower portion of the left frontal cortex could produce a profound disturbance of the ability to speak. He called this behavioral syndrome *aphemia* (literally, "no words," or more colloquially, "bad speech") to emphasize its motoric character, though the term *Broca's aphasia* came to replace Broca's own term. A little more than a decade later, in 1874, a young German physician name Carl Wernicke demonstrated that damage to the posterior part of the temporal cortex on the left side produced a different language problem: a profound disturbance of the ability to understand speech, along with a tendency to speak fluently but with anomalous words and word combinations.[1] This came to be termed *Wernicke's aphasia.* The corresponding brain regions that each of these physicians associated with their respective aphasias are now also referred to as Broca's and Wernicke's areas, respectively (see Figure 10.1).

Wernicke provided the first systemic analysis of the relationships between brain damage and aphasia. By organizing his theory around a diagrammatic analysis of cortical areas and putative connections with respect to input and output functions, he predicted a whole range of aphasic disorders defined by the possible ways that the physical connections between these presumed brain centers could be interrupted. Not only did he account for some of the prominent symptoms associated with auditory analysis and motor disturbance, but he predicted a number of additional syndromes that might result from damage to some of the presumed connections between these and

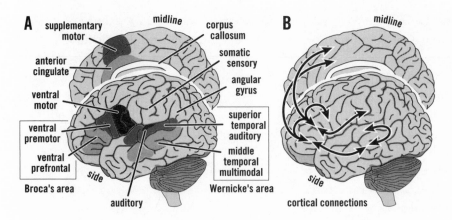

Figure 10.1 *Cerebral cortical areas most commonly associated with language functions in the left hemisphere along with a depiction of a few of the major cortical connectional pathways as identified in other primates.*

A. *The classic language areas cluster around the horizontal fold called the* Sylvian fissure, *which divides the temporal lobe from parietal and frontal lobes.* Broca's area *can roughly be identified with the ventral premotor and prefrontal areas and* Wernicke's area *can roughly be identified with the superior and middle temporal gyrus. Midline areas are involved in initiation of speech (supplementary motor cortex) and arousal-attentional control in comprehension and production of speech (anterior cingulate cortex). The angular gyrus was thought to be important for multimodal language processing, as in reading and naming objects. The corpus callosum is the fiber bundle that links the hemispheres.*

B. *The general pattern of major connections has been studied in other primates (but still cannot be determined in human brains). These cortical connections are reciprocal (though connections are not symmetric with respect to layers innervated) and form a tiered organization in each modality (adjacent tiers are connected by short U-connections). The tiered pattern is reinforced by long intermodality connections, which tend to link cortical areas at the same level across modalities. Connectional data simplified from Deacon (1988 and 1992b).*

other areas. Among these were predictions of "transcortical" aphasias due to interruption of links between input-output systems and "higher" cognitive centers.

More than a century has followed their ground-breaking discoveries, but even today, the syndromes associated with Broca and Wernicke provide the backbone of aphasia research. Yet a precise and broadly accepted interpretation of these disorders or exactly what part of the brain is responsible for the component symptoms remains controversial. What made Wernicke's analysis so compelling was both its categorical logic and its surprising predictive power. Though few details of this theoretical scheme persist unaltered in present analyses of aphasia, his focus on the importance of anatomical location and sensorimotor modality have continued to provide useful guides. Not surprisingly, in the more than a century since these men began to analyze the relationship between language breakdown and brain damage, we have come to appreciate that aphasias occur neither along either strictly linguistic nor along strictly sensory versus motor lines. They have a logic all their own, an uneasy hybrid logic that results from the post hoc evolutionary marriage between brain architecture mostly evolved in a world of concrete sensory and motor adaptations, prior to language, and a representational system that depends on a decidedly nonconcrete logic.

There are many problems with the enterprise of extrapolating from brain damage to brain function. Deterioration of a function is not just the reverse of development of that function, and loss of a brain region does not just result in a loss of function; the intact regions of a damaged brain are also changed in function by the loss. Nevertheless, aphasias and other pathologies affecting language due to local brain damage provide essential clues about what brain regions contain parts of the network that are associated with different language functions. Wernicke's genius was in recognizing that brains needed to be analyzed in terms of circuits, and not just collections of regions with distinct functions. In recent decades, this insight has been carried much further than Wernicke could have imagined. Impairment of a specific language function following brain damage to a particular region does not mean that those language functions are located in that region. *Some* function that is crucial to these linguistic operations has been disturbed by damage in these cases, but the correspondence may be quite indirect. The task facing the neurolinguist is to decipher the code by which language symptoms can be translated into brain processes, but it is clearly not a one-to-one mapping, especially for functions that cannot be specified in purely sensory or motor terms. The mismatch between linguistic and neural logic is probably exaggerated when dealing with such gross analytic distinctions as syntax and semantics. This difficulty is exemplified by the problems of explaining grammatical deficits that result from brain damage.

Agrammatism is the term applied to disturbances of a patient's ability to analyze grammatical relationships in ways that cannot be accounted for by a general lexical impairment alone. The history of the study of agrammatic disorders reflects some of the pendulum swings of neurological theory during the last century. Many theorists who divided language disorders into production versus comprehension difficulties tended to understand deficits in grammatical analysis in terms of lexical problems (therefore associated with Wernicke's area). Difficulties in stringing words together to make a sentence, or syntactic difficulties, were considered merely motor problems. These difficulties were characteristic of Broca's aphasia; and so, beginning with such influential physicians as Sigmund Freud, Pierre Marie, and Henry Head at the start of the twentieth century, it became common to consider Broca's aphasia not a true language loss, only a speech production problem. Many modern aphasiologists still concur with this assessment, and they additionally point to the fact that there seems to be a high incidence of recovery of function after damage to Broca's area, suggesting that there is no permanent loss of language knowledge per se. If language functions were located in discrete areas of cortex, then loss of the particular area that performed grammatical operations should produce a severe persistent agrammatism, just as damage to the primary visual cortex produces a permanent loss of vision in the quadrant of the visual field that corresponds to the damaged sector. Of course, this is a big assumption to make.

One of the most significant challenges to this simple dichotomy, which directly affected the interpretation of agrammatism, came with the discovery in the 1960s that many Broca's aphasic patients had problems with grammatical comprehension.[2] Though the analysis of the major content words of a sentence seemed relatively unimpaired, when the logical structure of the sentence could not easily be guessed from the content words alone (i.e., when grammatical function words or word-order clues were essential to understand how the nouns, verbs, adjectives, and adverbs referred to each other), the patients tended to have difficulties.

Of the many theories proposed to explain this difficulty, the one that has been most widely cited suggests that the patients behave as though they simply don't understand the grammatical function words. These include words like "was," "that," "who," "which," etc., that act like flags, signaling structural relationships within a sentence. Notably these sorts of words are by far the most frequent in any text or utterance, and form what amounts to a closed set within a language. Whereas new nouns, verbs, adjectives, and adverbs are added and invented all the time, and in this sense form an open class, these little function words don't seem to allow any other logical space within the language for additions. One interpretation, then, might be that the disorder could be described as a disturbance of the lexicon of closed-class words. Thus, two different lexical storage systems were invoked: one for closed-class words (associated with Broca's area) and one for open-class words (associated with Wernicke's area)—a challenge to the simple comprehension/production dichotomy.

The difficulties in interpreting agrammatism have only become harder as researchers have attempted to test these opposed hypotheses. For example, a simple lexical explanation (i.e., loss of the part of the mental dictionary that contains grammatical function words) does not appear to account for patients' problems in cases where manipulations of word order are central, as in embedded sentences ("The girl the boy liked lived in a red house") and only partially accounts for passive tense difficulties in English clauses ("The dog chased by the cat . . . "). Time also appears to be a critical factor. "On-line" analysis of spoken utterances tends to show significant impairments that seem to ameliorate if the patient is given written material or a longer period to reflect on the analysis. In other words, it appears that the information about grammar is not lost in these patients; it is just very much less accessible, and its use is more subject to error. It is, metaphorically, as though the main route to this information has been disrupted, forcing these patients to take more circuitous routes through the winding back roads of memory to connect the same end points. Again, it does not look as though grammar functions are "located" in Broca's area.

Some aphasia researchers have countered that what we are observing in the impermanence of this deficit is a takeover by the paired homologue to Broca's area on the right hemisphere, by other spared neighboring regions, or just recovery of the remainder of an incompletely damaged brain system. This interpretation, however, is harder to maintain in cases with more extensive or bilateral damage. It has also been implied that Broca's aphasia is just a motor deficit, and that with motor cortex damage there is a permanent loss of function in this domain. In this interpretation, the cognitive impairments might be blamed on motor interference effects. Even more serious challenges have been offered by comparisons of agrammatism in patients who are native speakers of very different languages. Broca's aphasics who speak languages that are highly inflected, for example, do not seem to be as agrammatic as their English-speaking counterparts. More will be said about this variability later.

All these challenges to the localization of grammatical functions to Broca's area are often couched in terms either of denying Broca's area status as a language area or of suggesting that other areas perform the functions that have traditionally been assigned to Broca's area. But there is a different way of approaching this question, and it depends on letting the notion of localization of function go while retaining something like the localization of computation. Another way to put this is that we need to stop conceiving of localization of *language* functions and instead try to understand how language functions map onto *brain* functions, which are likely to be organized to a very different logic. As with so many other body and brain functions, there may be a number of different ways of achieving the same goal. To offer a very simple analogy, throwing a ball at a target will utilize very different muscles and coordinated patterns of neuronal activity depending on whether it is thrown overhand, underhand, or backwards. The resultant flight of the ball and interaction with the target may be very similar, even though very different muscles and body movements produced them. Even recognizing that there are many commonalities (e.g., all may involve arm-muscle flexion, rapid ballistic movement, and visual orienting to a target), what ties them together are not these but rather the *requirements of the functional goal,* and this is something extrinsic to the anatomy.

I suggest that we need to think of language functions in a similar way, especially those that are imagined to be "deep" to the surface structure of speech. We need to treat them as composite behavioral products or logically defined outcomes, as opposed to neural operations. There need not be any specific association between a brain region or connection and a class of linguistic operations, and there may even be many alternative neurological means of achieving the same symbolic end. This does not assume that the human brain lacks local specializations for language, or that language abilities can be described in terms of general learning mechanisms alone. And it does not depend on any denial of area specialization within the cerebral cortex. It implies only that the neural distribution of language functions need not parallel a linguistic analysis of those same functions.

The assumption that there are distinct language areas in the brain that correspond to these language disorders (e.g., Broca's area and Wernicke's area) is an example of what, in another context, the philosopher Alfred North Whitehead called "misplaced concreteness." Abandoning the Procrustean enterprise implicit in correspondence theories of brain and language functions opens up a number of additional sources of information for understanding this relationship. One messy detail haunting classic theories was that Broca's and Wernicke's areas did not correspond to divisions of the cortex identified by distinctions of neural architecture, though many have insisted on trying to force a one-to-one correspondence. These areas even seem to vary in location from patient to patient. Not only are there no reliable microanatomical criteria for these areas, but the syndromes themselves are also not precisely delineated. They are not the same from person to person or language to language. Is it possible that there is no one ventral frontal region always associated with Broca's aphasia, nor any one posterior temporal region always associated with Wernicke's aphasia?

Data from brain-damaged patients provide insufficient resolving power to probe below the level of behavioral or linguistic generalities in order to carve the language problem at its natural neurological joints, but research tools with better resolving power for studying function within intact brains are now rapidly becoming available. These can help overcome many of the limitations of anatomical precision and functional dissection. One consequence is that we are being bombarded with a flood of new kinds of data that are often dissonant with existing paradigms. These data—derived from *in vivo* imaging techniques, electrophysiology, and numerous other approaches to brain function—often defy explanation along classic lines because they provide a view of these processes at a different level of structure and function from anything ever envisioned by classic theories. Classic high-level models of language functions were conceived in order to explain the large-scale features of language breakdown, what might be called *macrocognition.* We now must face up to the daunting task of analyzing the *microcognition* of language: analyzing phenomena at a level of scale where the functional categories often no longer correspond to any of the familiar behavioral and experiential phenomena we ultimately hope to explain.

The central problem faced by researchers studying the brain and language is that even the minutest divisions of cognitive function we hope to explain at the psychological level are ultimately products of the functioning of a *whole* brain—even if a damaged one—whereas the functions we must explain at a neurological level are the operations (or computations) of only a small fragment of this highly integrated and distributed network of structures. If there was ever a structure for which it makes sense to argue that the function of the whole is not the sum of the functions of its parts, the brain is that structure. The difficulty of penetrating very deeply into the logic of brain organization almost certainly reflects the fact that the brain has been designed according to a very different logic than is evident in its most elaborated behavioral and cognitive performances. This is precisely where the comparative and evolutionary approaches can provide their most crucial contribution.

Though other primate brains have not evolved regions that are specially used for language processes, those regions of the human brain that are did not arise *de novo.* The language areas are cortical regions that have been recruited for this new set of functions from among structures evolved for very different adaptations. They were selected during language evolution because what they already were doing offered the best fit to the new problems posed by language. Thus, we should stop thinking about Broca's and Wernicke's areas as "language areas." They are the areas that language most intensely utilizes. The question we should ask then is why? What is it about these areas that dictates the distribution of language functions that we observe?

The simple answer, proposed by the nineteenth-century aphasiologists, is that their location is a function of the predominant input and output routes: sound analysis and speech production. Broca's area is adjacent to the mouth-tongue-larynx region of the motor cortex and Wernicke's area is adjacent to the auditory cortex. In quite general terms, these are association areas for vocalization and sound analysis. But are there distinct anatomical structures that correspond to these operationally defined language structures? The answer probably is no. Broca's and Wernicke's areas represent what might be visualized as bottlenecks for information flow during language processing; weak links in a chain of processes. Once we abandon the idealization that language is plugged into the brain in modules, and recognize it as merely a new use of existing structures, there is no reason to expect that language functions should map in any direct way onto the structural-functional divisions of cortex. It is far more reasonable to expect language processes to be broken up into subfunctions that have more to do with neural logic than with linguistic logic.

The one fact that is clear is that language functions are dependent on the interactions among a number of separated regions within the brain. Languages themselves are heterogeneous structures, which often encode complementary functions in very different components of the speech signal. This probably aids the computational systems that support it by breaking it up in a way that most effectively allows the simultaneous processing of a number of parallel functions. The intense natural selection operating on language forms should have produced variants that map easily to these processing demands. Complementary linguistic operations should end up by being distributed to different brain structures, and differences in aphasic disturbances should reflect this. But language is fitted to an ape brain plan, and our ape brains, though modified in response as well, are by far the more inflexible element in this relationship. So what is the anatomical logic to which language has had to adapt?

The Illuminated Brain

The first source of brain-language data to offer a significantly different perspective from that provided by brain damage came from studies of the effects of electrical stimulation of the cortex in patients undergoing neurosurgery for unrelated pathologies. In the 1950s, a neurosurgeon by the name of Wilder Penfield perfected a technique for assessing functional localization by electrical stimulation of the exposed cortex of awake, locally anesthetized patients. By passing low-level electric current into the cortex near the presumed language areas of the left hemisphere, he found that he could selectively interfere with different language tests he had his patients perform. If he asked them to speak or name an object, he could selectively block speech, cause it to be distorted, make it difficult to come up with a name, and so on. Since neural impulses are also electrochemical processes, stimulation essentially bombards a region with very "loud" neural noise and thereby disrupts whatever function is mediated by that region. This effect is remarkably temporary and localized. It tends to cease immediately current is cut off, and stimulation of closely adjacent cortical surfaces (within a few millimeters) can produce quite different effects. By probing many areas, Penfield and others mapped the extent of cortical areas where electirical stimulation interfered with language functions. This allowed surgeons to identify those regions that would likely disturb language functions if removed during brain surgery to eradicate tumors or epileptic centers.

Penfield's findings were consistent with what was known from brain damage—and yet also strangely different in many ways. He found that stimulation of presumed language areas was indeed very likely to produce language disturbances, but these disturbances were not like aphasias. Moreover, he found both a much wider distribution of language-disturbance sites than that suggested by aphasia, and a symmetry of posterior and frontal functions that did not match the apparently dichotomous effects of damage. More recently, another neurosurgeon, George Ojemann, with his colleagues has carried this work much further, using more sophisticated neurolinguistic testing, stimulation, and recording. A composite summary diagram of many of these results is shown in Figure 10.2.

What these stimulation studies demonstrate is that the regions where stimulation disrupts language function fan out from the frontal mouth area into the prefrontal lobes, and from around the auditory area back into the temporal and parietal areas. Those regions where stimulation reliably disrupts the same language functions are organized in what appear to be tiers radiating outward from these two foci. Electrical stimulation of the regions closest to motor and auditory areas produces problems with phoneme identification and oral movements. Stimulation further out disrupts naming of familiar objects and grammatical assessments. And stimulation even further out appears to disrupt retention or recall of words. There is also a rough front-back mirror symmetry of these tiers, so that the very same responses are elicited by the second and third tiers both front and back. The language

Cortical Electrical Stimulation

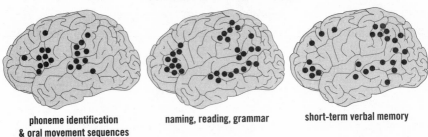

phoneme identification
& oral movement sequences

naming, reading, grammar

short-term verbal memory

Figure 10.2 *Electrical stimulation studies of selected language tasks indicate that language operations are fractionated and distributed in many regions of the left hemisphere. Electrical stimulation (indicated by dots) appears to indicate a tierlike pattern of functional organization fanning out from the classic language regions into the prefrontal, temporal, and parietal areas. Areas more removed from primary motor and auditory cortex appear to be involved in processes with greater temporal integration. The figure graphically summarizes data from Penfield and Roberts (1959); Ojemann (1983; 1991); and Ojemann and Mateer (1979).*

regions cover a large fraction of the entire left hemisphere, but stimulation within these regions does not produce uniform results. In fact, stimulation of most sites produces no language disturbance in a given individual, and there may be only a few sites that disrupt each sort of language task. The stimulation maps of language functions are therefore composite maps, demonstrating statistical consistencies between patients.

The most surprising pattern in these results is the mirror symmetry of anterior and posterior effects. This is an entirely different pattern from that derived from brain-damage studies. How does it change our understanding of brain-damage effects and their distribution? The key to interpreting these differences is that electrical stimulation is something added, whereas damage is something removed. Though cortical areas just next to a stimulation site might not be directly affected by the stimulation, areas some distance away may be, so noise from many sites may be able to feed into the stream of information supporting a given linguistic function. This should not be surprising, considering the high degree of interconnectivity of cerebral cortical areas. But the pattern of distribution and kinds of disturbance tell us more than just that there are lots of ways to mess up language signals.

First, language functions extend into all major lobes of the neocortex, including the temporal area (auditory), parietal area (tactile), and frontal cor-

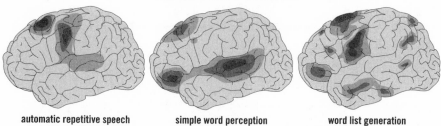

Regional Cerebral Blood Flow

automatic repetitive speech   simple word perception   word list generation

Figure 10.3 *Cerebral blood flow studies of selected language tasks indicate that language operations are also fractionated and distributed with respect to this assay of relative metabolic demand of the given tasks. In these drawings, blood flow activation is superimposed on outline drawings of human brains; darker areas are more activated by the specific task. Wernicke's area, the motor speech area, and ventral prefrontal areas associated with Broca's area are differentially recruited for different language tasks. Ventral prefrontal cortex is recruited for both word perception and word association tasks (e.g., generating word lists). The figure graphically summarizes data from Larsen, et al. (1978); Lassen, et al. (1978, 1980); and Roland (1985)*

tex (attention, working memory, planning) of the left hemisphere. This wide distribution suggests a far less localized language system than is suggested by brain-damage results. Though the language impairments due to parietal and prefrontal damage are often not as noticeable and devastating as left temporal damage, both can produce problems with speech fluency, word finding, and certain sorts of semantic analyses. Second, the stimulation sites that are located closer to the Sylvian fissure are mostly correlated with sensorimotor functions of language, while sites further out are correlated with higher-level linguistic and cognitive functions. This is consistent with the the fact that the innermost tiers are located adjacent to primary tactile, auditory, and motor areas, and the outer tiers are distributed within multimodal and association areas. One obvious interpretation of this hierarchic pattern is that it parallels the layout of what are sometimes called secondary and tertiary association areas in each modality. But notice also that the differences from tier to tier reflect different degrees of linguistic integration and different language-processing time scales. From inner to outer tiers there is a progression from parts of words, to words and phrases, to verbal short-term memory. Analyzed temporally, the inner tier of areas deals with events that take place on a time scale of tens to hundreds of milliseconds (phonemes), and the outer tier of areas deals with information that must be retained over many seconds (inter-sentential relationships, for example, must be analyzed at this rate).

Time is a critically important factor, especially in an information-processing device that tends to operate almost entirely in parallel (instead of funneling all operations through a single processing unit, one after the other, as do most desk top computers). Processes that utilize information presented at very different rates tend to become segregated within the brain (as they must be in many other systems, from mechanical devices to bureaucracies). Maintaining a signal within a circuit long enough to analyze its part in some extended pattern would tend to get in the way of processes that require rapid and precise timing. Slow neural signal transmission can also become a limiting factor in the brain, so that very rapid processes are best handled within a very localized region, whereas the accumulation of information over time might better be served by a more distributed and redundant organization that resists degradation. So it makes sense that for each modality, it might be advantageous to segregate its fast from its slow processes. This is probably one of the major factors that distinguishes the cortical subregions that we label "primary areas" (fleeting signals) and "association areas" (persisting signals). Separating processes according to a gradient of time integration may be every bit as important for handling potentially interfering functions as is the segregation of different modalities and submodalities.

The symmetry between anterior and posterior functional representation can probably best be understood in connectional terms. Studies of axonal connections between cortical areas in primates have demonstrated that the connectional logic parallels the tierlike organization of secondary and tertiary regions. Within a single modality, adjacent cortical areas tend to be more highly interconnected than to areas more distantly separated, but across modalities there appears to be a preference for connections at the same level in the sequence of areas progressively removed from primary input and output. Thus, the tierlike organization of stimulation effects in anterior and posterior cortex is probably a reflection of the direct interconnections between them, and of the comparability (possibly via synchrony) of the temporal-processing domains in each. The apparent lack of interference across levels may reflect the relative temporal isolation of processes at these different levels. To sum up, evidence provided by electrical stimulation suggests that there may be multiple areas in both the anterior and the posterior cortical regions that contribute to language processing at different levels of analysis and production, and that these may be linked in parallel in a temporal mapping relationship.

Once we abandon the reification of language areas as modular language algorithm computers plugged into an otherwise nonlinguistic brain, it becomes evident that language functions may be widely distributed and processed simultaneously in many places at once. They may also be distributed according to a computational logic that is not necessarily obvious from the apparent external speech signal. Our experience of the linearity of speech belies an internal parallelism and hierarchy of the underlying cognitive processes. Indeed, one possible interpretation of the hierarchic logic of grammatical and syntactic operations is that they evolved to accommodate this mismatch between cognitive processes and production constraints. This is important information to keep in mind for linguistic analysis. We may need to begin thinking of the various syntactic and morphological tricks used by languages to mark their logical and symbolic structure as markers that aid in rapidly breaking up the serial signal and distributing the processing of segments matched to different time domains to their most appropriate processors. For example, the small function words that are specially important for grammatical analysis signal the breaks between syntactic transitions, as do pauses and tonal/rhythmic changes of voice. The hierarchical complexity of sentence structure and the multiplication of tricks for compacting clauses into sentences with considerable analytic depth may not in fact add computational difficulty. They may instead be ways to distribute language processes more efficiently across many parallel and partially independent systems.

Another window on the brain's control of language processes is provided by techniques developed in the last decades for producing images of brain metabolism. The basic assumption of metabolic imaging is that regions of the brain that are more active will exhibit a greater turnover of glucose and will demand more blood than relatively quiescent areas. Thus, hot spots will show up in areas intensely involved in a given cognitive task. Three different metabolic-imaging techniques have been used to image functional changes during language tasks: regional cerebral blood flow (rCBF); positron emission tomography (PET); and functional magnetic resonance imaging (fMRI). For rCBF images, inert radioactive gas dissolved in blood indicates where blood flow has increased. For PET images, a radioactive analogue of glucose (the simple sugar that all neurons require for energy) is injected into the blood to measure cumulative metabolism in different brain regions.[3] For fMRI images, the response of water molecules within the brain to intense magnetic fields produces images of brain structures, and magnetic solutions (called "contrast agents") are injected to visualize blood flow changes.

In response to language tasks, the greatest increases in blood flow are seen in cortical regions that correspond with those damaged in acquired aphasias (Figure 10.3). The patterns for different tasks show that mouth movements of speech tend to activate motor cortex, and that listening to words tends to activate auditory regions of the left hemisphere. But there are other, more interesting patterns. Simply repeating words over and over (like counting to ten repeatedly) does not seem to activate the region we would call Broca's area. Instead, it activates mostly motor areas, and to a lesser extent auditory areas and a dorsal motor region known as the supplementary motor area.[4] These studies also showed that listening passively to (nonrepeating) words does not activate motor areas. It activates auditory cortex, including the area that probably corresponds to Wernicke's area, and also a ventral prefrontal region just in front of what most researchers would call Broca's area. More complicated language tasks, however, such as generating lists of words without repeating any (for example, naming objects that can be held in the hand), produce cerebral blood flow patterns characteristic of both speaking and listening. The ventral prefrontal area is active and the motor areas are active, including areas that would be called Broca's area (ahead of and below the motor mouth area). Uniquely recruited to this task, too, are regions of the temporal and parietal lobes that probably contain multimodal responsiveness (association areas).

These patterns reveal a number of features of functional organization that are not evident in either damage or stimulation effects. First, "mindless" repetition only minimally recruits classic language areas or association areas. Second, interpretation of words seems to involve both auditory and prefrontal regions, but apparently *not* what would have traditionally been called Broca's area.[5] This ventral prefrontal region of the cortex is, however, often partially damaged in Broca's aphasics with large lesions, as is motor cortex for the mouth region. Notice also that this ventral prefrontal region is associated with naming and short-term verbal memory as a result of stimulation and word list generation, though the degree of overlap or difference from task to task is unclear and suggests that different subdivisions may be involved in these tasks. The premotor part of Broca's area is most actively involved when both word analysis and speech are required.

Data from another metabolic imaging technique, positron emission tomography (PET scanning for short), produce similar results to those from rCBF studies (see Figure 10.4).[6] These data show that passively listening to novel words mostly produces auditory activity, with some ventral prefrontal cortex activity. Similarly, looking at novel written words involves visual cortex. Visual input does not, in this example, require significant auditory activity to be interpreted (this may have something to do with the simplicity of the task, as well). In the study summarized by the idealized drawings in Figure 10.4,[7] the experimenters augmented their analysis by using an interesting subtractive technique. The images that were produced from tasks that involved word perception were subtracted from images that were produced from tasks that also involved speech, and both of these images were then subtracted from images produced from tasks that additionally involved word association processes (generating a verb that is appropriate to a presented noun). Without this approach, the resulting images would have obscured the differences between more complicated tasks as they progressively occupied more and more of the brain. Subtracting images from one another isolates where in the brain activity is additionally recruited to handle specific increases in task complexity, though it doesn't indicate that regions activated by previous simpler tasks do not also contribute in unique ways to the computations of the higher-order task as well.

Several things are revealed by these subtractions, when compared to previous results. First, though different modalities of stimuli can be used (e.g., visual versus auditory word presentation), the differential activity patterns produced by the more complicated tasks using these inputs are very similar. Second, the ventral prefrontal area again seems activated when word analysis is required. Third, areas on the midline are active. Repeating words seems to involve the supplementary motor area, and word analysis seems to involve the anterior cingulate cortex. The cingulate cortex seems essential for most tasks that require intense attention, and so may not be uniquely part of linguistic processing (though essential). Fourth, both motor activity and word analysis independently produce intense activation of the cerebellum (on the opposite side of the brain, since the pathways linking cerebral cortex and cerebellum are crossed—see Figure 10.4, images 2 and 3).

Not surprisingly, when this task is practiced and familiar words are shown repeatedly, the task gets easier, and the activation level of all three of the differentially activated regions (left ventral prefrontal, anterior cingulate, and right cerebellar cortex) drops close to background levels. Other brain regions such as the insular cortex and striatum become slightly more active instead. The mental effort required for the task is thus reflected in the degree of differential activation of these structures.

It is surprising (from a classical perspective) that the cerebellum is recruited for this task, not primarily as a repository for automatic behavior, but for overcoming such tendencies in the service of producing alternatives.

Positron Emission Tomography (PET)

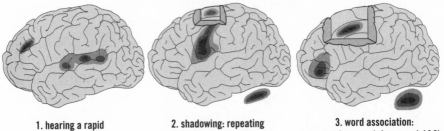

| 1. hearing a rapid sequence of words | 2. shadowing: repeating rapidly presented words (-1) | 3. word association: generating a verb for noun (-1&2) |

Figure 10.4 *Positron emission tomography (PET) studies of language tasks reveal similar patterns to those seen with cerebral blood flow imaging. In these drawings of PET image activation, patterns are superimposed on outline drawings of human brains; darker areas are more activated by the specific task. The three tasks are hierarchically constructed: 1 = Listening only; 2 = repeating the input word; 3 = producing a word associated with the word presented. The images to the right were produced by subtracting the results of those to the left to reveal differences. Notice the critical involvement of the ventral prefrontal cortex in the word association task. The activated area below the back of the brain indicates an additional intense involvement of the contralateral cerebellum (see also Fig 9.1). The figure graphically summarizes data from Petersen, et al. (1988); Posner, et al. (1988); and reviewed in Posner and Raichle (1994).*

Congenitally reduced cerebellar development thus not only should be expected to produce a variety of problems associated with the learning of skilled movements requiring precise timing and complex coordination; it should also remove an important support for a variety of prefrontal functions, particularly those demanding rapid production of novel associations. Since some of the most demanding language processes are likely to be those of generating the words that make up novel sentences we must produce in real time, this cerebellar-prefrontal link may play a critical role.

Finally, I want to draw attention to the contribution of the ventral prefrontal region that is specially activated in all these examples of word analysis. What is this region in anatomical terms? Evidence bearing on this question comes from my lab and others, and from experiments with monkeys, which have shown the connections that link the corresponding area with other cortical and subcortical regions.[8] This evidence shows that this region of the brain is the principal recipient and source of connections linking the auditory areas with other prefrontal and frontal areas (see Figure

10.1B). The areas just behind it—between the prefrontal cortex and motor cortex that have classically been linked to Broca's area—do not get auditory inputs, but rather supplementary motor inputs, and inputs carrying tactile information about the mouth region. The ventral prefrontal region also connects to the dorsal prefrontal areas (likely also those active in other word perception tasks) and with the anterior cingulate cortex, which also is highly active during this demanding word association task. As is suggested by electrical stimulation studies, systems that must be co-activated in a given language task often show evidence of linkage, and different tasks are segregated both in terms of modality and in terms of the temporal domain. For example, word association analysis in normal speech is required for interpreting the many relationships between words distributed throughout an utterance, perhaps even between sentences. Single word production or perception, on the other hand, involves articulatory precision at the level of individual phonemes, and the recognition of words requires little more than phoneme comparisons. These are distinguished by an order of magnitude difference in time, as well as by a totally different mode of analysis.

Comparing information from these very different methods for tracking the correspondence of language functions and brain regions makes it possible to begin to recognize repeating themes. The effects of ventral prefrontal stimulation on naming, grammatical decisions, and verbal short-term recall can help explain why Broca's aphasics often show poor fluency, relatively spared word meaning analysis, and difficulties with grammar where positional cues in a sentence are most important. Some less well understood aspects of Broca's aphasia are also given a new context of interpretation. For example, it turns out that Broca's aphasics demonstrate some interesting word association biases as well. They tend to have difficulty with word lists but they have an easier time producing lists that are within categories (animals) than those that involve shifts in the part of speech (i.e., noun-verb relationships). Additionally, Broca's aphasics have difficulty reconstructing hierarchical relationships among both words and objects. Finally, Broca's aphasics even exhibit word perception deficits, which is consistent with frontal activation in simple word perception tasks.

In summary, images of the working brain doing language tasks show a hierarchic organization, which correlates both with time and with the hierarchic-segmental organization of sentences, and a segregation with respect to the physical form, presentation, or manipulation of the signal. They also show that the classic language areas are not unitary modules, but rather complicated clusters of areas, each with different component functions. This is not consistent with the view that there is one self-contained "language organ" in the brain, even one dedicated only to grammar. If language had evolved as a consequence of the addition of a language module in the human brain, we should not expect such an extensive distribution of linguistic processes in diverse cerebral cortical areas. If a grammar module exists, it is not localized to one cortical region, since subfunctions associated with grammatical and syntactic processes are found in both the anterior and the posterior regions.

If we keep in mind that primate brains very much like our own have been around for tens of millions of years and that the mammal brain plan which our brain follows has probably been around for over 100 million years, it becomes evident that the logic of language is probably highly constrained to fit an ape brain logic. Though breaking up language analytically into such complementary domains as syntax and semantics, noun and verb, production and comprehension, can provide useful categories for the linguist, and breaking it up according to sensory and motor functions seems easier from a global neuronal viewpoint, we should not expect that the brain's handling of language follows the logic of either of these categorical distinctions. The patterns we observe probably reflect, in a very indirect sort of way, the processing problems produced by mapping a symbolic reference system encoded in a serially presented modality onto the processing logic of an ape brain.

If brain regions are recruited during language acquisition on the basis of how they map to implicit modally and temporally segregated, parallel computational domains, then we should also expect that the representation of language functions should be diversely and multiply implemented in different individuals' brains, and even within the same brain at different times in the life span or under other special conditions. Also, even within a single brain there may be some degree of redundancy of functional capacity, so that alternative structures, besides those best optimized for linguistic operations, could be recruited to handle linguistic tasks when these more optimal systems are unavailable (e.g., due to brain damage, to the interference of multilingual communication, or to pressure from simultaneous competing cognitive tasks). Such functional substitutions almost certainly would be associated with a significant decrement in efficiency, precisely because language is not processed by some general learning capacity, but by quite heterogeneous cognitive subsystems, none of which is a language processor by design.

Some of the most compelling theoretical models of language analysis argue that the incredible rate at which sentence analysis proceeds, and the relative automaticity and inaccessibility of our experience of these processes, can only be explained by postulating the existence of domain-specific modules that are closed to information from other systems. If speech analysis and sentence production are highly parallel processes mapped onto the serial medium of speech, then we should indeed expect that the multiple components of this process would run in comparative isolation from one another. The logic of distributed processing demands it, because significant levels of "cross-talk" between distinct processing domains would inject considerable interference. However, the logic of this sort of adaptive modularity is in many respects the reverse of that suggested by linguistic theories about grammatical processes. Indeed, grammatical cues, such as are embodied in small "function words," may be the primary agents for initially tagging and distributing sentence "chunks" to be separately processed. For this reason, it is precisely these features of language that need to be subject to minimal symbolic analysis. They serve a predominantly *indexical* function. And as we have seen, indices can be interpreted in isolation as automated, rote-learned skills.

In the interest of rapidly and efficiently organizing such highly distributed associative processes, some symbols had to be stripped of all but the vaguest symbolic content in order to provide a set of automatic "switches" for shuttling symbolic work to the appropriate regions of the brain. So, automatization of speech production and comprehension is accomplished by setting aside a small, closed set of symbols to be used as though they were indices. As a result of being stripped of semantic links, they can be learned by rote and implemented with minimal mnemonic search. And their representation within the brain can be highly local, even subcortical. The function of these modular operations is to implement grammatical rules, but the rules are implicitly symbolic and therefore distributed. These automated language functions are not grammar modules, but merely symptoms of the grammar, which is itself probably highly distributed.

Unlike closed modules, the separately processed levels of sentential information cannot be entirely "closed" to the information processed in others. Parallelism requires synchrony in order to keep the partially decoupled processes organized with respect to one another, and selective cross-talk so that the results of some processes can constrain the operation of others. This too can be facilitated by breaking up language processing according to temporal domains, because extended (and therefore more redundant) processes can serve as a frame within which many more rapid processes can be constrained; and conversely, the progression of a slower, more global associative process can serve as an integrator that helps overcome the intrinsic noisiness of the rapid processes, which must by design have minimal associative scope.

Where Symbols Aren't

So, where in the brain might one expect to find symbols represented? The answer appears to be that individual linguistic symbols are not exactly located anywhere, or rather that the brain structures necessary for their analysis seem to be distributed across many areas. The systemic nature of symbolic reference suggests that the representation of symbolic associations within the brain should be distributed in diverse brain regions, and yet that similar classes of words ought to share neural commonalities. Though words, as symbol tokens, may be encoded by specific sound patterns or visual inscriptions, the symbolic referential relationships are produced by a convergence of different neural codes from independent brain systems. Because they are symbolic, word comprehension and retrieval processes are the results of combinations of simpler associative processes in a number of quite separate domains, involving the recruitment of many separate brain regions. For this reason, they cannot be located in any single neural substrate; rather, as each supportive representational relationship is brought to play in the process of producing or comprehending a word, each corresponding neural substrate is activated in different phases of the process.

To the extent that each higher-order form of representational relationship must be constructed from or decomposed into lower levels of representation, we can expect that their neural representations will exhibit a similar nested hierarchic structure as well. There should be a sort of truncated recapitulation of this acquisition hierarchy, in opposite directions, depending on whether a symbolic relationship is being constructed or interpreted—from icon to index to symbol, or from symbol to index to icon, respectively. To predict how such a process might proceed within the brain, then, we need first to ask how iconic and indexical processes are likely to be represented.

The lowest level of iconic relationships inevitably should map to processes within single sensory modalities. For example, similarities recognized between phonemes in different speech contexts or between visual shapes in different visual experiences are based on assessments of only a few sensory dimensions in a single modality. Such simple within-modality iconic processes likely have a highly localized character, perhaps represented in the cerebral cortex by activity in a contiguous sensory or motor territory. Words and familiar objects often require compound iconic analyses involving more than one sensory dimension and sometimes more than one sensory modality, so depending on the object or relationship, the recognition process might be distributed across many areas. In most pragmatic contexts, however, only a few features need to be assessed (and the rest implied) in order to make a recognition decision, because the competing alternatives are limited. That the whole constellation of iconic criteria has been activated to a minimal degree in the process becomes apparent only when one of these incidentally is found to be missing in the stimulus (as in the example described in Chapter 3 of discovering a previously camouflaged moth as it flies off the bark that it resembles). This linkage between different iconic features is an indexical one, and reflects the intrinsic hierarchic relationship between these modes of representation. The mental representation of a complex object is based on a correlation of numerous iconic associations in different dimensions or modalities which predict one another's presence. The more complicated the object or relationship, the more numerous the iconic and indexical assessments that are required to recognize it. In this regard, there is only a difference in numbers and diversity of icons involved between learning to recognize objects, on the one hand, and learning to recognize relationships between objects or between events, on the other. Recognition and prediction processes can be streamlined as we learn to focus sensory analysis on only the assessment of the most relevant features.

The indexical associations between words and associated objects tend mostly to involve cross-modality relationships (e.g., sound and vision). Consequently, we should expect the substrates of word analysis to involve a highly distributed collection of brain regions, not even confined to a single modality. Recall that for symbolic reference to develop it is necessary (1) to establish a set of indexical associations between signs (e.g., words) and objects (things and events) in experience; (2) to establish a systematic set of indexical associations between different signs in the form of logical alternation and substitution correlates; and (3) to recognize correspondences (iconisms) between the combinatorial sign-sign relationships and the implicit relationships between the various objects to which the signs refer. When all these pieces of the symbolic puzzle come together, a referential shortcut becomes available: it is possible to bypass each of the indexical intermediaries and use the relationships implicit in combinations of signs (e.g., phrases and sentences) directly to refer to relationships between physical objects and events. As the learner comes to recognize this indirect mapping of sign relationships to object relationships, attention can be shifted away from the more concrete indexical associations (in effect unlearning them by diminishing the expectation of a physical correlation), allowing the more efficient and powerful combinatorial logic of between-sign relationships to provide the mnemonic support for retrieving and reconstructing them when needed.

The symbolic recoding of systems of iconic and indexical relationships is so useful because it ultimately allows us to ignore most of the vast web of word-object, word-word, and object-object indexical associations. The availability of this mnemonic shortcut makes possible the incredible acceleration and compression of information transmission and reception during language production and comprehension, as compared to most other forms of communication. We become lightning calculators of reference. These ignored indexical relationships are still the implicit grounding of word reference; it's just that these interpretive steps can be put off until it can be determined exactly which are relevant and which are not. The potential of being able to invoke these lower-order associations as elements in an interpretive process or use them to cue steps in symbol construction (e.g., sentence generation) is crucial. Symbolic reference is interpreted by supplying its indexical support, and these component indexical relationships are in turn interpreted by supplying their iconic support. This is a reductionistic process, like interpreting molecular interactions by considering the contributions of their atomic and then subatomic constituents. Symbolic interpretation requires a sort of idealized recapitulation of the indexical acquisition history that led up to the establishment of this referential relationship, which need not invoke anything but the most skeletal elements of the underlying indexical and iconic support—only what is essential to the immediate combinatorial and pragmatic context.

Historically, many neuropsychologists have suggested that word reference requires the involvement of special association areas or multimodal areas of the brain (particularly of the cerebral cortex). Studies of cross-modal transfer of information from discrimination learning in different sensory modalities demonstrated that, at least in monkey brains, there were multimodal parietal, temporal, and prefrontal cortical regions that were critical to the ability to transfer learned associations across sensory modalities, such as recognizing by sight an object previously known only by shape. More recent versions of these ideas have invoked a related concept—"convergence" zones in the cerebral cortex—in order to explain how distributed sensorimotor traces of experiences might be interlinked to support word meanings. This captures an essential aspect of word meaning, but not its symbolic aspect. Rather, these cross-modal associations between images and experiences on the one hand and their associations with particular word sounds on the other provide the indexical associations of words, but their symbolic association—what we call the meaning—involves these and something more. The something more includes both the associative relationships between words and the logic of how these map to the more concrete indexical relationships. These components have long been implicit in theories of brain processes involved with retrieving meaning.

Combining brain-imaging approaches with brain-damage approaches to language processing has made possible a renewed appreciation of the multifaceted nature of word association relationships and the corresponding distribution of their component processes within the brain. In general, studies of the mnemonic and lexical processing of words indicate that different brain regions may be differentially involved, depending both on the semantic features and on perceptual features engaged in the analysis. One demonstration of this dissociability of the different aspects of word analysis involves the left ventral prefrontal region of the brain, the same region that is preferentially engaged by verb-for-noun production tasks (discussed above). This same area also appears to play a critical role in retrieval of words representing actions. A similar region is activated during word completion tasks, as shown by fMRI. However, PET studies of subjects responding to nouns with verbs reveal patterns of cortical activation that only partially overlap with nearly identical tasks involving responding to nouns with other nouns, or analyzing grammatical and lexical features of words (see Fig. 10.5). This suggests that there are subtle differences in recruitment of the prefrontal areas as semantic relationships change. Moreover, the left ventral prefrontal region does not appear to be so involved in the naming of familiar objects. Retrieval of object words versus persons' names in response to visual cues has been shown to be differentially impaired in patients with damage to different sectors of the left middle and ventral temporal lobe, and this distinction also appears to be paralleled by imaging studies showing preferential activation of these areas in naming tasks (see Fig. 10.5).

Interestingly, these inferior and middle temporal regions are not associated with either auditory processing or Wernicke's aphasia. In fact, anatomical evidence suggests that these ventral temporal regions probably include cortical areas that are considered to be visual as well as auditory in function. This should not be entirely surprising because the stimuli were pictures. Yet it is important to recognize that this does not simply reflect picture recognition processes. In these PET studies, the brain activation pattern for the visual component of the task (as assessed by using data from parallel object recognition trials) was *subtracted* from the pattern exhibited in the picture–word association task, so the remaining pattern reflects some-

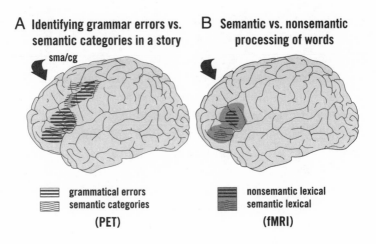

A Identifying grammar errors vs.
semantic categories in a story

sma/cg

B Semantic vs. nonsemantic
processing of words

▭ grammatical errors
▧ semantic categories
(PET)

▬ nonsemantic lexical
semantic lexical
(fMRI)

Figure 10.5 *Summary of brain-imaging studies investigating localization of semantic processing of words and sentences.*

A. *In this study, subjects were required to monitor a story either for grammatical errors or for the occurrence of words in a specific semantic category. The two tasks produced slightly different, but overlapping PET activation patterns in frontal motor-premotor areas and in ventral prefrontal cortex. Anterior cingulate gyrus (cg) cortex and supplementary motor area (sma) were also activated (arrow) (summarized from Nichelli, et al., 1995).*

B. *Functional magnetic resonance imaging (fMRI) used to distinguish between the effects of task difficulty of a lexical processing task and semantic processing of a similar task showing separate effects in adjacent ventral prefrontal areas (summarized from Demb, et al., 1995).*

C. *Combined data from PET studies and patients with focal brain damage were used to construct maps of areas critically recruited for three categories of picture naming. Different temporal cortical areas (outside traditional Wernicke's area) were*

thing else. An additional visual processing stage appears to come into play in order to generate the word (though it may involve some of the same visual areas as were required for the object recognition task). A different use of visually encoded information appears to be playing a critical role in this lexical (i.e., word-word) level of the task, a use that differs depending on the category of the named object. The patterns of cortical recruitment thus appear entirely dependent on the lexical demands of the task.

This differential recruitment of temporal-visual versus prefrontal regions offers support for a theoretical speculation first proposed by the famous MIT structural linguist Roman Jakobson many decades ago. He proposed that one could analyze word associations underlying language

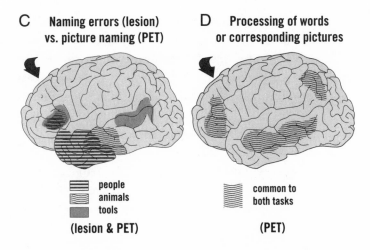

C   **Naming errors (lesion)
vs. picture naming (PET)**

D   **Processing of words
or corresponding pictures**

people
animals
tools
(lesion & PET)

common to
both tasks

(PET)

*recruited, depending on different lexical associative features of the words (person
names, animal names, tool names). Lesions were confined to the temporal lobes. PET
images additionally showed ventral frontal and cingulate cortex activation (summarized from Damasio, et al., 1996). Semantic processing appears to recruit different
ventral and lateral temporal lobe subdivisions depending on the semantic features
involved. (A similar pattern of anterior versus posterior temporal lobe recruitment
in processing names for inanimate versus animate objects was also demonstrated by
surface electrode recording in awake patients by Nobre, et al., 1994).*

    *D. Cortical areas that are activated during semantic judgments about either
words or pictures, identified using PET analysis of cerebral blood flow and subtracting the areas uniquely activated by each task alone (summarized from Vandenberghe, et al., 1996). All involve activation of ventral prefrontal cortex, as does
the verb-for-noun word association task depicted in Figure 10.3. Additional involvement of inferior temporal cortex demonstrates distributed recruitment of cortical areas in semantic processing.*

---

processes into two broad categories of associative operations that he termed
the *syntagmatic* and *paradigmatic* dimensions of word association,[9] and he
suggested that these processes should be distributed to frontal and posterior regions of cortex, respectively. Paradigmatic operations are reflected
in substitution relationships between words. Metaphors, anaphors, and pro-
nouns all serve this role. In the most general sense, all words of the same
part of speech are paradigmatic of each other to some degree since they
can substitute for one another. Words that serve the same function in a sentence tend not to co-occur in the same sentence or context, except in a sort
of renaming function. Syntagmatic operations are reflected in the complementary relationships between words from different parts of speech (e.g.,

nouns, verbs, adjectives, adverbs, or articles) and the way these different classes of words alternate in sequence in a sentence. Word association by metonymy, such as in the sequence "glass-water-thirsty-lunch," also employs a similar logic of complementarity. The content words that combine in sentences are almost invariably metonymically related to one another. In a sentence or in the developing context of a description or story, subsequent words operate on one another, expanding or contracting semantic relationships, introducing new topics, or simply producing a sentential function, such as indicating or requesting something. This metonymic function is not merely a shift of paradigm (in Jakobson's sense) but a shift that involves some ground for one word to operate on the other with respect to determining reference.

Producing a metaphoric association requires selecting words with common semantic features, whereas producing a metonymic association requires shifting attention to specifically alternative features. This is why there may be a posterior cortical bias to metaphoric operations and a prefrontal cortical bias to metonymic operations. Associations cued by attention to common perceptual features are analogous to perceptual recognition processes and so should recruit the function of the corresponding posterior cortical regions. However, associations that require a shift of attention in which some feature is used to generate its complement are likely to require prefrontal contributions. Metonymic word association thus offers a paradigm example of using information against itself to generate new complementary alternatives.

So, what parts of the brain should we expect grammatical and syntactical information processing to recruit? As with simple word association processes, the answer to this probably depends on processing demands, and not on some grammar-processing center. Syntactic operations and grammatical judgments can involve many different syntagmatic and paradigmatic processes, and these can differ from language to language. In languages like English and German, for example, word and phrase position in a sentence are used to determine many grammatical functions, such as relationships of possession or subordination, the difference between statements and questions, and certain changes in tense such as the passive tense. But in highly inflected languages, like Italian or Latin, affixes, suffixes or systematic changes in phonemes (as also in English verbs) tend to signal these functional roles. If grammatical operations were handled by some central processor we should not expect to observe neural variations correlated with language variations, but since grammatical relationships are symbolic relationships, they are probably no less distributed and task-dependent in their localization in the brain than are word retrieval processes. Even within bilingual individuals different languages may be organized differently and separately, sometimes in ways not even restricted to cortex. For example, bilingual patients with subcortical (striatal) damage are often paradoxically more impaired in their native language than in their secondary one.

The clearest demonstration of the variable relationship between language structures and brain structures has recently come from studies of acquired agrammatism: loss of grammatical analytic abilities due to focal brain damage in adults. Though it's long been unclear whether grammatical ability could be linked to specific brain damage, English speakers appear to be especially susceptible to disruption of grammatical abilities as a result of damage associated with Broca's area. Thus, patients who have significantly impaired speech fluency also tend to show difficulties interpreting sentences that depend critically on grammatical function words, and particular difficulties interpreting sentences that depend entirely on transformations of word order (like the passive tense in English). Such problems might suggest that this grammatical function is located in this part of the brain, and many have suggested just this. Curiously, however, a generalized grammatical deficit is not consistently associated with damage to Broca's area, and specifically, in speakers of highly inflected languages where word order is more free and where the passive tense is marked by grammatical words, morphemes, or inflections, there appears to be far less agrammatism associated with Broca's area damage. In these languages (such as Italian), Wernicke's aphasics, who also show disturbances of semantic analysis but not speech fluency, are more impaired in producing and analyzing the corresponding grammatical transformation than are patients with Broca's area damage (Figure 10.6). So if there is a grammar module, then the parts of this module map in very different ways to different grammatical operations, depending on the relative importance of positional or inflectional tricks for cuing grammatical decisions in different languages. This sort of module is a will-o'-the-wisp.

Probably the crucial factor behind this difference is the need to use very different sorts of neural computations and mnemonic tricks to analyze word order as opposed to individual words, suffixes, prefixes, and sound changes. Both coding methods offer viable means for marking the same grammatical distinctions, and are quite variably employed in different languages. Those languages that extensively utilize change in word form to mark grammatical functions tend correspondingly to allow considerable freedom of word order, and vice versa. English, for example, makes minimal use of inflection and extensive use of word order and special "function words" to

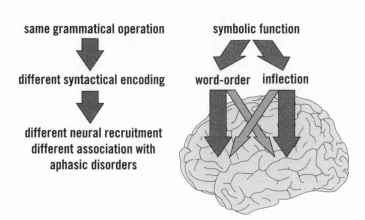

Figure 10.6 *Cross-linguistic comparison of aphasic patients is beginning to demonstrate that the same deep grammatical operation may recruit different cortical regions, depending on the way it is encoded syntactically. This is particularly evident when comparing languages that depend more on word-order cues to those that depend on morphological and inflectional cues to express the same grammatical function. The arrows pointing to the brain indicate that when word order is more critical, damage to ventral frontal regions produces greater grammatical impairment; when inflection and morphology are more critical, damage to temporal regions produces greater grammatical impairment (for details, see the discussions in Bates and Wulfeck, 1989; Bates and MacWhinney, 1991a).*

mark word-order interpretations. Italian and Latin, in contrast, rely almost exclusively on inflections and function words that change the grammatical function of content words they modify. If we took the view that a particular language function (e.g., the passive transformation) was a distinct primitive operation that was "computed" by brain regions specialized for it, then we would be forced to conclude that English and Italian speakers have different kinds of brains with different types of language regions that make this possible. This, of course, is absurd. What happens is that in the process of learning one of these languages, particular syntactic functions tend to demand most from areas of the brain that were previously specialized to perform similar manipulations of the signal (e.g., tracking order-dependency relationships). Over the course of maturation and in response to constant language use, the development of a certain degree of skill in this process is accompanied by progressive specialization of this region and progressive decline of contributions from other areas, as all become more specialized and differentiated in function.

Thus, a particular class of syntactical operation subserving a general grammatical function can come to "reside" somewhere in the brain, so to speak, and can be selectively lost due to focal damage, particularly in mature brains, and yet be located differently in the brains of speakers of different languages. Its localization is determined by the same competitive exclusion processes that hone all neural systems during development, in response to habitual computational demands, but these demands are imposed by the surface characteristics of the syntactic device employed in a given language and not by the underlying grammatical (symbolic) logic that it supports. As with a computer that is deaf and dumb to the content and function of the data entered into its memory, the physical characteristics of the data that the brain must handle in language—i.e., whether they are images, characters, or operations to perform on other bits of data—are a much better clue to where they will be processed than what they mean. The symbolic functions, the grammatical and representational relationships, are not processed in any one place in the brain, but arise as a collective result of processes distributed widely within the brain, as well as within the wider social community itself. Virtual reference is only virtually localized.

Taking Sides

It might seem odd that in a book on language and the brain, whole chapters haven't been devoted to lateralization—the difference in functional representation between the two cerebral hemispheres. This is in part merely a stylistic choice, a matter of emphasis on some otherwise less well known features of the brain-language relationship. But it is also a reflection of the fact that I consider this to be a side issue (bad pun) that is not an essential feature of language processing, only an incidental feature of the way human brains have adapted to the computation problems of language use. The fact that many specialized language functions are strongly lateralized to the left hemisphere in the vast majority of human brains has been a major impetus for theories suggesting that lateralization might have been a precondition for language evolution. Many argue that this robust side-to-side difference reflects some major organizational logic underlying language. But as with examples of the apparent preconditions for modern language abilities that were considered above, the logic of this argument is likely opposite to the evolutionary logic behind the correlation. Lateralization is almost certainly an effect and not a cause of brain-language co-evolution. Indeed, I think it is largely an effect of language development in an individual's lifetime. The structure of languages has probably evolved to take advantage of intrinsic subtle biases in developing brains to break up and distribute their component cognitive computations so that they can most easily be processed in parallel, and one important way this can be accomplished is by "assigning" functions to either side of the cerebral hemispheres.

Unfortunately, the study of lateralization has been afflicted with the problem of being an interesting topic for popular psychology, and of offering an attractive source of analogies for theorizing about almost every aspect of the mind. As a result, everyone's favorite complementary pair of mental functions can be mapped onto a brain whose functions differ on opposite sides. Since the middle of the nineteenth century, physicians and psychologists have argued over whether the left was female and the right male, the left verbal and the right nonverbal, the left linguistic and the right spatial, the left rational and the right irrational, the left differentiated and the right undifferentiated, the left localized and the right holistic, the left positive emotion and the right negative emotion, the left ego and the right id, the left dominant and the right subordinate, and even the left human cognition and the right primate cognition, to describe a few of the more prominent dichotomies. The attraction of discovering the most elegant way of dividing up the mind into two major complementary cognitive systems is almost irresistible.[10]

Although I wouldn't dispute the importance of lateralized function as an important clue about the cognitive demands of language, I want to emphasize that lateralization is probably a consequence and not a cause or even precondition for language evolution. In fact, it is a dynamic functional consequence of processing demands imposed by language performance during development. There is plenty of evidence to suggest that lateralization of similar functions occurs in some other species, including other primates. Because of this, it is likely that such phylogenetic biases were present in hominid brains prior to the evolution of language. But I want to make the point that although these preexisting biases might help us to understand why language could have become more commonly and more thoroughly dependent on the left and not the right hemisphere, these prior conditions do not *explain* lateralization. Lateralization is more an adaptation of the brain *to* language than an adaptation of the brain *for* language.

The starting point for thinking about lateralization is recognizing that although human brains are strongly biased for left-side representation of many language functions, this is neither universal nor irreversible, at least early in life. Somewhere under 10 percent of people are not left-lateralized in this way, some are reversed, others have ambiguous lateralization, and handedness and lateralization for language do not always coincide. Anatomical and maturational biases may stack the deck one way at birth and still not be sufficient to throw the game, so to speak. Cases of left hemispherectomy in early childhood with largely spared language later in life have also been cited as a demonstration of the right hemisphere potential. Though claims that these children had complete language recovery were probably an exaggeration, it must be remembered that single-hemisphere language learning lacks any option of separating specialized functions or relying on redundantly represented functions. The fact that hemispherectomized children can learn any aspect of language is a miracle, and as an aside, further demonstrates the limitations of brain size alone as a sufficient predictor of language and intellectual ability. Even a 700-gram brain (as might result from hemispherectomy), if organized appropriately, can acquire sophisticated language skills and pass as humanly intelligent.

Further evidence of the developmentally derived nature of language lateralization comes from a study of language development in children with focal brain damage. Bates, Thal, and colleagues (1994) report that right hemisphere damage as well as left can produce significant language development delays (discussed in Chapter 9). This developmental plasticity indicates that lateralization must be thought of as a dynamic process, driven by language and development of manual skills during childhood. Innate side biases are just that—biases in a competitive process that may involve both synaptic competition and competition for establishing memory traces of experiences and behaviors. The representation of language functions probably develops primarily in response to the need to perform simultaneous but competing operations when speaking or listening to speech. This is supported by the fact that lateralization is not so much a commitment of one side to language and the other not, but rather a segregation of component language functions, as well as many other competing functions, to the two sides. Lateralization is thus another reflection of the role of competitive processes in determining the ultimate representation of functions in the cerebral hemispheres during development (see the discussion in Chapter 6), a progressive streamlining of language operations, as competitive differentiation processes cause competing complementary functions to recruit corresponding opposite cortical areas on one side or the other. What is important for understanding the nature of language processing is not how the particular human bias came about, but rather what drives it in each individual's development.

Before we can begin to make sense of this functional competition, it is important to get one thing straight. The right hemisphere is not the non-language hemisphere. It is critically and intimately involved in language processing at many levels during both development and maturity. Perhaps most importantly, it is critical for the large-scale, semantic processing of language, not word meaning so much as the larger symbolic constructions that words and sentences contribute to: complex ideas, descriptions, narratives, and arguments. Symbol construction and analysis do not end with the end of a sentence, but in many regards begin there. The real power of symbolic communication lies in its creative and constructive power. Since symbolic representation is intrinsically compositional, there is no upper bound to the compositional complexity of a symbolic representation. A whole novel can be used to convey a sense of a unique life experience, a research paper full of equations can be used to present a subtle but elegant idea about physical relationships, a joke can offer a new and curious way to represent an old assumption, and so on. All of these are symbols in the same sense that a word or sentence is, but they require an extended effort at symbolic reference building.

The best evidence for this right hemisphere language involvement comes from analysis of how right hemisphere damage affects such abilities as story and joke comprehension.[11] Patients who have suffered extensive damage to their right but not left hemispheres are generally able to speak well, without any unusual increase in grammatical errors or mistakes in choice of words; but when required to follow and interpret a short narrative, they seem to fail to grasp the logic of the whole. For example, they do not recognize when important steps in a story have been left out or inappropriate or anomalous events have been included, though they can recount the details. They seem to be unaware of the constraints of the context. Jokes provide another window into this difficulty. Humor depends crucially on understanding both what should ordinarily follow, and how the insidious twist of logic of the punch line undermines our expectations. Assessment of what makes something funny depends on an awareness of two conflicting contexts: an expected, "appropriate" context, and a logically possible but very odd one. The aptness of the shift in contextual logic, the extent to which it effectively catches us off guard even when we know it is a joke, the way it caricatures what in a "straight" context might be serious or threatening, all these are the ingredients of good jokes. (How many scientists does it take to screw in a light bulb? Only one. It just takes him two years to get the grant support to do it.) Well, anyway, this poses a serious problem for someone unable to construct the appropriate narrative context in the first place. Patients with right hemisphere damage seem to rank jokes as funny based solely on the extent to which the punch line contains material that is different from what preceded it. (No, I do not have right hemisphere damage.)

Such inattention is also reflected in other aspects of their behavior. One of the more enigmatic symptoms of right hemisphere damage is a tendency to neglect or ignore things on the left side, including objects so placed, dressing or shaving on that side, even the left sides of objects. Asked to draw a clock, a right hemisphere patient might scrunch all the hands and numbers to the right side, or simply leave something out on the left side of the drawing. Another aspect of these patients' pattern of inattention is a disturbance of implicit learning. For example, when asked to keep track of how often an animal appears in a series of photographs of familiar objects, they do quite well. But when asked if there had been many pieces of furniture shown in that same same series of pictures, they will perform poorly. What was not being attended to did not register in background.

These are attentional functions that are crucial to symbol building, because the basis of shifting attention from the details to the implicit logic that organizes them requires an ability to be aware of more than one network of contextual relationships at a time. I do not mean to suggest that the right hemisphere plays some special role in this process during development or in evolution. But as language abilities become progressively more sophisticated with age and experience, the need to analyze symbolic relationships at many levels simultaneously grows. The highly automated interpretation of symbolic relationships encoded in word combinations and sentence structure requires a strategy of one rapid interpretation followed by another. It demands both rapid implementation and an ability to keep previous operations from interfering with subsequent operations. The same neural systems that subserve sentence-length analysis would probably also be critical for maintaining long-term mnemonic continuity of symbolic information. These simultaneous demands would thus likely conflict or interfere with one another, and so limit the efficiency of both processes. But because right and left brain structures are paired, it is possible to keep the processes from interfering with one another by compartmentalizing them to opposite hemispheres.

The right hemisphere also subserves another important language function that is nonsymbolic, but in terms of neural computations probably is competitive with phonological analysis and word processing. It is the processing of prosodic features of speech. Prosodic features are the rhythmic and pitch changes that we generally use to convey emotional tone, to direct the listener's attention to the more and less significant elements in a sentence, and in general to indicate how aroused we are about the contents of our speech. Not surprisingly, many aspects of this speech melody have been shown to have features in common with the innate vocalizations of primates.

These include the correlation of changing pitch, volume, and rate of production with the level of arousal; changes in the quality of vocalization as an indication of type of interaction (hostile, submissive, etc.); and the overall phrasing with respect to breath control. Though the extent of localization has been debated, there is a common association between both impairment in interpreting emotionality in speech and the production of prosodic speech in patients with right hemisphere damage.

Here again, language production and analysis effectively require that we implement two different modes of phonetic analysis and vocal control simultaneously: prosodic and phonemic processes. These tasks would tend to compete for recruitment of the same brain structures (probably the classic Broca's and Wernicke's areas), and as a result would probably interfere with each other. It would be far more inefficient to trade off use of the same cortical system for both. Like the monitoring of thematic context in background, the monitoring of prosodic information tends to operate against a foreground attention to specific words and phrases. Though we are aware of this peripherally, it tends to be attended to implicitly rather than explicitly. Exaggerating the representation of this background function to the right hemisphere, and phonemic and word analysis to the left during development, may similarly provide a means for processing these sources of information in parallel with minimal cross interference. Consequently, the right hemisphere may become more intimately associated with the midbrain homologues of innate call circuits that still exist in the human brain. Conversely, if there is a consistent developmental bias favoring such a preferential linkage, this could tip the scales toward this pattern of lateralization.

But the right hemisphere may be far more capable of full-scale linguistic functions than we normally imagine. Its abilities may be masked by this long developmental specialization, which actively reduces its roles in word- and phrase-level analysis and production processes in order to avoid cognitive conflicts. Could we find a way to observe right hemisphere abilities in conditions where there was no developmental selection to decrease its role in these processes? Data from schools that train simultaneous translators (such as are hired by the United Nations to listen to a speaker and simultaneously provide a moment-by-moment translation into a microphone for others to follow) suggest that under the special demands of this difficult language task, both hemispheres can to some extent become language hemispheres. The problem for the simultaneous translator is to keep the two languages from getting in each other's way. Listening to one while producing the other is like that old problem of patting your head and rubbing your stomach with opposite hands, and then reversing what each hand is doing but leaving them in place; or chewing gum while playing the drums or dancing or just walking out of sync with each chew. The direct competition of simultaneous similar language functions is often further coupled with a consistent asymmetry of auditory input: most translators develop an ear preference for listening to the source language. Studies before and after training demonstrate that most students begin with a right ear (left hemisphere) preference for both languages, but may develop an opposite ear advantage for each language by the end of their training.[12]

Thus, the two languages can come to be preferentially represented in opposite hemispheres. This is all the more remarkable since the shift from unilateral preference to bilateral segregation can be induced in young adults, not infants (there is undoubtedly self-selection in this population of students, so that such flexibility and language facility be the rule). This special case nonetheless demonstrate the general principle: when sensorimotor or cognitive operations tend to compete simultaneously for the same neural substrates, there is strong developmental selection pressure to segregate the competing operations to counterpart structures in the opposite hemispheres (Figure 10.7). Of course, any predispositions that contribute a bias that accelerates this process during development will be favored in evolution.

In general, then, it is misleading to think of language as though it is all in the left hemisphere. The right side is neither primitive nor mute. Both hemispheres contribute essential and complementary functions. These develop in tandem, and the biases for a particular pattern of asymmetry evolved with respect to this complementarity of functions. Lateralization is not so much an expression of evolutionary adaptation as of adaptation during one's lifetime, biased so as to minimize any neurological "indecisions" about what should go where.

The logic of the bias that leads most humans to develop the same pattern of lateralization may also reflect systematic temporal differences. Lateralization functions in both the linguistic and the manual domain seem to be segregated according to speed. The left hemisphere seems more often adept at ultra-rapid analysis of sound changes and control of rapid, precise, skilled movement sequences. Similar patterns seem to characterize visual analysis and manual skills as well. Thus, in the developmental competition for functional representation, a left bias for more rapid rate of processing might be sufficient consistently to tip the scales in one direction. Subtle maturational differences, myelination differences, cell-size differences, or differences in the proportions of cortical inputs from sensory systems on the two sides could provide such initial biases. In human brains, these subtle

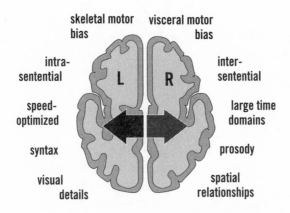

skeletal motor
bias

visceral motor
bias

intra-
sentential

inter-
sentential

L   R

speed-
optimized

large time
domains

syntax

prosody

visual
details

spatial
relationships

Figure 10.7 *Examples of segregation of competing language-related functions that have displaced each other into the cortex of opposite cerebral hemispheres.*

developmental biases may be amplified further by the cumulative statistical effects of large brain size (e.g., longer maturation process; longer distances producing greater conduction time differences; larger networks tending to break up more easily into functional subdivisions; and so on).

Similar biases are seen for other functions in species as diverse as birds, rodents, and primates. There are a number of crudely analogous patterns in other primates. Though there is no consistent species-handedness for fine-skilled manipulation, there does seem to be some consistent preference for left-side supportive functions. More relevant, however, for language evolution, there appears to be evidence for a left hemisphere advantage for processing rapid auditory stimuli, and conspecific vocalizations in some monkeys (also probably biased by rate of analysis effects). The advantages of increasing the innate bias toward one pattern probably can be explained by a kind of disruptive selection against unbiased brains. Lack of innate bias might produce a kind of physiological indecision in the developmental competition driving functional segregation. The hierarchic pattern of biases evolved to bias development is a consequence of the evolutionlike developmental wiring process. Like the disruptive evolution of different vervet monkey alarm calls, the intermediate unbiased state would be selected against. It didn't really matter which side "won" which function, just so long as the result was a decisive difference, and having both alternative outcomes as likely possibilities would increase the chances for stalemate. The prehominid biases, however slight, could have been enough to precipitate the evolution of further innate biasing in the same direction. But there may be an additional source of bias to consider.

A number of theorists suggest an evolutionary link between handedness and language lateralization. They argue that the left brain adaptation for more "dexterous" tool use might have paved the way for language evolution. The correlation between the asymmetry of language and of handedness no doubt indicates that the independent advantages of each may have influenced the other in their evolution. This is a reflection of "evolutionary overdetermination": the apparent convergence of many independent adaptive advantages that contribute to the same structural change. Because evolution is moved by patterns of biases, related biases from independent sources tend to reinforce each other over time. To the extent that either left hemisphere specialization for more precise manual skill or for more articulate verbal skill would have selected for similar biases in the neurological substrates, each would have increased the likelihood that the other function would co-localize in the same hemisphere. Lateralization for one or the other may have initially been slight, but their simultaneous co-evolution would have greatly amplified the resultant effect and linkage. One of these abilities need not be the evolutionary prerequisite of the other.

How old is lateralization for language? Lateralized biases in many functions appear to long antedate hominids. For example, corresponding lateralized biases in spatial and sensory processing have been identified in many mammals and birds. Though human lateralization can probably be traced to these roots, the consistency and almost completely dissociated extent of lateralization of language functions that is characteristic of most human brains has probably developed gradually over the entire period of brain-language co-evolution. Since the evolutionary course of the increase in vocal capabilities in hominids was probably protracted, beginning simply and taking more than 1 million years to become sophisticated, lateralized biases with respect to auditory processing have probably only recently reached their modern extremes. In contrast, as we will see in the next chapters, stone tool use was probably around since the very earliest stages of symbolic communication. If this was aided by lateralized biases for manual skills, then they may have been evolving in advance of vocal-auditory lateralization. But the evidence for any strong selective *disadvantage* of ambidextrous abilities seems weak. Indeed, athletic and manual skills seem at least as well developed in ambidextrous and left-handed individuals as in strongly right-handed individuals. But in language processing, an "ambiphasic" brain would have become increasingly disadvantageous as speech information transmission rates increased during evolution. Evidence for handedness in stone tool manufacture may thus tell us more about the lateralization advantages for language processing in these ancient toolmakers than anything about selection favoring handedness in their toolmaking. Even though lateralized tool manipulation could have offered an additional source of bias affecting language lateralization, it seems far more likely that the high incidence of right-handedness and left speech-processing biases in modern populations has been driven nearly to genetic fixation mostly by language.

PART THREE

Co-Evolution

And the Word Became Flesh

The human race took centuries or millennia to see through the mist of difficulties and paradoxes which instructors now invite us to solve in a few minutes.

—Lancelot Hogben

The Brain That Didn't Evolve

The phrase from the Bible (John 1:14) that I have quoted as the title of this chapter reflects an ancient mystical notion that certain words can invoke magical powers, and have a direct power to create or destroy. To know the "true name" of a thing was thought to be a source of power over it in many traditions. I want to invoke a similar sense of the autonomous power of words over things. I have borrowed this enigmatic biblical phrase out of context in order to describe an evolutionary process, not a divine miracle, but the process I describe is no less miraculous because it is explainable by science. The evolutionary miracle is the human brain. And what makes this extraordinary is not just that a flesh and blood computer is capable of producing a phenomenon as remarkable as a human mind, but that the changes in this organ responsible for this miracle were a direct conse-quence of the use of words. And I don't mean this in a figurative sense. I mean that the major structural and functional innovations that make human brains capable of unprecedented mental feats evolved in response to the use of something as abstract and virtual as the power of words. Or, to put this miracle in simple terms, I suggest that an idea changed the brain.

Now this may seem a rather mystical notion on its own, inverting our common sense notion of causality that physical changes require physical causes, but I assure you that it is not. I do not suggest that a disembodied thought acted to change the physical structure of our brains, as might a god in a mythical story, but I do suggest that the first use of symbolic reference by some distant ancestors changed how natural selection processes have af-fected hominid brain evolution ever since. So in a very real sense I mean that the physical changes that make us human are the incarnations, so to speak, of the process of using words. It is the purpose of the remainder of the book to make clear what I mean by this enigmatic pronouncement and to follow its implications.

A subtle modification of the Darwinian theory of natural selection, first outlined almost exactly a century ago by the American psychologist James Mark Baldwin,[1] is the key to understanding the process that could have pro-duced these changes. This variation on Darwinism is now often called "Baldwinian evolution," though there is nothing non-Darwinian about the process. Baldwin suggested that learning and behavioral flexibility can play a role in amplifying and biasing natural selection because these abilities en-able individuals to modify the context of natural selection that affects their future kin. Behavioral flexibility enables organisms to move into niches that differ from those their ancestors occupied, with the consequence that suc-ceeding generations will face a new set of selection pressures. For exam-ple, an ability to utilize resources from colder environments may initially be facilitated by seasonal migratory patterns, but if adaptation to this new niche becomes increasingly important, it will favor the preservation of any traits in subsequent generations that increase tolerance to cold, such as the deposition of subcutaneous fat, the growth of insulating hair, or the ability to hibernate during part of the year. In summary, Baldwin's theory explains how behaviors can affect evolution, but without the necessity of claiming that responses to environmental demands acquired during one's lifetime could be passed directly on to one's offspring (a discredited mechanism for evolutionary change proposed by the early nineteenth-century French nat-uralist Jean Baptiste Lamarck). Baldwin proposed that by temporarily ad-justing behaviors or physiological responses during its lifespan in response

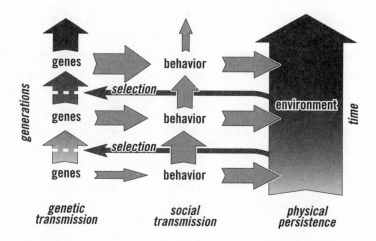

Figure 11.1 *Schematic diagram of processes underlying Baldwinian selection. Arrows pointing vertically depict three simultaneous transmission processes: genetic inheritance (left), social transmission by learning (middle), and persistence of physical changes in the environment produced by behavioral changes (right). Arrows pointing right indicate influences of genes on behavior and behaviors on the environment. Arrows pointing to the left indicate the effects of changed selection pressures on genes. The arrows for social transmission get thinner in each generation to indicate the reduced role of learning as a result of an increasing genetic influence to the behavior (indicated by arrows getting thicker from genes to behavior).*

to novel conditions, an animal could produce irreversible changes in the adaptive context of future generations. Though no new genetic change is immediately produced in the process, the change in conditions will alter which among the existing or subsequently modified genetic predispositions will be favored in the future (Figure 11.1).

Some of the best understood examples of Baldwinian evolution come from cases where human behaviors have changed natural selection in unexpected ways. The evolution of tolerance for the milk sugar, lactose, has long been seen as a case in point. Though the ability to utilize lactose is present in most mammals during infancy, the enzymes necessary to break it down are eventually down-regulated after weaning, when they become unnecessary. Most adult mammals are therefore unable to digest lactose and may develop digestive distress when it is broken down inappropriately by bacteria in the colon. This was probably the situation for most humans before the discovery of animal husbandry. It is no coincidence that the human populations with the highest percentage of lactose-tolerant adults are those where animals have been herded for the longest time, and those with the least lactose tolerance are those where herding was introduced most recently or not at all. Though many independent factors have also played a role in the evolution of this digestive ability,[2] it seems clear that the use of animal milk as a food source, despite digestive difficulties for some, favored the reproduction of those most tolerant to it.

The story is further complicated by the various adjustments that people around the world have made to lactose intolerance. For example, letting the milk sour as bacteria break down the milk sugar, or making it into cheese where molds can modify it and not incidentally aid in its preservation, may have eased the consequences of intolerance and served as a bridge allowing early herding populations to develop a greater dependence on milk. This allowed herding societies to move into more northerly and more mountainous climates inappropriate for agriculture, where the reduced sunlight exposure accidentally provided an additional, more ubiquitous selection for even a slight degree of tolerance to lactose: the advantages of providing vitamin D from milk to substitute for what is not produced endogenously due to reduced sunlight. Over time, large fractions of whole populations in Europe have become lactose-tolerant.

In the 1950s and 1960s a pioneering British evolutionary biologist at Edinburgh University, Conrad Waddington, described a related process he called "genetic assimilation." What he argued (and tried to demonstrate with breeding experiments using fruit flies) was that natural selection will tend to replace flexible adaptive responses to continuing environmental constraints with genetic predispositions. In his terms, the production of these adaptive responses will become progressively more "canalized." This means that after many generations their production during development will become ever more strongly biased by genetic inheritance and less dependent on external signals. Developmentally "anticipating" an important response by making its expression obligatory rather than contingent on external conditions provides a more efficient and less risky response than relying on an appropriate response being triggered by external signals. This account provides a more genetically and developmentally sophisticated description of an important class of Baldwinian processes. But Waddington was critical of overly deterministic models of genetic production of traits. "Canalization" refers to the more or less biased or constrained influences of genes on development of body structures and behavior. Genetic assimilation was not meant to imply the one-to-one replacement of learned adaptations with instinctual counterparts, but rather to describe an evolutionary trend toward increasing developmental constraint. In many ways this is antithetical to the idea that there could be any simple mapping of a once facultative response to genes that produce it like an internalized set of instructions, but it is consistent with an evolutionlike process of development.

The indirectness of genetic assimilation is borne out by the fact that there are seldom obvious links between behaviors induced by environmental changes and their long-term evolutionary consequences. A typical example is offered by the evolution of a variety of genetic blood disorders in response to malaria. The best known of these is sickle cell anemia. This deadly disease results from inheriting two copies of a mutated gene for one of the components of the oxygen-carrying molecule, hemoglobin. When the mutated hemoglobin gives up oxygen to the tissues of the body, it tends to precipitate into microcrystalline structures that can rupture the red blood cells that contain it. The damaged appearance of these cells gives the disorder its name. The key to its evolution is its characteristic distribution in different populations, specifically, its high incidence in populations indigenous to central Africa. A few decades ago, scientists discovered that people who had only one copy of the mutated gene thereby gained some resistance to malaria. This is because malaria is caused by a blood-borne parasite that reproduces within red blood cells during part of its life cycle within the host body. Having some sickle cell hemoglobin in one's blood cells (though not enough to produce rampant sickling of blood cells) interferes with the reproduction of the parasite. Sickle cell genes were thus selectively favored in a small proportion with respect to normal hemoglobin genes in the population because they conferred reproductive (survival) benefits for malaria. The cost of some individuals inheriting two such genes and dying from anemia was balanced by this benefit.

The sickle cell trait spread quite rapidly in Africa in recent prehistory because of human activity. The other host of the malaria parasite, by which it is passed from human host to human host, is the mosquito. Probably the critical historical event that catapulted malaria to an epidemic disease was the introduction of agriculture and animal husbandry into Africa, between five and ten thousand years ago. This culturally transmitted practice modified the tropical environment to create the perfect breeding ground for mosquitos. Slash-and-burn agriculture and animal husbandry replaced dense rain forests with land that tended to have lots of ponds and puddles of standing water, not to mention lots of human hosts, and mosquitos need both to breed. The introduction of agriculture thus changed natural selection affecting all three species, and the human population was thrust into a context in which powerful selection favored reproduction of any mutation that conferred some resistance to malaria. Human ideas, in the form of agricultural practices, initiated an indirect cycle of Baldwinian evolution that ended with a mutated blood protein that is potentially deadly, but it did not produce mosquito-resistant skin or an abhorrence to standing water. When such a complex system is modified, there is no telling what will happen. Evolution seldom follows straight lines. In fact, in the case of the adaptation to malaria (and to tropical agriculture), there appear to have been a number of simultaneous parallel molecular adaptations of blood proteins to inhibit the disease, all with their own unique costs.[3]

These two examples demonstrate how it is possible for human behaviors, in the form of agricultural practices, indirectly to affect the evolution of certain human genes and the physiological traits they contribute. But these examples also demonstrate that the route from behavior to evolutionary consequence may not be as simple as the analogy to Lamarckian evolution originally suggested, where what is accomplished by effort in one generation is accomplished by genetics in a later generation. Unlike the analogue to Lamarckian inheritance of traits acquired by habit, where adaptive responses are presumed to be mapped one-to-one from outside to inside the genome, *any* traits that ease the costs imposed by behavioral adaptation to new conditions will be selectively favored by Baldwinian selection and thereby subject to genetic assimilation. There is an inherent difficulty in tracing causes and consequences in this sort of process because it involves a one-to-many mapping. Inevitably, a given adaptive response can be biased in its probability of appearance during development in a very large number of diverse ways, and all may be simultaneously modified by Baldwinian selection.

Of all the forms of adaptation, the flexibility to learn new behavioral responses during one's lifetime can produce the most rapid and radical evolutionary consequences. Indeed, the ability to learn and thus inherit acquired behaviors may be one of the most powerful sources of evolutionary change. It provides an organism with access to a repertoire of *potential* adaptations, and so amplifies and extends the range of the behavioral predispositions that can be "sampled" by natural selection. A learned behavioral response can be genetically assimilated to become a behavioral predisposition by virtue of the costs that it imposes on the organism. There are costs in terms of learning time, costs for failing to learn or learning incorrectly, and costs for simply being inefficient. Individuals who, for any number of reasons, learn more quickly and reliably and implement this behavior more efficiently will benefit in reproductive terms. One of the features of this process is that *any* predisposition that even remotely contributes to producing a more reliable and efficient response will be positively selected. So again, unlike the Lamarckian caricature, what will tend to evolve is a constellation of many indirectly related contributory influences and biases, and not an innate replica of the prior facultative behavioral response.

Human brains have been changed by just such an evolutionary process. The question is just how specific and extensive these changes have been. Could it extend to such innate knowledge of grammar as innatists like Noam Chomsky envision? Though Chomsky himself has not suggested that grammatical knowledge could have been selected for during human evolution, others have. As discussed in chapters 1 and 4, the idea that an innate universal grammar could have evolved by Darwinian (and Baldwinian) processes has found its clearest explication in Steven Pinker's recent book *The Language Instinct*.[4] In general, what Pinker means by the somewhat archaic biological term *instinct* is some behavioral predisposition that is preformed and ready to operate prior to any experience. This is essentially equivalent to Noam Chomsky's notion of an innate competence or skill. Examples of such innate skills appear easy to recognize in other species because of the way they inevitably unfold as an animal develops, irrespective of experience or learning. For example, I am always amazed by the fact that young kittens can be so easily induced to use a box full of sand or Kitty Litter as a toilet, and inevitably scratch the dirt in the same stereotypic way. Even more surprising are nest-building behaviors in birds. An experienced bird watcher can tell what sort of bird built a particular nest because of the characteristic choice of materials and the specific way they are woven together. And there are instinctual skills that seem to extend across vast numbers of species, such as scratching and grooming behaviors. All these are what we commonly recognize as instinct. Clearly, even though many of these examples may involve some period of experience and shaping of the behavior, such predisposed skills have mostly been shaped by evolutionary processes. Wouldn't a biologist from another planet, observing the universality of human language development and the characteristic patterns by which the parts of a language are woven together to form a whole, consider it much as the bird biologist considers the structure of a bird nest: as an instinct? Couldn't a large part of language skill and knowledge have evolved to be built into the human brain from the start?

In Chapter 4 we found that it was possible to explain many of the remarkable feats of language acquisition in children by taking into account the fact that languages themselves have been shaped by a sort of cultural equivalent to natural selection, in which children's special learning predispositions have shaped language to fit; but clearly brains have been shaped to fit the demands of language as well. The middle part of the book systematically outlined some of the most robust of these changes in the brain. In this chapter we will explore the evolutionary processes that were responsible for such massive changes, and the first step is to ask just what features of brain and behavior are and are not susceptible to such a process. If evolution is capable of producing such massive changes in human brain structure, couldn't it also easily produce more subtle changes embodying the basic logic of a universal grammar?

Both Pinker and I argue that a very simple protolanguage could have evolved in an early hominid ancestor in the absence of any specific language adaptations of the brain, and that the adaptive advantages of language communication would have subsequently provided selection for progressively internalizing certain crucial features of language structure in order to make it more efficient and more easily acquired. Where we differ is in the description of what has and has not been internalized in this way, and more specifically in what theoretically could have been internalized. In formal terms, innate grammatical knowledge would aid language acquisition and could explain how an apparently unlearnable set of rules could nonetheless be mastered by children. It could also account for the apparently universal features of languages and the idiosyncrasies of language structure. The question is whether this is a biologically plausible explanation. Are there constraints on brain evolution that limit what knowledge can and cannot become internalized during evolution? I will suggest that although our brains and sensorimotor abilities exhibit many adaptations for language that together might be called an instinct, grammatical knowledge cannot be one of them.

To what extent could Baldwinian processes provide an explanation for the neurological basis for innate language abilities? This depends on the specific demands created by language use on human learning and behavior. Some behavioral adaptations may be able to contribute to Baldwinian evolution and others may not. The key determining factors are *the intensity of selection, the stability of the conditions being adapted to, and the invariant features of the adaptive response.* Whether Darwinian or Baldwinian, the evolution of genetically based adaptation is a function of the long-term invariance of conditions that affect reproductive success or failure with respect to some trait or traits. Depending on population size and the intensity of selection, the specific adaptive demands imposed by the environment must remain unchanged over hundreds or even thousands of generations in order to produce the levels of gene replacement necessary to cause a new trait to become a regular feature of a species. The question is whether such constant and consistent selection pressures are associated with language evolution.

For a language feature to have such an impact on brain evolution that all members of the species come to share it, it must remain invariable across *even the most drastic language change possible.* Though this might at first seem an overly restrictive constraint, it is an inevitable consequence of the very great difference in evolutionary rate between genetic evolution affecting brain function and the speed of language change. Most researchers would agree that language change is likely to be many orders of magnitude more rapid than genetic change. The relative slowness of evolutionary genetic change compared to language change guarantees that only the most invariant and general features of language will persist long enough to contribute any significant consistent effect on long-term brain evolution. The evidence from Indo-European language histories suggests that massive language changes can arise in just thousands of years, so even in the most minimal estimates of the age of modern languages, a vast range of tricks for mapping grammar to strings of words would have been sampled within every language lineage. Since the first languages emerged, innumerable variants of grammar and syntax could have arisen many times and faded again. Even in the extremely unlikely circumstance that the entire population of language-using hominids was at some point relatively small and localized, and happened to be using a single language (as might be imagined in an extreme species "bottleneck" where all but one local population went extinct), the time from the appearance of a beneficial mutation to its fixation within the population would still likely extend across a vast sampling of possible language structures.

For these reasons, there is little possibility for mental adaptations to specific syntactic structures. But there are many features of languages, from the presence of words and sentence units to the noun-part/verb-part distinction and many more subtle and idiosyncratic features, that are common to essentially every natural language. Universal features of language do not change even though their surface implementation does. So they could have persisted in all languages for hundreds of thousands of years. But unchanging persistence is only one facet of invariance. For genetic assimilation to take place, this persistent aspect of language must also impose *consistent invariant demands on neural processes*, and this is a more difficult criterion for language structures to satisfy. The problem with language, from an evolutionary perspective, is that what is relatively invariant across languages, what is often called the "deep structure" of grammar, following

Chomsky, only weakly constrains the highly variable surface structures that implement it. Only if these deep grammatical invariances correlate with invariant neural computational processes can there be any degree of genetic assimilation. Despite the fact that a noun is a noun and a change in tense is a change in tense irrespective of the words involved and the syntactical manipulations that encode these functions, there can be no assimilation of its function by evolutionary processes unless these functional distinctions are always processed in the same way *in all brains* under all conditions. The key requirement for genetic assimilation is the existence of some invariant sensorimotor features or invariant mnemonic features of the adaptation.

The impact of invariance of neurological representation can be illustrated by considering the evolution of vervet monkey alarm calls. Four stimulus-response parameters remain unchanged for each type of alarm call. The first invariant features are the motor programs and sensory templates of the vocalizations themselves. Both could be encoded in brains as specific neural network architectures. Not only are these essential structural features likely to be highly localized in the networks of auditory and respiratory-vocal muscle circuits, they are likely to consist in similar circuit details. Second are the invariant sensory characteristics of the stimuli. Though the classes of vervet predators are distinguished by a number of sensory attributes, including contextual cues, the visual appearances of individual eagles and leopards can be quite variable, and so only certain general sensory attributes will consistently distinguish the one predator from the other. Consequently, only those reliable sensory distinctions that are consistently associated with distinct neural circuits will be subject to any degree of genetic assimilation. Such general attributes as overall silhouette, size, movement patterns, and whether silently framed against the sky or associated with the rustle and movement of bushes all may form part of the sensory templates that distinguish these predators. In addition, there are a number of more general "egocentric" features that distinguish between these classes of predators. For example, eagles attack from the air, and invariably approach from above their prey. Leopards attack from the ground, and are usually not above their prey. So, direction of sensory orienting to prey might itself be a major component of an invariant discriminative feature (it would be informative to know what kind of call would be given if a vervet monkey on the ground was startled by a leopard in a tree). Automatic orienting responses tend to be controlled by circuits localized to specialized regions of the dorsal midbrain.

The third invariant feature is the emotional arousal state elicited by both calls and real predators. The startle and fright responses elicited by the sight of a predator are, of course, ancient and consistently localized features in the nervous systems of most species of vertebrates. So this component of the response would also be constant from one individual to the next, though not necessarily distinguishing one type of predator from the other. The fourth, and final factor is the invariant nature of any successful escape response. The direction of escape is invariably opposite to the predator's direction of approach (down in the case of eagles, and up in the case of leopards). Predators' distinctive attack modes also determine certain predictable locomotor requirements associated with the geometric features of trees, and so favor certain different predisposed vervet behaviors (e.g., a tendency to leap upward with arms extended versus crouch down and cling). Similar nervous systems will encode all these behavior patterns in similar ways.

In summary, a remarkable array of features implicit in the alarm-calling context can be correlated with consistent differences in neural architecture, either learned or inherited. The opposed and mutually exclusive nature of the best escape strategies, combined with the serious costs of failure to escape, can produce consistent and powerful selection for calls that can be unambiguously distinguished, and selection for consistent linkage of these calls with a set of highly specific sensory, motor, and attentional predispositions. The evolution of these alarm call systems created a kind of innate "foreknowledge" of useful stimulus-response associations appropriate to the environment. These are the sorts of built-in predispositions that we would feel comfortable calling "instincts," and they have many of the features that might lend themselves to a genetic assimilation interpretation that parallels Lamarckianism.

Genetic assimilation here is analogous to associative learning in a number of ways. Associative learning also depends on reinforcing responses that anticipate certain invariant links between events in the surrounding environment. But the differences are critical. The conditional relationships between stimulus parameters during genetic assimilation need to remain consistent across hundreds of generations, and different individuals must internalize these associations in identical ways. For this reason, the sorts of relationships that evolution can translate into neural architecture, and the level of detail to which they can be innately prespecified, are severely limited. This parallel between learning and evolution is particularly relevant for language, and the restrictions even more critical. The problem is this: the discontinuity between stimulus associations and symbolic reference associations, which is the basis of their function and which makes symbolic associations so difficult to learn, also makes them *impossible to assimilate genetically.*

The evolution of specific indexical referential signals, such as vervet monkey alarm calls, was made possible by features unavailable to the evolution of language. Languages exhibit minimal correlation between words and their references, and grammatical operations have even less of a correspondence to things in the world, imposing a very nonphysical logic (i.e., subjects/predicates, articles, adverbs, etc.). Thus, there has never been much of an opportunity for the evolution of innate language reference. We each have had to learn these symbolic correspondences from scratch. But are there underlying neural computational constancies involved with language processing? If the operation of determining the difference between a subject and predicate or of modifying tense of verbs in some way required the same patterns of neural activity in each and every person, there would be at least a possibility that this operation could have evolved an innate substrate. Like words, the surface markings and syntactic transformations that are used to represent different grammatical functions vary extensively from language to language. So this level of analysis does not exhibit sufficient constancy. But what about the underlying grammatical functions themselves? These are precisely what linguists have focused on as universal features of grammar. Are the deep grammatical functions that are ubiquitous in all human languages represented in brains by invariant neural processes? If so, they should be susceptible to Baldwinian evolution and would be candidates for genetic assimilation. If not, no level of adaptive importance, no amount of selection pressure, no extent of evolutionary time could contribute to instantiating them as innate neural predispositions.

This criterion of neural computational invariance does not require that there is some local structure in human brains in which this computation is performed. Sophisticated theories of innate language competence have come to avoid localizationist assumptions with respect to language acquisition "devices," etc. Many recent models instead appeal to computer software metaphors and invariant "data structures" to explain the nature of language universals. But there is an important sense in which brain evolution requires specificity of the neural substrate, not just of the "program." Although "location" in the brain can refer to multiple and distributed circuits, in order for innate information to be "instantiated" in the brain it must utilize certain neural circuits in repeatable and consistent ways from one individual to another across a long stretch of time, so that selection repeatedly operates on the same morphogenetic processes generation after generation. The operations themselves need not be localized, but how they are distributed within different brain structures should be invariant.

Evidence that such constancy is not characteristic of language-specific functions was provided by the neural variability of agrammatical deficits described in the previous chapter. Speaking a highly inflected language means that a different part of the brain is more critical to a particular grammatical function than it is in speaking a relatively uninflected language. The implication for constancy of neural representation is that the same grammatical operation, when represented by very different surface features of language structure, may also be represented by very different brain regions as well. Though the underlying logical-symbolic operation that is performed by these different syntactic strategies is the same one, this is not what determines which brain structures perform the operation. Rather, it is the ways that the surface operations on the words are carried out (i.e, the analyses of the physical signals themselves) that determine what part of the brain gets involved. This has important and little appreciated implications for evolution. The majority of the deep structures of grammar that have been proposed as universal are logical operations that have quite variable surface implementations from one language to another. The ways that questions are derived, the subject/predicate distinction, the marking of tenses or moods, the many other grammatical distinctions that must be encoded in words and their positions within sentences, and the deeper logical rules that govern the relationships between them are all subject to this variable link with surface attributes of language structures. This hierarchic relationship is crucial to their symbolic functions. But it poses an interesting riddle. The most universal attributes of language structure are by their nature the most variable in surface representation, variably mapped to processing tasks, and poorly localizable within the brain between individuals or even within individuals. Therefore, they are the *least* likely features of language to have evolved specific neural supports. Those aspects of language that many linguists would rank most likely to be part of a Universal Grammar are precisely those that are ineligible to participate in Baldwinian evolution! If there are innate rules of grammar in the minds of human infants, then they could not have gotten there by genetic assimilation, only by miraculous accident.

Where does this leave us? Are there any grammatical and syntactic universals of language that satisfy the criteria that would enable Baldwinian evolution to take place? What about the most general and universal principles? Consider, for example, the subject-predicate distinction. If any categorical feature of grammar should correlate with a neural substrate distinction, the segregation of syntactic and semantic analyses according to some version of these complementary propositional functions should do so. It is always a necessary grammatical distinction, irrespective of the language, and without its being assumed, few other grammatical categories and functions can be defined. For languages to have served as a means of indicating, commanding, and seeking additional information (have pragmatic or propositional functions as opposed to just a marking function), they would have needed something akin to what in modern languages we recognize as this distinction. Though the terms *subject* and *predicate* do not capture the full diversity of such complementary functional roles and phrase components of language (other variants include topic-comment, agent-action, operator-operand), they can serve to identify this core combined symbolic function present in all.

The earliest symbolic systems would necessarily have been combinatorial and would have exhibited something like this operator-operand structure (and probably subject-predicate structure) right from the start. This is the minimum requirement to make the transition from indexical to symbolic reference. In other words, some form of grammar and syntax has been around since the dawn of symbolic communication. There was never a protolanguage that lacked these and yet possessed words or their equivalents. This satisfies the first requirement, consistency across all languages throughout time. But satisfying the second is much more difficult. Are the grammatical and syntactical processes supporting the complementary subject/predicate function always carried out the same way and utilizing the same brain systems, irrespective of language differences? In terms of sensorimotor invariants, there is probably no invariant feature in the speech signal that can be relied on to mark noun and verb phrases. There are no universal words or sounds that mark them, not even any reliable intonation contour or ordering rule that invariably points to the elements in a speech stream that must be collected into each of these grammatical functional units. The theory of deep structures paints itself into an evolutionary corner, so to speak, by recognizing the logical independence of universal features from surface features. In this regard, Chomsky's abandonment of Darwinian explanations for innate language knowledge is at least consistent. So what is left?

Language Adaptations

In the beginning chapters of this book we identified one unprecedented cognitive computational demand associated with language that is ubiquitously associated with all symbolic activities. An unusual sort of contortion of normal learning processes is required simply to overcome the mnemonic and attentional threshold that tends to prevent sets of rote indexical associations from becoming recoded into symbolic systems of association. This one cognitive demand would introduce an incessant selection pressure in a society of hominids habitually dependent on symbolic communication, in whatever form this symbolic communication took. As was explained in Chapter 9, the particular neural computations that are required to surmount this mnemonic-attentional threshold largely depend on processes that are carried out in the prefrontal cortex. Thus, the neural computations associated with symbol acquisition were unavoidably required by all languagelike behavior; they imposed a significant demand on a comparatively underdeveloped cognitive process; they were invariant across a wide range of sensorimotor applications; and they depended on a specific common neural substrate in all brains. This is a recipe for a powerful Baldwinian selection process.

Did it have to be language that drove this process? Could language have emerged later, after this neural shift in emphasis arose for other reasons? In other words, could some other social or ecological demand have selected for this particular learning bias and thus prepared the way for symbolic learning? To some extent, selection for this cognitive function is implicit in the fact that prefrontal cortex is present in all mammal brains, and is particularly well developed in primates. There are a number of behavioral and learning contexts that require maintaining attention on something in short-term memory in order to do something opposite or something complementary. Foraging for a resource like fruit is one example, but there are likely many social contexts for which learning complicated conditional associations are relevant, as well. However, these almost certainly comprise the minor exceptions in a world where most adaptational contingencies can be based on immediate context, and thus correlational, indexical learning tends to far overshadow these more indirect modes. But it is not the presence of conditional learning/unlearning strategies that needs to be explained in the case of human evolution; some such ability is essential for many complicated animal behaviors. What is unusual in humans is the radical shift in the balance between attention to higher-order recoding possibilities, and thus unlearning, as compared to more typical first-order learning processes which are more appropriate to the vast majority of physical and even social adaptations.

What other adaptive demand could account for a such an exaggerated predisposition for this rarely needed mode of learning? There certainly are few if any spontaneous analogues to the symbol-learning problem in nature. In order for a set of objects to serve as symbol tokens, they must be susceptible to being recoded in a consistent way. In other words, they need to be associated with one another in a pattern that maps to a closed logical system. Such lucky coincidences of object relationships very rarely occur by chance or by physical constraint. Indexical information is sufficient to adapt to the majority of complex social relationships, since most are dependent on making highly reliable assessments of others' behavioral propensities, anticipated by subtle behavioral symptoms. In fact, it appears that our pet dogs often do a more accurate job of reading a person's behavioral predispositions than we do, precisely because our far less reliable predisposition to rely on others' words and our own rationalized predictions gets in the way of recognizing nonverbal signals. In summary, I believe symbolic reference itself is the only conceivable selection pressure for such an extensive and otherwise counterproductive shift in learning emphasis. Symbol use itself must have been the prime mover for the prefrontalization of the brain in hominid evolution. Language has given rise to a brain which is strongly biased to employ the one mode of associative learning that is most critical to it.

Language adaptations don't end with symbolic cognition. There is a lot more to language than its representational and grammatical logic, and there are many other attributes that fill the criteria for being subject to genetic assimilation. However, these are global perceptual, motor, and mnemonic regularities rather than universal logical regularities. They include a great many attributes of language that are associated with surface structures, such as those tricks used to map strings of sounds onto symbolic relationships and logical operations between symbols, and those needed to extract symbolic information from them in the very limited time that the flood of speech information provides. Speech places particularly heavy demands on the employment of these general features by auditory and oral-vocal systems. Modern languages depend on production and analysis of dozens of phonemes (units of sound distinction or "sound-gestures"), produced at a rate of well over ten per second in the thousands of distinct combinations that make up words. This rate of production and analysis exceeds any possibility of doing the perceptual analysis or sound production movements in a one-at-a-time, phoneme-by-phoneme fashion. Not only must we be quick learners and skilled articulators, we must be able to "offload" a significant part of the lower-level analysis and production of speech to some remarkably facile automatic systems. Like the requirements for the genetic assimilation of defense strategies in alarm calls, the requirements for efficient speech analysis and production have all the qualities to become progressively internalized in patterns of neural architecture. It is even conceivable that there could be some degree of genetic assimilation of specific, highly regularized phoneme-distinction mechanisms.

Use of language also places inevitable demands on neural systems not directly involved in the production or perception of speech. The amount and rate of the information that is presented during linguistic communication, and the requirement that symbolic reference must be constructed combining many component symbols, place special demands on short-term memory and attentional processes. For example, there seems to be a contiguity requirement for the application of rules between words, and for marking phrase relationships, that probably reflects invariant mnemonic requirements for speech processing. Though within the predicate part of a sentence there can be separation of aspects of a complex verb (as in the German past perfect tense, where the verb for "to be" heads the verb phrase and the past participle ends it), there do not appear to be languages that chop up noun and verb phrases and intermix their parts. For the most part, the hierarchic structure of sentences derives from building higher-order units from contiguous parts. Here we may be able to extract an invariant feature of the input-output signal; but interestingly, this universal is precisely one that may have far more to do with selection pressures that sensory analysis and mnemonic processes place on language evolution, rather than the other way around. Though adjacency dictates some bias in whether elements are analyzed as parts of a larger grammatical unit, there is no constancy in what sorts of operations are governed by adjacency constraints, nor what markers indicate whether perceptually adjacent elements are or are not within the same phrase. It might even be because of this ambiguity that languages inevitably evolve a small, closed class of marker words to indicate phrase boundaries.

Another aspect of language learning that we tend to take for granted is the ability and the predisposition to mimic others' speech sounds. Though other species are known to mimic behaviors produced by parents or peers, there are few that equal the readiness that human toddlers show to mimic speech. Unlike the evolution of specialized modular abilities, such as those involved in speech sound analysis, the evolution of learning predispositions is a more subtle and indirect result of Baldwinian processes. A predisposition to learn some particular class of associations, or to attend more closely to certain stimulus features during learning, or merely a bias in motivation toward certain realms of learning, all may be selectively favored by evolution. Unlike neural processes that map to specific perceptual or behavioral templates, however, differences in learning are probably not so readily localizable, because learning is ultimately based on links and biases among many separate neural systems. A learning predisposition may nevertheless be a consequence of genetic assimilation under circumstances where the demands on particular aspects of learning are consistent over time, and where the neural circuits that are involved are invariant from individual to individual. One of the challenges facing us is to try to understand how specific learning biases might be represented in brains.

The relationship between the evolution of learning predispositions and of perceptual predispositions can be illustrated by considering another aspect of vervet monkey alarm calling. Evidence for which distinctive sensory attributes of predators are innately prespecified in the brains of these monkeys comes from observing alarm calling in young animals. Though juveniles begin life with an alarm-calling ability, they initially overgeneralize their alarm calling to many nonpredators. For example, they have been reported to give eagle calls in response to a variety of startling disturbances occurring overhead, and to give leopard calls in response to a variety of startling ground-level disturbances. This suggests that the stimulus features that have been the most completely genetically assimilated are those associated with orientation, and that as they mature, young monkeys learn to associate these responses with a more limited subset of stimulus objects which share these characteristics. The features of the objects of alarm calling, though innately predisposed, have been incompletely "assimilated" into neural mechanisms. Learning must still bridge the gap between what has been internalized and what has not, possibly because the variety of possible predators makes genetic assimilation of any more specific predator identifiers unlikely. One could imagine that future vervet evolution might lead to more completely prespecified sensory templates for predators, if the same few species were always involved, though there will always be some level of genetic assimilation beyond which selection cannot push because of competing pressures and variable conditions. But the role of learning in bridging the gap between outside and inside is also subject to genetic assimilation effects. If there are constant attributes of the learning context, including social attributes, then the learning process by which these young monkeys come to respond selectively to these stimuli may be enhanced by the evolution of attentional and mnemonic biases of a slightly more general sort.

So the ability to learn is not some generic function, equally applicable to all stimuli. Learning biases are both cause and consequence of Baldwinian evolution. Whereas sensory and behavioral invariance with respect to some adaptive context may be a source for genetic assimilation of neural templates that correspond to the critical features of that context, invariant features of the learning context itself are also susceptible to genetic assimilation. Besides considering the unchanging sensorimotor attributes of language use, then, we must also consider the possibility that there are invariants in the language-learning context that have become internalized in language evolution. This almost certainly also includes adaptive biases to invariants in the social context of language development. It is not that all facets of our language adaptation are either innately predisposed or learned; some are strongly predetermined before language experience, some involve extensive interaction between innate biases and experiences, and some are almost entirely unconstrained by human predispositions. The point is that it should be possible to predict which aspects of our language adaptations are more or less susceptible to genetic assimilation, by virtue of the invariances they demonstrate. Universality is not, in itself, a reliable indicator of what evolution has built into human brains.

In summary, only certain structural universal features of language could have become internalized as part of a "language instinct," and these turn out *not* to be those that are most often cited as the core of a Universal Grammar. Instead, the best candidates for innate language adaptations turn out to be some very general structural characteristics of the primary language medium itself, speech, and the computational demands this medium imposes when it comes to symbolic analysis. Whatever learning predispositions are responsible for the unprecedented human facility with language, they specifically cannot depend on innate *symbolic* information. No innate rules, no innate general principles, no innate symbolic categories can be built in by evolution. A number of linguistic functions undoubtedly depend more on evolutionary specializations of certain brain areas than others, and are more available for genetic assimilation than others. But the evidence suggests that the deep logic of grammatical rules lacks the invariant characteristics that could allow them to be subject to natural selection. We must conclude that few, if any, aspects of the deep grammatical logic of language could have been prewired by natural selection in response to the demands of language use. The very abstraction from the surface implementation of morphology and syntax that provides the grammars with their generative power also shields them from the reach of natural selection. The noncorrelative nature of symbolic reference has cut language off from forces that shape biological evolution, and instead has shifted the burden of adaptation to a new level of information transmission. So, to expand on the argument in Chapter 4, the co-evolution of languages with respect to human neurological biases may not just be a plausible source for emergent universals of grammar, it may be the only plausible source.

The remarkable expansion of the brain that took place in human evolution, and indirectly produced prefrontal expansion, was not the cause of symbolic language but a consequence of it. As experiments with chimpanzees demonstrate, under optimal training conditions they are capable of learning to use a simple symbol system. So, it is not inconceivable that the first step across the symbolic threshold was made by an australopithecine with roughly the cognitive capabilities of a modern chimpanzee, and that this initiated a complicated history of back-and-forth escalations in which symbol use selected for greater prefrontalization, more efficient articulatory and auditory capacities, and probably a suite of other ancillary capacities and predispositions which eased the acquisition and use of this new tool of communication and thought. Each assimilated change enabled even more complex symbol systems to be acquired and used, and in turn selected for greater prefrontalization, and so on. In other words, the computational demands of symbolization not only are the major source of the selection pressures that could have produced the peculiar restructuring of our brains, they are likely also the indirect source for the selection pressures that initiated and drove the prolonged evolution of an entire suite of capacities and propensities that now constitute our language "instinct."

Homo symbolicus

Though the analysis of brain organization and function in modern species can tell us a great deal about the human/nonhuman difference with respect to language, it leaves open questions concerning the evolutionary transition from nonlanguage to language communication. It does, however, offer some new tools for approaching these old questions about language origins. When did languagelike communication first appear? Which hominid species first began to use it? What was it like? Did speech and language complexity develop all at once or gradually? There is little physical evidence to support any but the most generic answers to such questions, but the brain structure/symbol-learning connection offers some important constraints which can guide our speculations concerning this critical phase in our prehistory, and which can help place language evolution on a timeline defined by other more robust sources of evidence about our fossil predecessors.

For heuristic purposes, let's invent a new species designation: *Homo symbolicus*. This name, based on one trait, would apply to all hominid symbol users. The first appearance of this species would correspond to the first hominids who habitually used symbolic communication. It thus refers to a sort of virtual species, not a genetic species, because it is based on something other than just genetic or morphological features. Its members are defined by a dual inheritance. The designation might, then, seem like an exercise that demonstrates the dangers of using behavioral traits as a basis for classification, because they are too prone to the influence of parallelism and convergence to be useful for tracing lineage relationships.

But I don't propose this as an idle terminological exercise. Biological species are defined in terms of their ability to reproduce viable offspring, that is, to trace their genes from and contribute them to a common closed genetic pool. This genetic criterion has a clear semiotic counterpart. All symbolizing hominids are linked via a common pool of symbolic information, one that is as inaccessible to other species as are human genes. We are all heirs of symbolic forms that were passed from one generation to the next and from one group to another, forming a single unbroken tradition. We derive all our symbolic "traits" from this common pool and contribute to its promulgation. Being a part of this symbolic information lineage is in many respects a more diagnostic trait for "humanness" than any physical trait. Evolutionary phylogenies are defined in terms of inheritance of information, but not all the information that determines a species' defining characteristics is coded in genes. So, to the extent that there are partially independent forms of information transmission that are crucial for determining a species' attributes, they too can determine potentially independent descent relationships. *Homo symbolicus* might thus be termed a *nöo*-species designation (in which "nöo," from the Greek νόoζ, refers to "mind") as distinct from a biological or zöo-species designation. It is nonetheless a biological classification. Without considering this lineage, the most distinctive physiological feature of the human species, a uniquely modified brain, would lack an evolutionary explanation.

The phylogenetic history of *Homo symbolicus* may cut across fossil phylogenies in more profound ways than just including an ascending series of ancestral species. A number of paleospecies may be entirely contained within this superordinate nöospecies, and others may be subdivided by it, with some individuals within each species included and others excluded. Also, because symbolic communication can be present without specific biological correlates, we shouldn't expect that its first appearance corresponds with any fossil species transition, so that we cannot rely on biological markers to identify members of this nöospecies. Nevertheless, there are clear biological correlates to long-term symbolic communication produced by its selection on brain traits, specifically brain size and the correlated changes in internal organization that are so critical to symbol learning. These effects are evident in fossil skulls. They do not allow us to claim that a given species lacks symbolic abilities; rather, they can be used to identify whether symbolic activities have been present for some time and have provided selection pressures on brain organization. This can tell us which species must be included in *Homo symbolicus,* but not which can be excluded. Though the size of fossil hominid brains provides only very crude information about this most complex organ, and so should not be overinterpreted, it provides sufficient evidence to answer a correspondingly broad question: Was languagelike communication present in a given fossil species?

But aren't there distinctive anatomical features in brains that are capable of language? What about the evolution of the language areas of the brain? Might we be able to identify the appearance of language-specialized brain regions in fossil species if we could actually see what their brains looked like? Indeed, it is possible to get a fair idea of the outward appearance of fossil hominid brains. Paleoneurologists have long hoped that the patterns of bumps and grooves that are barely exhibited on the surface of fossil cranial endocasts[5] might provide critical evidence for language origins. Two modern endocast researchers, Philip Tobias and Dean Falk,[6] have each concluded that the patterns appearing on endocasts of *Homo habilis* specimens probably reflect the presence of language abilities.[7] More specifically, they argue that one habiline brain exhibits distinctive sulcal patterns that mark the location of Broca's area in modern brains (two sulci that ascend into the frontal lobe from the tip of the Sylvian fissure). These folds can be recognized in the lower-left frontal lobe of *Homo habilis* endocasts, but not in australopithecine endocasts. Could this indicate the first appearance or the special enlargement of a unique and critical language area?

Though Broca's area is indeed a language area that is important for speech in most humans, these surface landmarks are only incidentally relevant. Language areas are simply not the repositories of linguistic skill and knowledge that they were once thought to be. Though damage to the cortex near this location was classically associated with Broca's aphasia, even in modern brains this is a variable correlation, and, as we have seen, distribution of functions, individual differences, and language differences in aphasia effects do not support any simple localization of language functions to this ventral-frontal brain region. More important, comparative evidence demonstrates corresponding brain regions in other primates, and developmental evidence makes assumptions about localized additions or expansions untenable. As other paleoneurologists have noted, the appearance of additional folding in this region is probably an inevitable correlate of overall expansion in brain size.[8]

The presence or absence of putative language-specialized areas is not the critical factor. They do not represent what is unavailable in other species. Though such surface details are not terribly informative about brain reorganization, the far more general and unambiguous changes in relative brain size may be able to provide more specific evidence of reorganization. This is because the critical difference is not the addition of some essential brain structure, but the quantitative modification of relationships within the whole brain, particularly the relative contribution of the prefrontal cortex.

Developmental analysis indicated that relative prefrontal size is determined by competitive processes during development, which involve interactions converging from widely separated brain regions. The bias in favor of prefrontal circuits in modern brains is a function of disproportions between the developing cerebral cortex and other brain and peripheral nervous structures. So, as the cerebral cortex became proportionately larger during hominid evolution, the connecting systems that determine prefrontal cortex recruited a greater fraction of potential targets. This means that the relationship between hominid brain and body size in fossil species can be used as an index of the degree of prefrontalization in their brains. The developmental constraints on brain structure allometries provide a means for unambiguously predicting innumerable details of neuroanatomical organization from an easily estimated feature that is discernible in fossil materials: brain size with respect to body size. We are no longer in the dark about what was inside the brains of our fossil ancestors.

The main trends in the fossil record of brain size are clear (see Figure 11.2). The size of the hominid brain increased almost threefold from African ape averages, which were similar to those of our australopithecine precursors. There is little disagreement among paleontologists that hominid brains first began to enlarge significantly in comparison to body size approximately 2 million years ago, with the appearance of the paleospecies designated *Homo habilis*. This habiline increase was as much as 150 percent above australopithecine values, from roughly 500 to 750 cc. The transition was not quite this simple, because an increase in stature made the net increase somewhat less, and because there is a considerable range in brain sizes among fossil specimens that have been categorized as *Homo habilis*. Subsequent brain expansion took place incrementally up until the near present. *Homo erectus* fossils, which are dated as early as 1.8 million years ago and as recent as 350,000 years ago, overlap with the high end of *Homo habilis* and the low end of modern *Homo sapiens* brain volumes, that is, from

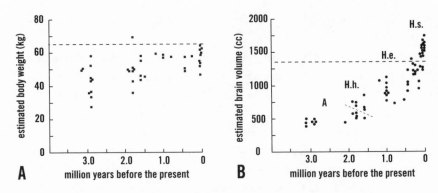

Figure 11.2 *Change in body size (A) and brain size (B) in hominid fossils from 3 million years ago to the present. Dates and brain volumes are taken from published estimates; body sizes are estimated using published long bone and vertebral measurements by regressing on modern human long bone and vertebral measurements. Though the data are from heterogeneous sources that may carry different biases, they provide a graphic demonstration of the dramatic increase in brain size during a period of relative stasis in body size. Dashed lines indicate modern mean values; letters only approximately indicate which points are from which species: A = australopithecines (all groups); H.h = Homo habilis; H.e. = Homo erectus; H.s. = Homo sapiens.*

about 800 to over 1,000 cc. Though it is conceivable to see these shifts as categorical *between* species,[9] it is clear that within the *erectus* and *sapiens* lineages there are also brain size trends: a slight increase over time for *erectus,* and a slight decrease in recent times for *sapiens,* finally settling to about 1,350 cc. (Figure 11.2).

From these trends we can predict a correlated trend of internal brain structures: an increasing prefrontalization, with a corresponding shift in learning predispositions. It cannot be doubted that such a robust and persistent trend in brain structure evolution reflects forces of natural selection acting on major brain functions, and it is hardly coincidental that these most salient changes in human brain structure correlate with the development of a unique human facility for symbols. For this reason, the increase in brain size in hominid evolution is an important record both of the relative ease with which symbols were able to be acquired by a given fossil species, and of the prior effects of selection on this ability.

Though the evolution of most traits in most species can be explained in simple Darwinian terms, attempting to apply this model to the central features of human evolution can lead to confusions about cause and effect. The arguments presented above suggest that human behavioral and brain evolution in particular can only be adequately explained in terms of Baldwinian evolutionary processes.[10] In general, behavioral adaptations tend to precede and condition the major biological changes evident in human evolution because they are far more facile and responsive than genetic and morphological changes. But once some useful behavior spreads within a population and becomes important for subsistence, it will generate selection pressures on genetic traits that support its propagation. More than any other group of species, hominids' behavioral adaptations have determined the course of their physical evolution, rather than vice versa. Stone and symbolic tools, which were initially acquired with the aid of flexible ape-learning abilities, ultimately turned the tables on their users and forced them to adapt to a new niche opened by these technologies. Rather than being just useful tricks, these behavioral prostheses for obtaining food and organizing social behaviors became indispensable elements in a new adaptive complex. The origin of "humanness" can be defined as that point in our evolution where these tools became the principal source of selection on our bodies and brains. It is the diagnostic trait of *Homo symbolicus*.

Thinking of human evolution in these terms goes against some basic prejudices in our thinking about causes and effects. Learning abilities and behavioral flexibility in general complicate the analysis of causes and effects in evolution because they loosen the linkage between structure and function. This can significantly alter the rate and direction of evolutionary trends, both by buffering old adaptations from the influence of new selection pressures, and by serving as a bridge between radically different adaptive responses. But many theories attempting to account for behavioral adaptations still implicitly assume a more rigid structure-function correlation, i.e., that a difference in behavioral adaptation must be predicated on some prior evolutionary modification of the brain. This is clearly reflected in assumptions about the beginnings of tool use, language, and brain expansion in hominid evolution.

Most scientists agree that the first significant increase in brain size should signal the origin of the genus *Homo*, since this trait is specially characteristic of our species. The once widely accepted assignment of the earliest hominids to exhibit significantly enlarged brains to the species *Homo habilis* reflects this conventional wisdom. Not coincidentally, this period in time is also marked by the beginnings of a stone tool record. The first stone choppers appear in the fossil record some time between 2 and 2.5 million years ago. These tools were fabricated with a few sharp blows of stone on stone to create a sharp but not particularly refined cutting edge. They were almost certainly hand-held, and used for cutting flesh and chopping bone to get at the marrow. It is generally assumed that the first tools were manufactured by these first large-brained hominids. Together, these two lines of evidence are usually interpreted to demonstrate how increased intelligence made this new kind of adaptation possible. Though this seems to offer an obvious interpretation of such an association, the association is not so clear as it might appear. In examining some of the unresolved details of this transition, we begin to expose some weaknesses in the assumptions behind this view.

First, there is considerable controversy about the associations between particular fossil species and stone tools from this time period. Part of this stems from the difficulty of finding clearly associated skeletal remains with the earliest stone tools. Second, the identification of fossil species from this period and of their associations with earlier and later hominid lineages is also in a state of flux. This too is a reflection of the fragmentary nature of the evidence, but also a result of the apparent diversity of concurrent forms. Depending on which expert is consulted, there are two or three potential candidates for the "true" ancestor to the subsequent *Homo erectus* lineage, and each has traits that might be interpreted as ancestral to derived traits found in *Homo erectus,* including reduced dentition, reduced jaw and face structure, increased relative brain size, changes in the opposability of the thumb and fingers, and so on.

The question of identifying the first toolmakers is not merely a matter of establishing a species-artifact correlation. Behavioral adaptations need not be uniquely linked to one species or one lineage, and because toolmaking and tool use are behavioral adaptations, there may not even be any biological features that distinguish the fossils of the first stone toolmakers from those who were not making and using tools. Stone toolmaking is not a physical trait like opposable thumbs or small canines. It is not passed on genetically. It is a learned skill, passed from individual to individual. As a result, it is not impossible that one species might under certain circumstances "inherit" a behavioral adaptation previously used by another, not genetically but behaviorally. Human history provides ample evidence of the dissociability of socially inherited traditions and biological genealogies. Acquired behavioral adaptations have frequently passed from one population to another, either as a result of retrieval of abandoned artifacts or by direct mimickry. The initial population to "discover" an adaptation may even subsequently become extinct, and yet the adaptations they originated may survive in another. Indeed, many animals have been taught to make and use simple human tools, including stone tools.[11]

Due to the limited interactions between separate species, it is probably exceedingly rare for animals' behavioral adaptations to be transmitted even beyond a single lineage; but genetic continuity is not implicit. Species barriers to information transmission are not absolute, and in the case of artifacts such as stone tools, which can persist longer than any specific use and can be retrieved and examined independent of another's use of them, there may be many opportunities for information transfer. The transmission of tools could even be aided by living out on the open savannah, where visibility and group mobility might increase the likelihood of intergroup interactions and observation. Although I pose this more as a possibility than as a serious hypothesis, it cannot be easily discarded as irrelevant. We cannot assume that all tool users were our ancestors. Paleontologists have even noted that later australopithecines had hands well equipped for simple tool use, with widened fingertips and more opposable thumbs than their forebears.

With this caveat in mind, let's reconsider the fossil origins of brain size increase and stone tools. The first stone tools slightly predate the first appearance of the genus *Homo,* as defined in part by its expanded brain. But this difference in time may not only be due to their numbers and better preservation. The Baldwinian perspective suggests another possibility: that the first stone tools were manufactured by australopithecines, and that the transition into *Homo* was in part a consequence rather than the cause of the foraging innovation they introduced.

For most of their evolution, australopithecines were clearly not symbol users. There is no evidence of the changes in brain structure we can now associate with this function during their presence for millions of years before any such "humanlike" neural adaptations appeared. If some troop of australopithecines hit upon this communication trick, it did not take hold—except for once. But when it finally took hold, perhaps as much as 2.5 million years ago, it introduced selection for very different learning abilities than affected prior species. During this brief epoch, when the first symbolic systems flourished among populations of australopithecines, they were being learned and used by individuals who had brains with an internal organization roughly comparable to that of the modern apes—brains, in other words, that were not well adapted for symbol learning.

Earlier, we considered the problem of teaching symbols to apes and other species with brains that are not predisposed by evolution for symbol learning. The results of these animal "language-training" studies show that considerable external social support is necessary to attain minimal symbol learning. Nevertheless, under special circumstances, at least one species of ape (and possibly many other bird and mammal species) appears capable of making the conceptual leap necessary to support a basic symbol system, if provided with the necessary supports. This is probably a reasonably accurate analogue of the symbolic abilities of the first symbol users, as well as of their needs for external support. During the initial phase of the symbolic adaptation, considerable external support must have been required to back up even a very basic symbol system. As a result, these first symbolic abilities were likely dependent on fragile social adaptations that were subject to periodic failure. But as a consequence, any source of support that could be recruited to help overcome these handicaps and reinforce these fragile conditions of transmission would have been favored by evolution. The expansion of the forebrain, and specifically the prefrontal cortex, in *Homo habilis* reflects only the core adaptation, supplemented over time with an increasing number of other diverse language supports. The introduction of stone tools and the ecological adaptation they indicate also marks the presence of a socio-ecological predicament that demands a symbolic solution (as we'll consider in Chapter 12). Stone tools and symbols must both, then, be the architects of the *Australopithecus-Homo* transition, and not its consequences. The large brains, stone tools, reduction in dentition, better opposability of thumb and fingers, and more complete bipedality found in post-australopithecine hominids are the physical echoes of a threshold already crossed.

One of the complications of dual inheritance is that independent trends in morphological evolution can be driven along parallel lines by common *non*genetic inheritance. Parallelism in genetic evolution is exemplified by dolphins, seals, and manatees. As their ancestors adapted to the aquatic environment, each independently evolved reduced, flattened limbs and streamlining in response to the locomotor requirements of swimming. The environment created by the introduction of stone tools and symbols would have likewise contributed parallel selection pressures on whatever lineages used them, and would have selected for the same changes in body and brain organization, even if they remained genetically isolated. This may be relevant to the problem of determining which, among the many potential fossil species present 2 million years ago, was the ancestor of *Homo erectus*, and ultimately ourselves. It now seems clear that there were at least two biological species that could be considered intermediate in time and morphology between australopithecines and *Homo erectus*. One had a large brain but australopithecine-like facial and teeth features, and the other had a small brain but other features more like *Homo erectus*. Which of these was our biological ancestor? In terms of our mental evolution it will not be easy to determine, and may not much matter. Either could have been.

The dawn of stone tool use and symbolic social communication among late australopithecines could have precipitated the diversification of many fossil lineages. In the period from the first appearance of stone tools, separate lineages that were each inheritors of this adaptive behavioral complex would likely have responded slightly differently to its effects depending on different initial conditions and environmental contingencies. But over time, we should expect these differences to converge. The many features that we might attempt to trace from this epoch to later lineages in order to establish phylogeny, such as reduced dentition, loss of sexual dimorphism, more efficient bipedalism, more complete precision grip, and increased brain size, could thus have been represented separately in isolated populations of early stone tool users. Even so, whichever lineage managed to survive would have eventually evolved parallel versions of these previously separate traits in response to the selection pressures imposed by this common behavioral adaptation. Which transitional fossil lineage represented our true ancestor may not be as important as the fact that implicit in their stone tools and social-ecological adaptations were the seeds of future human characteristics. Ultimately, all these curious physical traits that distinguish modern human bodies and brains were caused by *ideas* shared down the generations.

The Co-Evolutionary Net

Although the threshold that separates linguistic from nonlinguistic communication is not the complexity or efficiency of language, human adaptations for language that enable it to be so complex and yet so effective were also produced by selection on language abilities. Once symbolic communication became even slightly elaborated in early hominid societies, its unique representational functions and open-ended flexibility would have led to its use for innumerable purposes with equally powerful reproductive consequences. The multitiered structure of living languages and our remarkably facile use of speech are both features that can only be explained as consequences of this secondary selection, produced by social functions that recruited symbolic processes after they were first introduced. These are secondary insofar as they became selection pressures only after symbolic communication was well established in other realms. They are, however, the primary causes for the extensive specialization that characterizes spoken languages and for the great gulf that now separates our abilities in these realms from those of other species.

The evolutionary dynamic between social and biological processes was the architect of modern human brains, and it is the key to understanding the subsequent evolution of an array of unprecedented adaptations for language. This is an important shift in emphasis away from what might be called "monolithic innatism," that is, the view that the "instinct" that humans have for language is some unitary and modular function: a language acquisition device (LAD). Co-evolutionary processes instead have produced an extensive array of perceptual, motor, learning, and even emotional predispositions, each of which in some slight way decreases the probability of failing at the language game. Each by itself might appear to be a rather subtle shift from typical ape predispositions; but taken together, these many language adaptations (LAs) inexorably guide the learning process, the way carefully planted tidbits of misinformation can dupe a victim to fall prey to con artists. Though no one may be either indispensable or sufficient, together they guarantee the replication of language.

Seen in hindsight, the myriad of spin-off uses of symbolic communication—from actively passing knowledge acquired in one generation to the next, to manipulating and negotiating with others about every aspect of social life—each can appear as independently converging selection pressures for language. Many competing scenarios of language origins differ in their focus on some particular social advantages deemed most likely to have been the source of selection for language abilities. Was language selected because of its importance in supporting close cooperation in mother-infant relationships, passing on tricks for extractive foraging, organizing hunts, manipulating reproductive competitors, attracting mates, recruiting groups for warfare and collective defense, or providing a sort of efficient social glue by which individuals could continually assess common interests and support networks, as grooming does for so many other primates? The answer is that all are probably significant sources of selection, not so much for language origin, but certainly for its progressive specialization and elaboration. In this regard, they do not offer competing, mutually exclusive hypotheses, but rather a list of domains into which symbolic communication has been successively introduced. The value of each of these uses, in turn, would have contributed new selection pressures that further supported and elaborated symbolic abilities.

This spreading of selection pressures to support diverse adaptations is an inevitable outcome of the Baldwinian recruitment process. It is analogous to the way a new technological innovation, developed for one application, can become the device of choice for a vast number of totally unexpected applications. Any one or more of these spin-off uses could eventually become the primary support for its continued production. Yet in hindsight, historians who were not privy to the initial events might view any one of these ongoing uses as a reasonable candidate for the original use, or else could argue that all these needs played some role prompting the initial invention. A vast range of applications may be relevant to the *persistence* of Velcro in late twentieth-century societies, for example, but have little or nothing to do with its invention (the achievement of a Swiss engineer, George de Mestrel, attempting to mimic the clinging trick used by cockleburs). Many inventions were originally created for purposes for which they were soon found to be inapplicable, but they persisted because of a more relevant spin-off use.[12] Building adaptational stories in hindsight, extrapolating from current adaptational evidence to past initial conditions, floods us with false leads and a multitude of independently plausible explanations. Because evolution is a historical process, it is like entropy in the sense that it cannot be analyzed backwards, the way a logical proof can be analyzed. Many parallel threads get ensnared in the net of natural selection and weave together as though they belonged together from the start.

The biological version of this process has been termed *exaptation* by Stephen J. Gould and E. Vrba.[13] In most cases of exaptation, whether in biological or technical realms, there is probably neither complete abandonment of the original adaptive function nor takeover by a single new function. As the adaptive contributions of a novel structure become more diverse, the influence of selection becomes distributed over its parallel useful consequences, and this increases the likelihood that it will be replicated in some form in future generations, whether or not any one function persists. A new innovation that is adaptive in one realm will tend to be applied to an ever-widening range of additional uses as time goes by. Consider, for example, the evolution of feathers. What may at first have been an adaptation for heat retention has become the paramount component in an adaptation for bird flight. This exapted use eventually became "the tail wagging the dog," and played a prominent role in the evolution of feather shape and distribution. Both functions are still contributors to the persistence of feathers; but in addition, many tertiary adaptations have subsequently arisen, including such functions as nest bedding, flotation, and mating display, to name a few.

Of course, the more intrinsically flexible or general the adaptation, the more potential there is for diverse involvement in secondary functions. For this reason, neurological adaptations may be among the most highly susceptible to this spreading exaptation process. An important consequence of this is that neural adaptations should tend to evolve away from strict "domain specificity" toward functions that can be recruited for multiple uses, so long as the original domain-specific function is not thereby compromised. This tendency should be even further exaggerated for neural adaptations that are one or two steps removed from perceptual and motor domains, as are various learning predispositions. In this regard, symbolic strategies for communication and mnemonic support are minimally constrained. Once symbolic communication became essential for one critical social function, it also became available for recruitment to support dozens of other functions as well. As more functions came to depend on symbolic communication, it would have become an indispensable adaptation. And the more indispensable it became, the more this would have raised the reproductive costs of failed symbol learning and awkward symbol use.

Some sort of positive feedback process like this has been invoked by most theories of human cognitive evolution. And yet few have considered the consequences of as much as 2 million years of such intense selection on the many diverse neural and body systems that are involved in the analysis and production of languagelike symbolic communication. All aspects of sensory, motor, and cognitive function that are involved in language learning and use eventually may have been subject to adaptive changes to one degree or another. All would have left their marks in brains, minds, bodies, and even human social institutions. However, because the ability to understand symbolic relationships underlies all, selection for anything that benefited this prerequisite function would have been constant and intense throughout. This probably accounts for the increases in brain size (and prefrontalization) that have continued for 2 million years since the introduction of symbolic communication.

But any source of support that could be recruited to help overcome handicaps to efficient symbolic communication would have been favored by evolution. Prefrontalization is only one facet of this process. Throughout the last 2 million years, there must have been many other aspects of language structure that evolved to maximize language reproduction by minimizing the impact of the limitations of language users (this process and some examples are depicted in Figure 11.3). Effects would have been exhibited not only in the vocabulary, grammar, and syntax of languages but also in the medium of communication. Their descendant features in modern languages are still at least partially evident as the fragmentary "fossils" of the earlier stages of this process. Though most details of these prior language adaptations have been worn away by the refining hand of selection, a number of major features of modern language have preserved a significant impression of their past selection pressures, and these allow us some basis for speculative reconstructions.

If the primary mystery of language is the origin of symbolic abilities, the second mystery is how most symbolic communication became dependent

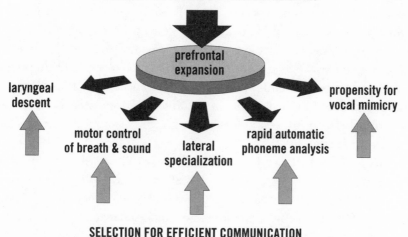

Figure 11.3 *Diagram indicating how selection for the core symbolic function distributed selection to a wide variety of supportive adaptations that became significant only once this core function was established. These in turn distributed selection pressure back on the core function as they became entrenched in later epochs of the process. As a result, prefrontal cortex became additionally recruited for other supportive functions as well, as did the other systems. Numerous serendipitous "spin-offs" would thus become coopted, or exapted, by the growing cluster of language adaptations.*

on one highly elaborated medium: speech. We employ symbolic representation in many realms of social and internal communication, yet the medium of speech has become its overwhelmingly most important conduit. Spoken language is the vehicle that first introduces the power of symbols to children, and the principal means by which symbolic markers are insinuated into most human endeavors. But it was probably not always this way. This is a snapshot of only the most recent epoch in a long process of condensation and streamlining, in which one medium of symbolization has progressively taken on an increasing fraction of the load of symbol transmission.

This process is evolution's analogue of monopolization. Initially, it is likely that the "language" of the earliest symbolizers was far more multimodal than are modern languages, and that only later did spoken language become the relatively independent and closed system we would describe as a language today. In the 2 million years since the introduction of symbolic communication, alternative modes for expressing symbolic information have been in competition with one another, much in the way that alternative pronunciations of words compete for representation in future generations. Over time, vocal symbolic forms progressively took over functions previously served by a variety of nonvocal symbolic forms, so that presently it appears nearly autonomous and complete in itself. Indeed, this apparent self-sufficiency is the implicit assumption of linguistic theories that endeavor to provide complete rule-governed accounts of language functions. But this is an idealization of a communicative system that is still not completely freed from complementary nonverbal supports, despite a million years of consolidation.

Much of the evidence for the progressive assimilation of nonvocal symbolic functions into vocal form still exists in cryptic form in modern languages. The structure of syntax often only vaguely conceals its pragmatic roots in pointing gestures, manipulation and exchange of physical objects, spatial and temporal relationships, and so on. For example, it is not uncommon for languages to demonstrate number, intensity, importance, possession, etc., by corresponding conventional iconicity of repetition, inflection, adjacency, and so on, in their syntactic forms. Though children's language development likely does not recapitulate language evolution in most respects (because neither immature brains nor children's partial mapping of adult modern languages are comparable to the mature brains and adult languages of any ancestor), we can nevertheless observe a progressive assimilation of nonverbal language supports to more flexible and efficient vocal forms as their language abilities develop.

The idea that the removal of vocal limitations released untapped linguistic abilities has been a major theme of a number of language origins theories (most notably argued by Philip Lieberman, in a number of influential books and articles).[14] Though some have imagined that spoken languages could have emerged rather suddenly with the removal of constraints on vocalization, it would be an oversimplification of language processes to suggest that this was the major constraint limiting language evolution, and it would also be an overinterpretation of the fragmentary fossil evidence to suggest that there were sudden discontinuous changes in these capacities. Nevertheless, increasing articulatory skill has probably been an important pacemaker for the evolutionary assimilation of symbolic functions to a single medium, because vocal ability imposes significant constraints on the capacity and flexibility of the medium. Both the neurological basis for vocal learning and the anatomical basis for vocal sound production have changed radically with respect to the likely ancestral condition, but the development of skilled vocal ability was almost certainly a protracted process in hominid evolution, not a sudden shift. The incremental increases in brain size over the last 2 million years progressively increased cortical control over the larynx (see Chapter 8), and this was almost certainly both a cause and a consequence of the increasing use of vocal symbolization. Though the record for vocal tract evolution is far less complete, comprised mostly of australopithecine data at one extreme and *Homo sapiens* (including Neanderthal) data at the other extreme, the greater use of vocalization during successive epochs of brain evolution would inevitably have imposed selection on vocal tract structure to increase its manipulability during this same period. It is probably safe to conclude that although fully modern vocal abilities were not available to earlier *Homo* lineages, their vocal abilities were nowhere near as limited as those of other apes.

In experiments begun in the 1950s and 1960s investigating the language abilities of apes, it was soon realized that despite their otherwise precocious and humanlike learning abilities, chimpanzees (and indeed all the great apes) were extremely limited in their vocal-learning abilities. Even with very intense efforts at language training and efforts physically to assist in mouth shaping with a home-reared chimp named Vicki, only a couple of simple words could be trained. A breakthrough came in the late sixties, beginning with the work of Alan and Beatrice Gardner, who showed that, provided she was allowed to use a different expressive medium—manual signing—their chimp, Washoe, seemed able to acquire a large "vocabulary" of American Sign Language (ASL) signs. Thus, freed of this one limitation there was a superficial appearance that language learning could be considerable in nonhuman primates.

This suggested to many that signing might be the missing link in the language origins story. The idea that human ancestors might have once communicated in hand signs had been suggested many times in earlier speculative accounts of language origins. This history has been reviewed by one of its modern proponents, Gordon Hewes.[15] Such "gestural origins" theories offered a way to bridge the gap between the inarticulateness of apes and the fluency of humans. If the first languages were manual languages, they might have provided an intermediate stage between the ape and modern human conditions. But great apes are not like persons mute from birth, nor are the signed languages used by modern deaf communities a useful model for the communication of our less vocal ancestors. The vocal medium was different, not unavailable to our early ancestors; auditory perception was not limited; and symbolic-learning abilities were significantly advanced over those of other apes but not so well supported as in modern brains. These facts suggest that gesture likely comprised a significant part of early symbolic communication, but that it existed side by side with vocal com-

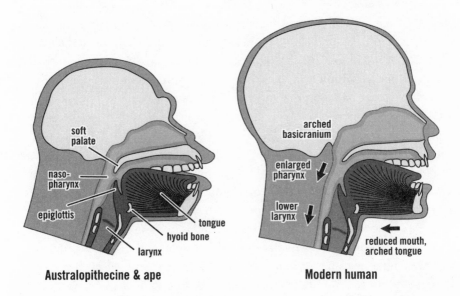

Australopithecine & ape
Modern human

Figure 11.4 *Descent of the larynx in human evolution. Relative positions of the major structures of the vocal tract as in apes and australopithecines* (left) *and modern* Homo sapiens (right) *depicting how relative reduction of the mouth and face and relative expansion of the cranium in human evolution has resulted in a correlated descent of the larynx and epiglottis lower into the throat, enlargement of the pharynx, and increase in the role of the tongue in modulating the shape of the pharyngeal and oral cavities. This has significantly increased the range of sounds, especially vowel sounds, that can be produced and decreased the degree of nasality of speech sounds. The evolutionary time course of this shift in vocal tract anatomy is still a matter of considerable debate, but most researchers would agree that* Homo erectus' *anatomy was more or less intermediate between that ancestral ape state and the modern state.*

munication for most of the last 2 million years. Rather than substitutes for one another, gesture and speech have co-evolved complex interrelationships throughout their long and changing partnership. Both should therefore exhibit evidence of this prolonged evolution in some near-universal interdependencies between them, as well as in predispositions that reflect the comparative recency or antiquity of these abilities. The current "symbiosis" between speech and gestural communication is abundantly reflected in culture-specific gesturing that accompanies most conversation. It is not difficult to imagine communication that is a much more complete hybrid of speech and gesture.

Probably the last feature of the language adaptation complex to be in-

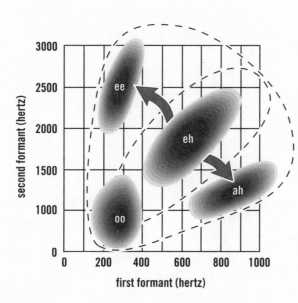

Figure 11.5 *Graph of the first and second formant ranges (major harmonic resonant frequencies of vocalizations) for three extreme vowel sounds* (gray areas). *The region encircled by the inner dashed region is a restricted range, where the two formants are relatively more linked, such as might be characteristic of the higher laryngeal position in early hominid ancestors. The arrows indicate the relative increases of range as a result of greater independence, as well as of the range of pharyngeal and oral cavity volume changes that occurred during later hominid evolution.*

corporated into the mix was the position of the larynx. Analyses of the comparative laryngeal anatomy of many vertebrates (particularly by Jan Wind)[16] and reconstructions of fossil hominid vocal tracts (by Phil Lieberman, Jeff Laitman, and Ed Crelin)[17] have demonstrated that modern humans are unusual in exhibiting a comparatively low position of the larynx in the throat (see Figure 11.4). Lieberman has emphasized how this peculiar anatomical feature may have significantly influenced the range and distinguishability of vocal sounds that earlier hominids were able to produce. The higher vocal tract of other primates and mammals significantly reduces the range and flexibility of sound production. In contrast, the low position of the human larynx increases the range of sounds that can be produced both by allowing greater changes in the volume of the resonant chamber formed by the mouth and pharynx, and by shifting sound to the mouth and away from the nasal cavities. The major result is that the sounds that comprise the vowel components of speech are much more variable than could be produced by any other ape, and include such extremes of resonance combinations as the "ee" sound in "tree" and the "ah" sound in "flaw," both of which require the pharyngeal space (behind the mouth) to be relatively enlarged. Of equal importance is the way this maximizes the ability of the mouth and tongue to modify vocal sounds. It is unlikely that the increase in vocal flexibility provided by this shift of laryngeal position was merely coincidental. This shift also comes with certain costs—easier choking—which might otherwise have selected against it (Figure 11.5).

But the cause-and-effect relationships with respect to language evolution aren't obvious. It is not at all clear that a reduction in vowel range and an increased nasality of speech, such as in other apes, would by itself be a major impediment to articulate vocal communication. Though reducing both the range and distinctiveness of speech sounds, it would not eliminate the most critical sound elements of speech: the sounds made by partially or entirely interrupting the voice. These make up the majority of phonemes that we identify with consonant sounds. Thus, even australopithecines would have vocal tracts capable of this degree of articulation if only they had had the necessary degree of vocal motor control. However, their brain sizes suggest that they were not in fact able to exert such control over the muscles of the tongue and larynx. Our more recent larger-brained ancestors, *Homo habilis* and *Homo erectus,* would have had greater motor control, and probably also exhibited some intermediate degree of laryngeal descent as well. *Homo erectus'* speech might have been somewhat less distinctive as well as slower than modern speech, and the speech of *Homo habilis* would have been even more limited. So, although their speech would not have had either the speed, range, or flexibility of today, it would have at least possessed many of the consonantal features also found in modern speech.

These predictions have some interesting implications. First, they suggest that some nearly universal aspects of modern spoken language (e.g., constants of consonant articulation) may have appeared as early as 2 million years ago. Second, they suggest that vocal articulation was not likely to have been the sole, or even the major vehicle, for the earliest forms of symbolic communication by the immediate australopithecine predecessors of *Homo habilis.* Third, they suggest that there were many vocal abilities that were relatively late to appear in evolution, some possibly as recent as early *Homo sapiens,* and so could not have been subject to selection for any significant period of time. Fourth, they suggest that the precision and speed of many aspects of vocal articulation were in continual development during most of human evolution, probably reflected in relative brain size.

Considering the incredible extent of vocal abilities in modern humans as compared to any other mammal, and the intimate relationship between syntax and speech, it should not surprise us that vocal speech was in continual development for a significant fraction of human prehistory. The pace of evolutionary change would hardly suggest that such an unprecedented, well-integrated, and highly efficient medium could have arisen without a long exposure to the influence of natural selection. But if the use of speech is as much as 2 million years old, then it would have been evolving through most of its prehistory in the context of a somewhat limited vocal capacity. It is during this period that most predispositions for language processing would have arisen via Baldwinian evolution. This has very significant implications for the sorts of speech adaptations that are present in modern humans.

By studying the ways people analyze and produce speech sounds, Al Liberman and his colleagues at the Haskins Lab pioneered research demonstrating human beings' unique adaptations for speech. This work provided some interesting clues to exactly which aspects of speech may have been subject to some degree of genetic assimilation over the course of human evolution.[18] To summarize decades of research from the Haskins Lab and others in a single phrase: "speech is special." Humans utilize a remarkably efficient and automatized system for compressing and decompressing the information of the sound stream of speech, because speech perception mechanisms are largely automatic modular responses that preprocess many sounds below the reach of conscious awareness. Surprisingly, unlike many other aspects of auditory perception, auditory processing of speech sounds does not appear to be based on extracting basic acoustic parameters of the signal, as a scientist might design a computer to do, before mapping them onto word sounds. Speech analysis appears designed instead to predict which oral-vocal movements produced them and to ignore the rest.

Because the speech signal originates from a rather limited source, a human oral cavity, a full acoustic analysis is unnecessary; only the linguistically relevant information about what movement generated the sounds needs to be separated from the sound. What is important is the speaker's *intended* words, which are reflected in intended movements of oral and vocal muscles. All else can be ignored for the sake of word analysis (though other slower-paced prosodic cues are simultaneously present), including size of vocal tract, rate of speech, failure to achieve intended acoustic targets, and so on. This is further aided by the social evolution of many speech sound distinctions to occupy ends of phonetic spectra where differences are not fuzzy. We tend to perceive speech sounds in terms of "articulatory gestures," whose boundaries and distinctions correspond to articulatory (i.e., somato-motor) features, not just sound features, and this perceptual process is mostly automatic, taking place beyond the access of conscious analysis.

Almost certainly, oral-aural adaptations have been fine-tuned for deal-ing with the intense processing demands of spoken language during human evolution; however, the extent to which these are unprecedented in other species and language-specific is not clear. Given the incredibly rapid and facile nature of sound change in language (especially when compared to change of brain structures in evolution), and the powerful social evolution selection pressures that articulatory constraints and auditory-processing bi-ases would contribute to such sound change, it should not be surprising to find that the articulatory distinctions recruited for indicating the units of speech make the best use of predispositions that were already present be-fore the evolution of speech.

An example of such a perceptual predisposition is provided by voice-onset time, which is a measure of the time from the beginning of conso-nant sound production to the point when vocal sound is begun. Consonants such as "*p*" and "*b*" are distinguished by differences in voice-onset time. When the "*p*" sound is produced, it is characterized by an initial pressure release at the lips in the absence of vocal sound. When it is combined with other sounds, as in "*pa*," there is a slight delay before this is followed by the vocalization we associate with the vowel. This is the "voice-onset time." In contrast, in the syllable "*ba*," vocal sound coincides or slightly precedes pres-sure release at the lips, so the two phonemes are distinguished by their voice-onset time. What is curious about this unit-of-time delay is that it is highly predictable from phoneme to phoneme and language to language. Moreover, it seems to correspond to a perceptual categorical boundary. In a famous experiment, in which people were presented with artificial speech where the entire spectrum of possible voice-onset time distinctions was pre-sented, they did not hear the sounds as grading into one another, but rather as one of two distinct forms (e.g., either "*p*" or "*b*"), and made the transi-tion between the two in a surprisingly narrow band of delays, between 0 and 0.05 seconds delay. Greater than 0.05 seconds delay and one always hears "*p*," and less than 0 (i.e., laryngeal sound production starts before the lips open) and one hears "*b*" in all normal circumstances. One might be tempted to treat this phonemic universal as a special, genetically assimilated, perceptual-motor adaptation for language, were it not for the fact that this same temporal delay has also been found to provide a categorical percep-tual boundary in other species as well.[20] This does not imply that genetic assimilation was totally uninvolved. Rather, it indicates that languages have taken advantage of predispositions for sound analysis already present in the nervous system, recruiting them for this special purpose. Because of this habitual new use, the importance of this predisposition has increased, and Baldwinian selection has undoubtedly enhanced both its analysis and production.

We not only hear this sound boundary, we produce speech sounds that quite distinctly fall to one side or the other, and avoid the intermediates. This subtle and facile skill has no clear animal precedent. Baldwinian evolution has probably caused articulatory predispositions to evolve to complement those of sound analysis, but the selection pressures of sound analysis have also constrained and guided the social evolution of speech sounds in every language. Evidence has been accumulating for many other parallel examples of a sort of "preestablished harmony" between articulation patterns and sound processing used in language. In each case, a coevolutionary convergence has probably taken place, aided by the rapid adjustments of sound change in language to evolve toward optimal patterns of sound contrasts between alternative phonetic units.

The quite extensive evidence of auditory specialization and automatization is thus a sign of a longtime use of the corresponding articulatory gestures by our ancestors. Oral-aural specializations that are extensively constrained and invariable from society to society likely have a comparatively ancient origin; those that are less specialized and more variable probably have become important more recently. Thus, differences in specializations for different aspects of speech production and analysis offer evidence for differences in the age of their incorporation into regular use in speech. The ranges and variability of features in modern languages, then, may be useful clues to their antiquity, and to the structure of truly ancient languages.

For example, nearly all of the categorical production and perception specializations that have been demonstrated for speech seem to involve consonants, not vowels. Vowel sounds grade easily into one another in both production and perception, they were among the last speech sound distinctions to get encoded into written phonetic languages, they are comparatively free to vary in some languages, and they are highly susceptible to change in language histories, dialects, and regional accents. This is consistent with the independent evidence that changes in the position of the larynx in the throat have only quite recently allowed the full range of vowel production in hominid evolution. In the vocal repertoires of *Homo erectus*, vowel variations were likely insufficient to serve as reliable features for distinguishing between words, and so language changes during that phase of prehistory would have continually selected against their use and would have provided minimal selection for any auditory specializations to treat them categorically as well. But the range of nearly universal consonant distinctions and the corresponding perceptual biases that appear to have evolved as a result provide evidence for ancient speakers who used an extensive array of these oral gestures to communicate.

If something analogous to American Sign Language long predated spoken languages and served as the bridge linking the communication processes of our relatively inarticulate early ancestors, then we should expect that a considerable period of Baldwinian evolution would have specialized both the production and the perception of manual gestures. Clearly, there are some nearly universal gestures associated with pointing, begging, threatening, and so on, but these more closely resemble the nonlinguistic gestural communications of other primates both in their indexical functions and in the sorts of social relationships they encode, rather than anything linguistic or symbolic. The absence of other similarly categorical and modularized gestural predispositions suggests, in comparison to speech specializations, that the vast majority of Baldwinian evolution for language has taken place with respect to speech. Pointing may be the exception that proves the rule. This universal gesture exhibits many features which suggest that its production and interpretation are subject to innate predispositions. The fact that it appears prior to language as a powerful form of social communication in children (but not in other primates), and subsequently plays a very powerful role in children's language development, is particularly relevant evidence that it traced a complementary evolutionary path with the evolution of speech. The way it is recruited into manually signed languages and recoded symbolically for use as a pronominal marker with respect to signing position in space (among other uses distinct from its nonlinguistic indexical role) demonstrates how such a predisposition might have enhanced the ease with which it entered symbolic communication in the past in languages that were somewhat less verbally facile.

Early symbolic communication would not have been just a simpler language, it would have been different in many respects as a result of the state of vocal abilities. A number of writers have speculated that early languages would have been like strings of words without regular grammar or syntax. Some have hypothesized that the labored, poorly articulated, and telegraphic speech of Broca's aphasics might be a good model for early language. Paradoxically, however, the limitations of a less facile articulatory ability might instead have been a major cause of structural complexity in early languages. With less articulatory skill and a reduced phonetic range as compared to modern humans, our ancestors would have had fewer phonemic differences available to mark word differences, and these would have been less distinctive. This has two important consequences. First, it would increase the chances of misinterpretation due to miscategorizing the sounds and confusing similar words. Second, possessing a smaller number of sound elements would probably, on average, lead to using longer strings of sounds to form each word and longer strings of words for each proposition.[21] Thus, it would tend to be more error-prone and less efficient than modern languages. A greater utterance length might even have been further complicated by a slightly slower articulation rate as well. All this would make interpretation more prone to another limitation, short-term memory. Increasing complexity of another sort might have helped.

Consider the fact that the best way to compensate for noise or error-proneness in communication is redundancy. We tend to repeat things, spell out important words, say the same thing in different ways, or add gestures and exaggerated tonality and volume in order to overcome the vicissitudes imposed by noisy rooms, distractions, inept listeners, or otherwise difficult-to-convey messages. Redundancy is implicitly built into language structure as well. Highly predictable phonetic elements, grammatical markers that all must agree within a sentence, and predictable word-order constraints can help one anticipate what is coming. These are all direct contributors to redundancy, while regular intonation patterns, breaks and pauses, nods and gestures are all parallel signals that provide additional clues to help minimize interpretive errors. The more predictable the category of each word in an utterance, whether due to additional cues or to constrained phrase structure, the fewer the options that have to be distinguished from one another.

Thus, we should expect that early speech was constrained in word order, structured in small chunks, and embedded in a richer nonspoken matrix of gestures and exaggerated intonation. Just the opposite of the flat, unmarked, unstructured, and somewhat irregular speech of Broca's aphasics. This would be a better description of "motherese," the label often given to the exaggerated speech (usually richly embellished with gestures and embedded in almost ritualistic schemas) that adults often spontaneously employ to communicate with young children. Social habits which enable one to distribute the interpretive workload redundantly across many communication channels at once, and embed small completed chunks of sentences within other chunks, lessen the communicative demands placed on short-term memory and articulatory skill. Conversations today are inevitably embedded in rather ritualized markers for greetings, turn-taking, demonstrating assent or dissent, and indicating objects. It seems reasonable that such language rituals would have been far more prominent during the early phases of brain-language co-evolution, not due to any greater innate predisposition but in response to intense social selection on communicative habits.

With the final achievement of fully articulate speech, possibly as recently as the appearance of anatomically modern *Homo sapiens* just 100,000 to 200,000 years ago, many early adaptations that once were essential to successful vocal communication would have lost their urgency. Vestiges of these once-critical supports likely now constitute the many near-universal gestural and prosodic companions to normal conversation. Their persistence may be facilitated by long-evolved predispositions, but they have now become somewhat free of these demands and have been recruited variably and to a much wider array of linguistic and paralinguistic adaptations in modern societies. Rather than being abandoned, many of these previously critical predispositions may have contributed to the increased effectiveness and efficiency of language communication by offering new options. It may well be that with the advent of modern vocal abilities, languages for the first time fully assimilated many functions that were previously supported by nonvocal counterparts. With the added power of increased distinctiveness and more rapid articulation, more information could be packed into the same short-term memory time bins, and the full range of pragmatic functions, previously supported by links to other simultaneous communications, was finally encoded as a series of syntactic operations.

The one class of paralinguistic functions which probably had the longest co-evolutionary relationship with language is speech prosody: the changes in rhythm, volume, and tonality of speech that are used both to direct attention to what the speaker deems to be more salient elements and to communicate correlated emotional tone. The evolution of this system of indices must also have been tightly linked to the evolution of speaking abilities because they are effectively opposite sides of the same neurological coin. As we discovered in Chapter 7, these prosodic elements of vocalization resemble stereotypic call systems in their use of intensity, frequency, and rhythmic modulation of vocal sound, and in their likely involvement of limbic and midbrain vocalization circuits. Since human cortical control of these same output systems is achieved by a more direct set of projections, which bypasses these emotional-arousal pathways, the two systems are in this respect parallel and complementary to one another anatomically as well as functionally. To some extent, they also operate on different though synchronized time scales. Word articulation takes place within fractions of a second, whereas prosodic changes unfold over seconds across whole utterances, punctuated by breath cycle. Their seamless complementarity is a sign of a long and consistent selection history. This may indeed be an additional clue to the origins of consistent laterality of language-related functions. As noted in Chapter 8, sites of brain damage that produce aprosodic speech or difficulties in analyzing emotional information encoded in prosody often involve the right cerebral cortex, opposite the "language-dominant" hemisphere. There may even be a sort of complementary mirror image location of Broca-like and Wernicke-like aprosodias. The significance of this functional segregation could be as a response to the potential for competitive interference between these systems, since they share both a common output pathway (speech) and a common feedback pathway (audition). An evolutionarily predisposed bias, perhaps in the form of differential right/left proportions of limbic versus nonlimbic connectivity in the immature brain, might support a more reliable developmental segregation of these functions and minimize interference effects.

Thus, as was also suggested for symbolic analysis, lateralization may not so much reflect the need to put speech under the control of only one hemisphere—it clearly is not—but rather the advantages of isolating parallel and potentially competing speech functions from one another. The outputs from the right hemisphere, producing the slower modulations of background subglottal pressure and vocal muscle tension, and from the left hemisphere, producing the rapid-fire articulatory movements, can be superimposed without conflict on the final common motor pathway analogous to the way a sound wave can be superimposed on a much faster carrier wave in a radio transmission. Rather than degenerating in the face of the takeover by cortical motor projections in human evolution, the systems involved in linking arousal and emotional state to vocalization have become more elaborated along with speech. This too provides strong evidence for the long use of vocal articulation in hominid evolution.

The Writing on the Wall

Our prehistoric ancestors used languages that we will never hear and communicated with other symbols that have not survived the selective sieve of fossilization. In their absence, it is tempting to use the minimal evidence of prehistoric material culture as an index of the mental and linguistic complexity of our ancestors. It is almost certainly a reliable expectation that a society which constructs complex tools also has a correspondingly sophisticated symbolic infrastructure, and that where tools and other artifacts differ from region to region, there is likely to be a corresponding diversity in traditions of tool use, resource production, and social organization. Moreover, a society that leaves behind evidence of permanent external symbolization in the form of paintings, carvings, or just highly conventional doodlings likely includes a social function for this activity. In short, archeological artifacts are one of the few windows through which we can glimpse the workings of the "mental" activity of a prehistoric society.

The problem is that this logic is not valid in reverse. Though we can deduce from artifacts what minimal social conditions are required to produce them, we cannot so easily deduce from their absence that certain social conditions are missing, and we certainly cannot deduce that the potential to produce them is lacking. This is evident from a comparative look at modern human societies. Prior to the twentieth century many believed that people in societies that still used stone tool technologies for hunting, farming and warfare were also more primitive in biological terms than were European or Asian people. This was an integral part of the Eurocentric conceit during the colonial era, and was used to justify numerous injustices, from genocide to slavery. Anthropologists in the early twentieth century quickly realized that the technological status of a society was no predictor of the complexity of its language or the symbolic richness of its traditions. In general, the average member of any society probably has roughly the same amount of linguistic and cultural information "in his or her head." There are no living *primitive* languages, in the sense of being more simple and elementary in structure, not even in ancient written texts. Children who have come from societies that still utilize stone tool technologies can adapt to modern industrial society and absorb its intellectual tradition as easily as those born within it. Not only does a lack of preserved archeological material say little about the likely extent of "perishable" culture; it says nothing about the capacity for it. It is instead a marker for a variety of ecological and historical variables that have selected for the social value of producing highly durable symbolic objects.

There is another related problem that is particularly troublesome for stone age archeology. The artifacts we find are a very biased sampling of a culture's physical traces. Only certain kinds of objects in certain environments get preserved. The primary factors determining what gets found and what disappears after tens of thousands or millions of years are the perishability or fragility of the material and the nature of the environments in which they have been left. In simple terms, this means that stone tools are preserved in the fossil record whereas wood and leather tools are not; or that paintings on cave walls and stone or ivory carvings are preserved whereas body decoration, clothing, wooden sculptures, and so on, are not.

Most of the symbol use in a society, even excluding language, is not even embodied in any material, but only in ceremonies, habits, and rules that govern everyday life. And what we know about most artifacts is that the vast majority are made of perishable materials. This is particularly true of foraging peoples, who move continuously from place to place. Like the words that prehistoric people spoke, the vast majority of their creative efforts would have produced results that vanished with them, or shortly afterwards. As in so many scientific enterprises, we must always keep in mind the caveat that "absence of proof is not proof of absence."

Imagine the impoverished view of African Pygmy societies we would construct from their bones and stone artifacts alone (excluding those things borrowed from outside societies). We would miss the incredible richness of their ritual life, not know of their complex languages, be unaware of their use of natural poisons in hunting, and not realize that they had developed elaborate musical instruments and musical traditions, among the many other things of which we would be ignorant. The simplicity of their tool kit as compared to agricultural and metal-using societies might suggest to some future archeologist that a great mental gulf separated these people from other contemporary peoples. Since there is good reason to suspect that this small population could well disappear as a distinct people in the very near future, we can imagine this future archeologist interpreting their disappearance, in the context of their few artefacts and smaller brains, to mean that their mental inferiority caused them to be supplanted by biologically more advanced hominids.

We know, in this case, that such conclusions would be entirely erroneous. The Pygmy people of the forests of Central Africa are every bit our physiological and intellectual equals. The differences in nonperishable material culture are simply the result of a sampling bias. This particular cultural adaptation might even be argued to be optimal to the extent that, unlike the destructive use of this habitat that has come in the name of "development," the Pygmy adaptation seems on the whole to be more in balance with the ecosystem and therefore more sustainable.

The significance of this argument for interpreting archeological data is that artifacts are not reliable indicators of mental abilities, and that lack of artifacts does not indicate a lack of the potential to produce them. Inferring intellectual and cultural complexity from preserved artifacts in a very different species, living a lifestyle that probably has no clear counterpart in recorded anthropology, should be even more prone to error than is suggested by this Pygmy thought experiment. We should need to be even more restrained in our tendency to interpret what we don't see. Careful archeology adheres to the maxim that the only supportable speculations are those that proceed from existing artifacts to what these require in the way of supportive contexts (both physical and social) and what they alone imply about their possible uses. Reasoning from absence of artifacts can only be allowed in very special contexts.

In the case of pre-agricultural peoples, we can be pretty sure about the introduction of behaviors associated with specialized tools, like fish hooks and harpoons, but can't know whether or not cave painting and ivory sculpture were the earliest forms of representational art. The date of the introduction of hunting or scavenging of meat cannot be definitely determined, but the point at which we can be sure it already exists can be inferred from the earliest appearance of tools designed for cutting flesh: sharpened stone tools. We can also argue that before these tools were available such activities were less efficient, and by further extrapolation, that they probably were a less critical component in the subsistence of earlier hominids (if they took place at all).

These caveats are often ignored when it comes to the consideration of the evolution of language and intelligence in our ancestors. The assumption that "simple tools mean simple minds" is often the tacit operating principle. As a result, technological change and variety are often interpreted as evidence of cultural and biological advance, and, conversely, lack of technological change is often interpreted to reflect lack of advance, which from a modern, progress-centered viewpoint is given the negative connotation of stagnation.

This problem has followed many efforts to extrapolate from the physical cultural evidence left by our ancestors to their language abilities. For example, *Homo erectus* can be traced through a fossil record spanning 1.5 million years, but during this time archeologists have identified a relatively stable array of tool types and tool manufacture techniques across a vast range of dates and habitats throughout the Old World. In contrast to this apparent stability, within the last one hundred thousand years of *Homo sapiens'* presence on earth there were increasingly rapid advances in tool-manufacturing tricks and a growing diversity of tool types. This has often been interpreted as evidence of a sort of cognitive stagnation in the *erectus* epoch of hominid evolution, replaced by a subsequent rapid progress in *Homo sapiens*. Part of this view may be the result of a misdating problem. Tools found in Southeast Asia, and originally thought to date to late stage *erectus* populations, reflected a pattern typical of the earliest stages in Africa. At this same later time in Africa there appear to be more sophisticated tools. However, recently new dating of the Asian fossils has indicated that they are as ancient as the earliest African *erectus* fossils. Members of *Homo erectus* apparently spanned the extent of the Old World, from Africa to Java, within as little as a hundred thousand years after their first appearance in Africa. When these Asian *erectus* populations were thought to be more modern than their African counterparts, the apparently primitive tools found in Asia appeared to suggest a lack of technological innovation over a very long time. Some later *erectus* tools are made of volcanic glass and shaped with a chipping logic that involved multiple stages and chipping tools, and this is quite different from those found at the beginning of this species' evolution. There were also other technological adaptations that appear only in later stages of *erectus* evolution. It seems probable from the appearance of hearthlike fire sites discovered in Africa, Europe, and China that later members of *Homo erectus* also used fire. Thus, reports of *erectus'* mental stagnation are likely exaggerated.

The pace of change and diversification of *erectus'* artifacts may pale in comparison to that which followed associated with groups of *Homo sapiens,* but this is not sufficient evidence to suggest that *erectus* had reached some plateau or "grade" of mental evolution, which was only eclipsed by later *Homo sapiens. Homo erectus* achieved a highly successful adaptation. Almost certainly it was supported by a complex symbolic culture that allowed them to organize reproduction, work, and social negotiations with an efficiency and a flexibility unknown in any previous species. Symbolic communication would have been essential for passing on information about survival details that were specific to the incredible range of habitats they conquered. Their stone tool technology was a very successful adaptation, in part, because it needed little modification to succeed in very different habitats, from subtropical to subarctic climates. Hunting and butchery do not require entirely different tools for different habitats or different species of prey animal. What kills or cuts flesh in one place will do so in another. If it works, it doesn't need fixing. This common tool technology could have been associated with considerable diversity of cultural and linguistic traditions. During their tenure on earth, members of *Homo erectus* used their symbolic and stone technologies to adapt to a diverse set of habitats and evolved brain sizes that match those of some normal humans living today. Now that more accurate dating has placed some of the Asian members of this species earlier in time, it also appears that there was a significant trend toward brain size expansion from the earliest to the most recent *erectus* populations. Their brains and their symbolic forms of communication were undoubtedly co-evolving together, even if the tools they were using were not progressing at a comparable pace. Though the forms of symbolic communication used by these people may also not have been improving in any simple linear sense, it is likely that there was both considerable diversity from population to population and considerable change from early to late *erectus*. Would we have called their symbolic communication "language"? Was it spoken? The answer to these questions is almost certainly both yes and no, because symbolic communication would not have been confined to one facet of social communication, any more than it is in modern societies. Regional variations may have favored an emphasis on vocal symbols in one population and an emphasis on gestural forms in another, but as we have seen there are a number of reasons to suspect that vocal communication was not as autonomous and elaborate in *erectus'* societies as in modern societies.

Another example of overzealous extrapolation from human artifacts to human minds has played a major role in theories stressing a relatively recent origin of speech. Many scenarios for the origins of language have focused on the middle and later stages of the Upper Paleolithic period, between about 40,000 and 15,000 years ago, as a source of evidence for the first use of symbols. The paleocultures of this period represent a major change in the kinds and varieties of technological and cultural artifacts. They also include the first surviving evidence of something akin to representational art. Taken together, these two indications of cultural sophistication of a sort more recognizable by modern standards have been cited as signatures of a major revolution in the symbolic capacity of these people—a revolution created by the "discovery" of spoken language.

Assuming a late discovery of speech also fit with early claims that Neanderthal populations lacked modern vocal abilities (and particularly some extreme claims that Neanderthals would have been incapable of language). Circumstantial evidence of vocal tract structure in Neanderthals had suggested to many researchers that their speech abilities might have been far more limited than those of modern humans. Because the major explosion of late Stone Age culture occurred only at the end of the Neanderthal occupation of Europe, and seemed largely to be confined to the more modern-looking *Homo sapiens*, it seemed reasonable to see this transition from one population to another, from one tool technology to another, and from a period without evidence of representational art to one with it as driven by some major change in information transmission. However, the timing and location of the transition and its association with changes in hominid populations have also undergone recent revision. Modern humans did not evolve from Neanderthals, and the association of variable and sophisticated tools with anatomically modern humans and simpler stereotypic tools with more archaic humans is not so neat and simple. The classic image of the dumb and brutish Neanderthal pitted against a vocally and culturally sophisticated Cro-Magnon is not well supported by the evidence.

More recent finds and more accurate dating of sites have demonstrated that the fossils of anatomically modern humans predate many Neanderthal fossils, though not in Europe. Evidence of more anatomically modern facial structure is found in fossils from caves in Africa that date to between 75,000 and 115,000 years ago. This puts them in the same time range as some of the oldest Neanderthal fossils. In the Middle East, anatomically modern remains found in a cave at Qafzeh are probably at least 90,000 years old. This is particularly devastating to the view of Neanderthals as our ancestors because nearby Neanderthal remains, in Kebara Cave, are probably no more than 60,000 years old. In other words, both types of recent humans occupied the same region of the Middle East, and neither were using Upper Paleolithic tools or producing paintings and sculptures.

The demise of the Neanderthals between 30,000 and 35,000 years ago is often portrayed in terms of classical survival-of-the-fittest competition: the replacement of an inferior subspecies by a superior one. The anatomically modern subspecies is assumed to have been intellectually superior, and endowed with complex language abilities and more sophisticated tools. The Neanderthals are instead presumed to have been adapted to the rugged subarctic climate of the recent Ice Ages, where strength and endurance were thought to be more critical than wit and communication. In this scenario, the anatomically modern populations presumably triumphed in an ecological competition with the Neanderthals because their superior ability to communicate made them able more efficiently to adapt to the changing climate of Europe as the ice sheets receded. There is considerable intuitive appeal to the view that Neanderthals were undone because they were mentally inferior, lacked modern language, and so had simpler tools; a smug comfort in imagining that the culturally and intellectually superior Cro-Magnons vanquished the dumb brutes—that living by one's wits won over living by strength and endurance. This is the sort of interpretation that prompts Misia Landau to warn of our tendency to read the fossil evidence through the lens of a hero myth.[22]

For the period that Neanderthals occupied Europe there are mostly tools from what is called the Mousterian stone tool industry, one that was shared with earlier anatomically modern humans as well in other parts of the world, including the Middle East. After the Neanderthals are gone, there is indeed a development of more sophisticated types of tools and the first appearance of durable art, but is this an incidental temporal sequence or did this difference make the difference? These more complicated and

varied tools appeared first in North Africa, tens of thousands of years earlier, and only spread across Europe in an east-to-west direction with anatomically modern populations and as Neanderthals were on the wane. It is
unclear whether this technological change was a cause or merely a correlate of the disappearance of the Neanderthal populations, and Neanderthals
were not left entirely out of the loop of this cultural modernization. There
is now clear evidence linking a tool and artifact period called the Chatelper-
ronian with the last of the Neanderthals in France, approximately 34,000
years ago. These artifacts include diverse tool types, a rich bone and ivory
industry, and artistically crafted adornments.[23] All of this makes it difficult
to argue that better tools and a more sophisticated culture distinguished one
of these two near-modern human groups and made the difference between
its expansion and the other's demise, and it particularly calls into question
the assumption that this replacement can be blamed on a hypothetical inferiority in Neanderthals' language abilities.

In neurological terms, it seems likely that Neanderthals were fully modern and our mental equals. They had a brain size slightly above modern values, and a slightly smaller stature, and so we can extrapolate that the internal
proportions of their brain structures were consistent with a symbolic capacity equal to anatomically modern humans.[24] The argument that the Neanderthal vocal tract was not well suited to produce modern-sounding
speech may also have been exaggerated, if we cannot rely on the assumption of their primitiveness and some questionable extrapolations from reconstructed skulls. The discovery of a Neanderthal hyoid bone and an
analysis of the attachment facets (that should indicate its relative position
in the throat) have provided evidence that they may have had a more modern vocal apparatus than previously suspected.[25] But even if we accept the
more extreme reconstructions that place the Neanderthal larynx very high
in the throat, we can hardly doubt that they possessed a symbolic communication system every bit as sophisticated as their anatomically modern
contemporaries.

Neanderthals were a local variant population of archaic *Homo sapiens*,
isolated by geography and inbred due to their small numbers. From this perspective, there are obvious similarities between the Neanderthal demise and
the demise of many historically recent indigenous populations during the
history of conquests in the Old World and the recent colonial expansions
into the New World and the tropics. Perhaps the closest parallel is to the
even more rapid decline of native populations in the Americas after Colum-
bus. Although the popular conception is of Conquistadores killing off the
indigenous peoples in warfare, this actually accounted for only a fraction of

the massive genocide that ensued. The real culprit, right from the beginning, was disease. Within the first two centuries after contact, it is estimated that 80–90 percent of the native populations had been eliminated by diseases introduced from Europe to which these populations had no immunities. The pattern was repeated throughout the globe during the age of colonization by European powers, and took a particularly high toll on island populations, such as that of Hawaii, because of their relative reproductive isolation and inbreeding.

In the circum-Mediterranean world, there had been millennia of migrations, wars of conquest, long-distance trade relationships, and repeated convulsions of epidemics. As a result, Europeans had experienced millennia of selection for resistance to diseases imported one after another from many continents, and causing widespread epidemics before burning themselves out. The buildup of resistance to these diseases in European descendants was not just a result of elimination of those lacking resistance, but it was also supported by the genetic variety provided by extensive interpopulation gene flow. However, when these same diseases were brought, *en masse,* so to speak, to smaller, more isolated and inbred populations, these many diseases hit all at once, like plagues upon plagues, and found hosts that were minimally resistant. It was not the superiority of the European mind, or even of European technology, that cleared the way for the Europeanization of the New World, but their demographic history as part of a larger Old World pandemic and pangenic system.

This historical tragedy bears many similarities to the interaction between Neanderthals and the encroachment of anatomically modern "colonists" from the South and East. The Neanderthals were scattered in small populations and were probably genetically isolated from the remainder of *Homo sapiens* for as much as a hundred thousand years, before this second contact in Europe. They were surrounded by an anatomically modern population that extended from South Africa to the Near East and parts of Asia. And during a comparatively brief period these outsiders began to move into Europe. Whether or not warfare, resource competition, trade, peaceful coexistence, or even mixing and interbreeding characterized their interaction, it seems almost certain that the Neanderthal gene pool would have been doomed from the moment of first contact. No special story about cultural and technical superiority or the origins of language need be invoked to account for it.

Though these comparatively recent events represent the most rapid and radical changes observed in the entire Paleolithic, there is no clear biological transition to correlate with it. It occurs in the context of populations in

Europe apparently adapting to somewhat more specialized hunting and gathering niches—beginning to utilize animal sources other than just the herds of ungulates that fed hominids since the dawn of stone tool technology, including fish, seals, and mammoths among others—and perhaps experiencing the consequences of their success as their efficient hunting progressively depleted precisely those resources they had specialized to exploit. There is, in short, another set of possible engines underlying rapid technological change and variation: changes in ecology, both as a result of climate change and as a result of human activity itself. The latter of the two has almost certainly become the more dominant force from that point on. It can hardly be coincidental that this expansion of anatomically modern populations heralds a recent epoch of large animals' extinction which claimed such Ice Age giants as the mammoth and the giant sloth. These transitions and the subsequent transition to agriculture in the Middle East probably both reflect a similar dynamic: the need to adapt to irreversible changes in the environment brought on in part by prior human adaptations. This is a view of "progress" that is not so much improvement as irreversibility, adaptations that are so successful that they are self-undermining, another twist on the adage, "Necessity is the mother of invention." In general, we should not invoke biological evolution as a cause of cultural-technological innovation where demographic and ecological processes can suffice to explain the changes. Conversely, we should not assume that relative stability of technological adaptations precludes continued biological evolution.

The first cave paintings and carvings that emerged from this period do give us the first direct expression of a symbolizing mind. They are the first irrefutable expressions of a symbolic process that is capable of conveying a rich cultural heritage of images and probably stories from generation to generation. And they are the first concrete evidence of the storage of such symbolic information outside of a human brain. They mark a change in the structure of human cultures at least to the extent that they are evidence of the use of media that have persisted to the present, but they don't correlate with any "advance" in human biology or neurology, and they probably don't demonstrate the origins of symbolic communication or even spoken language. Perhaps the effectiveness of new tools depleted a once-sustainable food source and forced these people to discover radical alternatives supported by new tool innovations; perhaps a changing foraging ecology altered group size and/or residence patterns, allowing people to accumulate and pass down objects other than just tools; perhaps the interaction of long-isolated African and European populations spurred change in response to exchange of information and technology; perhaps the presence of a medium like ivory or the regular use of caves in a cooler climate provided a hitherto unavailable permanent medium for expression; or perhaps an abundant source of game simply led to a people with more time on their hands.

The point of mentioning these alternatives is not to prove one or the other of these to be relevant, but to serve as a reminder that our ancestors were not inevitably headed toward painting, sculpture, and more varied and or-namented artifacts. The appearance of these things just a few dozen thou-sand years ago is more likely a reflection of incidental ecological changes and access to more durable media than it is an indication of some revolu-tion in communication. Indeed, probably many of the events we view as ad-vances, from our hindsight perspective, were desperate responses to the environmental degradation that human foraging success itself brought on. Nevertheless, though the invention of durable icons may not indicate any revolution in human biology, it was the beginning of a new phase of cul-tural evolution—one that is much more independent of individual human brains and speech, and one that has led to a modern runaway process which may very well prove to be unsustainable into the distant future. Whether or not it will be viewed in future hindsight as progress or just another short-term, irreversible, self-undermining trend in hominid evolution cannot yet be predicted. That we consider this self-undermining process advance-ment, and refer to the stable, successful, and until just recently, sustainable foraging adaptation of *Homo erectus* as "stagnation," may be a final irony to be played out by future evolution.

Symbolic Origins

. . . the human understanding is greatly indebted to the passions.

—Jean Jacques Rousseau, *Discourse on Inequality*

A Passion to Communicate

We have some idea what happened to the brain and body in the process of language evolution, and we have an estimate of when this began and how it developed in our prehistory, but none of this answers a more nagging question. Why? What was the spark that kindled the evolution of symbolic communication? If symbolic communication did not arise due to a "hopeful monster" mutation of the brain, it must have been selected for. But by what factors of hominid life? How can we discover the context of this initial push into such a novel form of communication? Can the forces favoring such an unprecedented evolutionary adaptation be reconstructed in hindsight? Or are the conditions underlying the critical transition from indexical to symbolic communication, and from animal to human minds, forever lost to analysis along with with the flesh, blood, and day-to-day behaviors of our fossil ancestors?

On the cover of the *Cambridge Encyclopedia of Human Evolution* and also on the cover of the book *Ape Man,* a companion volume to the television series by the same name (produced for the *A&E* Television network), there appear images of "morphed" faces that are not quite ape and not quite human.[1] "Morphing" is a computer image manipulation whereby one generates a spectrum of intermediates between different images that, if animated, smoothly transform the one into the other. These two cover illustrations depict half-morphed stages between an ape and a human face. The eerie realism of these images resonates with a deep predisposition for what we might call "reconstructive hindsight." In thinking about evolution, we naturally imagine a sort of morphing from one species to another. But we should be wary of letting our images of evolutionary change direct our thinking about it, because hominid evolution doesn't resemble a 5-million-year morphing from a chimplike species to *Homo sapiens*. There never was a creature who was half-ape, half-human, who was a bit more dull than most people but slightly more savvy than most chimps, and who communicated with a half-language. We can't just extrapolate from present species and their adaptations back to a state before language and culture. We must instead try to reconstruct the adaptations of our ancestral species from principles of behavioral ecology, physiology, and information about brain structure and function.

From the perspective of hindsight, almost everything looks as though it might be relevant for explaining the language adaptation. Looking for the adaptive benefits of language is like picking only one dessert in your favorite bakery: there are too many compelling options to choose from. What aspects of human social organization and adaptation wouldn't benefit from the evolution of language? From this vantage point, symbolic communication appears "overdetermined." It is as though everything points to it. A plausible story could be woven from almost any of the myriad of advantages that better communication could offer: organizing hunts, sharing food, communicating about distributed food sources, planning warfare and defense, passing on toolmaking skills, sharing important past experiences, establishing social bonds between individuals, manipulating potential sexual competitors or mates, caring for and training young, and on and on. This apparent lack of constraint on what constitutes a plausible hypothesis is one source of the frustration that caused past researchers to be less than charitable toward language origins scenarios. Could *any* significant reproductive advantage conferred by language be considered a plausible candidate for its prime mover? What would make some more plausible, more supportable, or more falsifiable than others? Though many of the myriad uses for symbolic communication are probably sufficient to provide selection for enhancing symbolic learning and associated speech abilities, they already assume a form of symbolic communication that is intrinsically superior to all preceding forms of social communication. This is precisely what we cannot assume to explain the origin of symbolic communication.

Though modern languages exhibit these assumed communication advantages, the first symbolic systems were almost certainly not full-blown languages, to say the least. We would probably not even recognize them as languages if we encountered them today, though we would recognize them as different in striking ways from the communication of other species. In their earliest forms, it is likely that they lacked both the efficiency and the flexibility that we attribute to modern language. Indeed, I think it is far more realistic to assume that the first symbolic systems would have paled in efficiency and flexibility in comparison to the rich and complex endowment of vocal calls and nonverbal, nonsymbolic gestural displays exhibited by any of our primate cousins. The first symbol learners probably still carried on most of their social communication through call-and-display behaviors much like those of modern apes and monkeys. Symbolic communication was likely only a small part of social communication.

But that's only half the problem. As we have seen, even learning the simplest symbolic relationships places heavy demands on a rather questionable learning bias. In this trade-off lies the explanation for the failure of languagelike, symbolic communication to evolve in all but one species. The cognitive requirements for efficient associative learning are in many ways in conflict with those that would enhance symbol learning. Attention to higher-order, more distributed associations and away from those based on temporal-spatial correlations may render these other forms of learning somewhat less efficient. And to learn symbols it's necessary to invest immense effort in learning associations that aren't much use until the whole system of interdependent associations is sorted out. In other words, for a long time in this symbol-learning process nothing useful can come of it. Only after a complete group (in the logical sense) of interdefined symbols is assembled can any one be used symbolically. To approach most learning problems with the expectations and biases that would aid symbol learning would be very inefficient for most species. Both the difficulty and the costs of learning symbols have kept other species from evolving symbolic abilities. Only if there were significant advantages to symbolic communication, that greatly outweighed these costs, could there be selection that both favored the unusual learning strategy and compensated for the decline in effectiveness of others.

Seen in this light, the problem of explaining symbolic origins is much more challenging. Even a small, inefficient, and inflexible symbol system is very difficult to acquire, depends on significant external social support in order to be learned, and forces one to employ very counterintuitive learning strategies that may interfere with most nonsymbolic learning processes. The first symbol systems were also likely fragile modes of communication: difficult to learn, inefficient, slow, inflexible, and probably applied to a very limited communicative domain. And we must assume that the vast majority of social communication was mediated by a more or less typical primate repertoire of nonsymbolic vocal, olfactory, and gestural displays. If the initial symbolic adaptation was neither more efficient nor more flexible than preexisting forms of communication, then the incredible power that modern languages confer on their users cannot be invoked to explain the origin of language. Neurologically and semiotically, symbolic abilities do not necessarily represent more efficient communication, but instead represent a radical *shift in communicative strategy*. It is this shift, not any improvements, that we eventually need to explain.

This form of communication must have provided a significant selective advantage to the majority of those who employed it. In what ways would symbolic communication, even if very rudimentary and inefficient, provide a means of adaptation that would be unavailable using other forms of communication, even if far more sophisticated and efficient in comparison? A generally less efficient form of communication could only have gained a foothold if it provided something different, a communicative function that was not available even in a much-elaborated system of vocal and gestural indices. Given these disadvantages, what other possible selective advantage of symbolizing could possibly have led a group of hominids to incur such costs? What difference could have offset the enormous costs in cognitive effort, time, social organization, and reduced efficiency? To answer these questions, we need to know something about the context in which the first symbolic communication evolved.

Before we can begin to understand the circumstances that provided the selection pressures to favor such a radical shift in communicative strategy, we need to understand more generally what circumstances tend to produce significant evolutionary changes in communication in other species. In general, this occurs in the context of intense sexual selection. In the classical terminology of ethology, the process whereby some behavior became progressively modified and specialized for its communicative function was called "ritualization," by analogy to human ritual and ceremonial embell-ishments of communication. The elaborate tail feather display of a male pea-cock, the hanging-upside-down display of a male sulfur-crested cockatoo, or the straw archway elaborately decorated with shiny stones and bright-colored objects pilfered from human societies by a male bower bird, all are elaborations of communication that have been driven by sexual competi-tion between males in order to attract females. But it is not only males that may be affected. Consider also the incredible part-swimming, part-flying, part-walking display of male and female grebes as they run in paired for-mation across the surface of a lake, or the head-bobbing conversations of mated penguins meeting on a crowded rookery. These are also forms of sex-ual communication that evolved to negotiate mate choice and pair-bond maintenance. But in these cases, the behaviors are not just used by males to attract prospective mates, they are used by both sexes to assess each other. In all these examples, we see the evolutionary elaboration of behaviors whose only significant function is communication. They occur at these times because communication is somehow more important for these con-texts. How do the specific requirements of assessing prospective mates and choosing mates drive these evolutionary trends?

In evolution, the transmission of genes is the bottom line, and the only way to do this is to reproduce or to help close relatives reproduce. But re-production in complex animals often involves a lot more than just produc-ing a fertilized ovum. This is particularly true for birds and mammals, where parental care can be a crucial part of reproducing. Not only must an ani-mal grow and mature to the point that it is physiologically competent to re-produce; it also may need considerable social experience to be successful. When caring for offspring demands a contribution from both sexes, com-peting for access to many mates or finding mates with the best physical traits has less of an impact on reproductive output than does direct care of one's mate and offspring. The need to exchange information about parenting abil-ity is similar in each sex, whereas obtaining information about desirable physical traits is not crucial. Even in species where only one sex is neces-sary to care for the offspring (typically, but not always, the female), as is the case in the majority of mammals, the opposite sex (typically males) is not freed from the need for maturity and experience. Natural selection for basic survival abilities is amplified and supplemented by sexual selection for "desirable" traits in these cases. If one sex is essentially freed from the need to care for offspring, its energy can be redirected to maximizing copulatory access to the opposite sex by competing for optimal feeding territories or directly competing over mates. This can have significant reproductive consequences, and in such a case the information-gathering needs of males and females may be quite different.

The most common pattern of reproductive social behavior among mammals is polygynous mating. In most mammals, the immense physiological and time investments of pregnancy and lactation make females and their access to food the limiting resources. Females make the most of their reproductive effort, not by producing as many babies as possible, but by directing as much support as possible to each infant. For a male in such a species, however, the number of fertile females that can be impregnated determines his reproductive potential. From the female's perspective, since only one male is likely to impregnate her at any one time, and all her efforts will be directed toward his progeny, she must be much more selective in who she accepts as a mate than is the male. This tends to promote the evolution of complex means of extracting information about prospective male mates, and of means for producing information that can influence and otherwise manipulate the female's choice. In other words, males try to communicate information that favors them, and females try to receive information that gives them a basis to choose between the males and cut through any potential misinformation provided by them.

Often, this transmission and receipt of information is "incidental" to other activities, to the extent that communication is not the sole function. This is the case for fighting behavior. Where extensive competition exists among males for access to fertile females, threats and fighting behavior are common between males. Such species are often said to engage in "tournament" behaviors. Although the point of male aggression in these conditions is to displace an adversary who is also vying for the same territory or the same mates, females can learn a good deal about the physical and mental abilities of the competing males by observing their confrontation, and thus make their choice based on performance. The success of one male or his ability to exclude all others from a valued territory may even be information enough. If the competition accomplishes the assessment for the female, the details of what ensued in the process may add little other useful information. The relative fighting succes of males is a representation—specifically, an index—of their comparable differences. As a result, the direct communication between male and female in such highly combative polygynous species may be minimal. If a female has sufficient information to determine that a male has been the most successful defender of a territory or a group of females (something that may be implicit in the social context), then little else about his state may require cautious interpretation. There will be little selection pressure for females to develop special means of getting more information from the males, or for the males to provide the females with any additional information about their physical state. Their communication mostly involves coordinating copulatory behavior—something that can be accomplished with a minimum of assessment and analysis, and just a few relatively automatic responses. This is why, from a human perspective, sexual communication and copulation in highly polygynous species often appear as perfunctory acts with few preparatory or follow-up interactions.

In contrast to the minimal communication between males and females in tournament systems, the communication between competing males under these circumstances may be extensive and highly elaborate. If a male in a highly polygynous and aggressive species misjudges another's physical prowess or willingness to risk physical harm, the consequences can be serious. One possible outcome is getting needlessly injured. If losing and being harmed are inevitable, it's better not to have engaged in conflict in the first place, since this may diminish future reproductive chances. Alternatively, if overcoming one's opponent is very likely and the chance of harm slight, then failure to engage is costly in direct reproductive terms. For these reasons, it may also be advantageous to provide misinformation to an opponent, to lead him to make a mistake in judgment; and inversely, it may be advantageous to check the information again and again in different ways, for signs of inconsistency.

When the reproductive and health stakes are both high, there is intense selection pressure to provide high levels of disinformation to competitors and to elaborate and prolong the assessment process before aggressive engagement in order to maximize the opportunity of seeing through the other's deception to gain a reliable estimate of the likely outcome. These complementary pressures cause both deception and detection adaptations to escalate over the course of evolution. This can produce a kind of short-term "runaway" process that is finally stopped when the other costs associated with continued elaboration of communication begin to diminish any value of doing a little better than one's competitors. In other words, spending too much time and energy deciding whether or not to attack or defend can be just as costly in reproductive terms as making periodic mistakes.

Very different constraints and forces of sexual selection influence the evolution of communicative behaviors in species where both males and females must care for and defend their offspring. Because of their common reproductive interests, dual-parenting species tend to form cooperative pairs with strong exclusive (emotional) attachments to one another—pair bonding. Nevertheless, the underlying dynamic of the evolution of communication is the same: where the potential for inaccurate assessment of the other is high, and the potential costs of misjudgment or the advantages of deception are also high (in reproductive terms), communication will tend to be more elaborate and more complex. Pair-bonding species are under pressure to assess the other's physical condition, resource defense capabilities, care-giving abilities, and likely fidelity.

However, the interests of prospective mates are not entirely symmetric. A male can still improve his reproduction by cuckolding another male (who will consequently raise young that are not his own), and a female might be able to maximize both care-giving and genetically valued traits by soliciting copulations from multiple males, so long as at least one remains to care for her young. From the other's perspective, however, either deception is reproductively costly. A philandering male is more likely to abandon one female for another, and a solicitous female is likely to cause her mate to spend time and energy supporting other males' offspring. Any tendencies that allow these deceptions to slip by unchecked will be strongly selected out. Evolution will therefore favor mechanisms for avoiding getting stuck with an incompetent, unreliable, or unfaithful mate. This requires that one check out prospective mates very carefully ahead of time, and continually recheck them throughout the parenting relationship. Of course, prevention is more advantageous than post facto reliance on punitive responses, once the damage is already done, and so courtship is the most important period in the relationship for accurate assessment.

Thus, it is at the point in the life cycle where choice of mate takes place that evolutionary theory predicts we should find the greatest elaboration of communicative behaviors and psychological mechanisms in both pair-bonding species and polygynous species, though the communicators and the messages may differ significantly in these two extremes. Between these extremes there are many more complex mixtures of reproductive social arrangements that add new possibilities and uncertainties, and thus further intensify selection on the production and assessment of signals. This is especially obvious in primate species living in large multi-male/multi-female groups, such as chimpanzees and baboons. In these species, communication about mate access and mate choice additionally involves group dynamics such as developing and sustaining coalitions between cooperating males or females, even as other aspects of physiology (for example, the highly arousing sexual swelling and solicitous behavior of females in estrus) act to undermine these relationships.

Why Human Societies Shouldn't Work

Though, on the surface, we might characterize the mating pattern of *Homo sapiens* as pair bonding, this is a misleading oversimplification. Every fan of human drama—from Sophocles to Shakespeare to television soap opera—knows implicitly that the relationship between the sexes in human societies is at the mercy of powerful social and sexual undercurrents, which both form and destroy human reproductive bonds and so constantly threaten the facade of social stability. The inevitable conflict between sexuality and the constraints of social cooperation has led to elaborate means of regulating reproduction, and has produced the great variety of sanctioned reproductive arrangements in present-day human societies. This aspect of social life is where I suggest we focus our initial efforts at identifying the selection pressures that led to the evolution of such an unlikely and novel means of social communication as the use of symbols. Understanding how we differ in the social negotiation of reproductive activities is likely to provide the most important clue to the forces that shaped our social communication.

Is there anything in the social context of hominid reproductive choices as unusual, compared to other species, as our distinctive mode of communicating? We are forced to turn to modern humans for the first clues, since the evidence of social behavior in our ancestors is only circumstantial at best. But despite the potential for getting misled by our own anthropocentric biases, this may not be a bad place to start, because it is likely that whatever first selected for this otherwise anomalous form of communication is still with us in some form. What trick of social engineering could have been so powerful and so peculiar in the history of animal communication that it required a totally novel form of communication? Such a trick—one that no other species has hit upon—is not likely to have been cast away too quickly in the subsequent evolution of our species.

What do we see when we take an "outsider's" view of human reproduction and sexual selection? We have a tendency to view other animals' social and reproductive relationships in the image of our own, and therefore to perceive what we do as normal reproductive behavior; but such comparisons are almost universally erroneous. We humans engage in some quite anomalous patterns of social interaction around the problem of sexual reproduction when compared to other species. This has important implications for that other human anomaly, language. To see how unusual we are in this regard, we need to look into that favorite distinction of anthropologists, the difference between mating and marriage. Consider three of the most consistent facts about human reproductive patterns:

1. Both males and females usually contribute effort toward the rearing of their offspring, though often to differing extents and in very different ways.
2. In all societies, the great majority of the adult males and females are bound by long-term, exclusive sexual access rights and prohibitions to particular individuals of the opposite sex.
3. They maintain these exclusive sexual relationships while living in modest to large-sized, multi-male, multi-female, cooperative social groups.

This pattern, of course, refers to marriage in its most general sense, though I do not mean to imply that monogamy is the rule—far from it— rather only that adult males and females are assigned (sometimes by their kin and sometimes by their own choosing, with the consent of the larger social group) to specific mates, often for life, and that this entails explicit exclusion of sexual access by other group members.[2] Exactly who is included or excluded may differ from society to society, particularly along the polygynous dimension (e.g., one male, many female mates), but marriages everywhere have reproductive rights and obligations as their central content, and so specify the reproductive status of the marriage partners within the parameters of the wider community, both family and nonfamily. Marriage is more than a reproductive arrangement, because it additionally establishes new rights and obligations for the larger kin groups to which marrying individuals belong.

There are explicit rules about who can and can't marry, and these rules have some highly predictable general patterns. In the vast majority of societies, incestuous marriage is prohibited, though what constitutes incest differs somewhat from society to society. Also, in all but a very few societies in the world, two males almost never have simultaneous sexual access to the same reproducing female.[3] A major factor contributing to differences in marriage patterns in different societies is the comparative difficulty of obtaining and defending reproductive resources, including women in many societies. Thus, tacit and explicit marriage agreements are about reproductive access not only in the direct sense, but also in the broader sense in which kin group cooperation, property rights, and promises of mutual aid and defense must be understood as factors critical to reproductive success.

In societies that rely mostly on foraged foods, as did all of our prehistoric ancestors until just ten or fifteen thousand years ago, the opportunities to accumulate, appropriate, and defend resources are minimal. Thus, group warfare, competition for polygyny, consolidation of wealth or political power, and so on, were unlikely to have played any significant role in the early stages of symbolic communication. There is one feature of human foraging ecologies that has always stood out as specially related to human origins, because it is uncharacteristic of most other primate societies and because it is associated so closely with many indices of the rise of peculiarly humanlike behaviors. This is the use of meat. Recently, the classic "man the hunter" theories of human origins have been strongly criticized as too narrowly focused on the activities of less than half the members of a society, and because there is a compelling argument to be made for scavenging as the earliest hominid meat-use adaptation. Yet the importance of this radical shift in foraging ecology cannot be ignored. The appearance of the first stone tools nearly 2.5 million years ago almost certainly correlates with a radical shift in foraging behavior in order to gain access to meat. And this clearly marks the beginnings of the shift in selection pressures associated with changes in the brain relevant for symbolic communication. Something about this complex of adaptations is central to the rise of *Homo symbolicus*. But what? The key to unlocking this mystery, I think, does not have to do with the cognitive difficulties of hunting or the use of communication or increased intelligence for toolmaking. What is important about this shift to a novel food source is the unprecedented demands it placed on the whole fabric of social group organization.

Women in foraging societies provide at least as large a fraction of the calories in the diet as do the men, but a mother carrying a dependent infant makes a comparatively poor scavenger and an even poorer hunter. She suffers from reduced mobility and the difficulty of employing stealth with a young child or infant in tow. Even more limiting is the threat from competing predators and scavengers, who are also attracted by a kill, and might easily turn their attentions to a relatively poorly defended infant. Consequently, men can provide access to a resource that is otherwise unavailable to women and children. Why should this introduce any special conditions into hominid social evolution?

If meat provides a necessary component of the diet of foraging people, even if just to get through otherwise lean times when vegetable sources are poor, a woman foraging with one or more children would be unable to provide adequate essential resources. Under these circumstances, mothers with infants are dependent on hunting by the males for an important and concentrated food source to supplement other gathered foods. But in any species where males provide a significant contribution of resources to help raise infants, there are special selection pressures on certainty of sexual exclusivity and predictability of male provisioning. A female who can't count on at least one male will have a high probability of losing her children to starvation and disease, and a male who can't rely on exclusive sexual access to at least some female will have a high probability of supporting the genetic fitness of other males.

In most mammalian species, sexual access is either determined by rank and ongoing competition, and results in polygyny; or else it is a result of a courtship process in which two individuals become "attached" to one another, and then isolate themselves from other members of their species. In order to maximize paternity and minimize cuckoldry in competitive polygynous social groups, males will threaten and fight with potential sexual rivals. Rival males display and threaten the dominant individuals, and will engage in direct physical contests or fights in efforts to depose them and gain access to females. In pair-bonding species, aggression is also a means of enforcing sexual exclusion, but it is often employed by both sexes to keep same-sex interlopers away from their shared territory and their mate. Under these conditions, long-term—even life-long—sexual exclusion can be common. Thus, the social mechanisms for maintaining sexual exclusivity in other mammals are ubiquitous, and exclusive sexual access is often maintained by the threat of physical harm. Almost certainly, such sexually motivated defenses or attacks are among the most intense emotional events in an animal's life.

Human reliance on resources that are relatively unavailable to females with infants selects not only for cooperation between a child's father and mother but also for the cooperation of other relatives and friends, including elderly individuals and juveniles, who can be relied upon for assistance. The special demands of acquiring meat and caring for infants in our own evolution together contribute the underlying impetus for the third characteristic feature of human reproductive patterns: cooperative group living.

Group living is not uncommon in the rest of the primates and in other mammals generally, but it is almost exclusively associated with polygynous reproductive patterns, or very special contexts in which nest sites or breeding grounds are a very limiting resource.[4] Reproductive access and reproductive exclusion are determined by ongoing competition in a social group. Males are able successfully to exclude others from sexual access for only a short time, when they are in their prime, so reproductive exclusivity is inevitably transient and unpredictable in mixed social groups. This is one reason why, under conditions where males must put a significant part of their energy into caring for offspring, male and female pairs tend to become isolated. The other reason is that a male contribution to the offspring is more critical in niches where resources are scarce and a larger social group would not have enough to go around. In human foraging societies, these conditions are not linked and pair bonding occurs in the context of group living. Resources are scarce enough that females can only rear their offspring with male support, and meat can only be acquired by groups of men. This pits two critical reproductive problems against one another: the importance of pair isolation to maximize the probability of sexual fidelity, and the importance of group size for access to a critical resource.

From foraging societies to agricultural societies to industrial societies, the same general sexual exclusion and residential patterns persist, despite shifts in the probability of polygyny. What is common to all is something that is exceedingly rare in other species: cooperative, mixed-sex social groups, with significant male care and provisioning of offspring, and relatively stable patterns of reproductive exclusion, mostly in the form of monogamous relationships. Reproductive pairing is not found in exactly this pattern in any other species. Why not? What's so special or peculiar about this social arrangement? I think that the answer to this question offers important hints about the initial impetus for language evolution, though probably not in ways that anyone might expect. This pattern of social-sexual organization is rare because it tends to undermine itself in the course of evolution. The combination of provisioning and social cooperation produces a highly volatile social structure that is highly susceptible to disintegration.

In evolutionary terms, a male who tends to invest significant time and energy in caring for and providing food for an infant must have a high probability of being its father, otherwise his expenditure of time and energy will benefit the genes of another male. As a result, indiscriminate protection and provisioning of infants will not persist in a social group when there are other reproducing males around who do not provision, but instead direct all their efforts toward copulation. Group size is an important factor because the chance of cuckoldry increases with increasing numbers, even when most are pair-bonded. The probability of cuckoldry should translate into the male's probability of abandoning a given female, so the complementary reproductive dilemma faced by females in these conditions is equally vexing.

In human foraging societies, meat offers a concentrated food source that is available far more reliably than fruit from season to season; but precisely when it should be most important for a female, when she is nursing an infant, she is least able to get it on her own. In order to be able to rely on this food source, she needs reliable provisioning from an individual or individuals not encumbered with an infant; a male or males. But a male, predisposed to provision females with whom he has copulated, will be an unreliable provider the more sexual access he has to other females. Cooperative hunting and scavenging by groups of males who each are provisioning one or more females and infants inevitably requires that groups of females live and forage cooperatively as well. Group living is a disadvantage both because of the proximity of many other sexually active females and because of their competing demands for the fruits of the hunt. In the face of these imminent reproductive threats, a female must find ways to ensure that some male will reliably provide her with meat and to minimize the probability of his opportunities for philandery.

These problems are further amplified in a context where males and females forage apart a significant fraction of the time. This is inevitable if hunting and scavenging are an important method of procuring food. Males pursuing meat on the savannah can't defend against cuckoldry, can't protect females from abduction, and can't protect their infants from infanticide by outside males. Females can't ensure that the males they depend on for meat are not copulating with females from another group and giving them meat that would otherwise have supported their own offspring. Hunting and provisioning go together, but they produce an inevitable evolutionary tension that is inherently unstable, especially in the context of group living. Besides ourselves, only social carnivores seem to live this way.

Most carnivores undertake significant food provisioning. But because hunting is a skill that requires a relatively mature body and takes a long time to learn effectively, there is a considerable period in the lifetime of carnivores when the young are too old to survive on mother's milk and too young to hunt for themselves. Many species cache their babies in dens and hunt for food that they will bring back in some form to feed to the young. In some ways this is analogous to the situation of nesting birds. In some species of carnivores, males and females will pair-bond, and one will stay in the den (more often the female) while the other hunts and brings back some of the meat for the remaining mate and offspring. In other cases, where young can be safely cached, a single female can even raise her young without additional male investment. Such "single parenting" is common in some of the larger cats, such as the leopard and the cheetah. But there are also many carnivore species that engage in cooperative group hunting. Hunting in groups makes it possible to employ a wider range of hunting strategies and to prey on a wider class of animals, including those much larger than the predator itself. Social carnivores include wild dogs, wolves, hyenas, lions, and meercats, among others. Certain reproductive and ecological conditions must be met for cooperative hunting with provisioning to be a mutually reinforcing aspect of an evolutionarily stable reproductive pattern. A few examples can help to demonstrate this.

Lions are a well-known species of social hunters, but the cooperation and provisioning that take place largely occur among females within a pride who are likely all closely related (sisters, half sisters, aunts, nieces). Two and rarely three males may cooperate to take over a pride of females and their young from other males, and will defend it against intruders, but they play little role in provisioning the young. The females, however, may take turns remaining with cubs while the remainder of the adult females hunt cooperatively. Although the spoils of the hunt may be made available to the offspring of another female, she is a close relative. This relationship is also eventually reciprocated, since all adult females are reproducing, and maternity certainty is not in question. Because of the females' kin-based common reproductive interests, their lifetime continuity of membership in a pride, their reciprocally shared defense and feeding of offspring, the males are largely irrelevant to offspring care, except in one way.

Male lions spend a good deal of time and risk their lives to defend the pride against other males, because the usurpers will likely kill the cubs. This infanticide brings lactating females back into estrus sooner and guarantees that the males will not be defending cubs who are not their own. Because males do not provide any significant provisioning to the cubs and have a common interest in female defense, cooperation between two or even three males in pride defense is also favored in evolutionary terms, so long as each has a roughly equal opportunity to mate. Common reproductive interests promote the evolution of cooperative behaviors between males and between females, for different reasons. The ability of the females to provision cubs without male contribution eliminates the value of paternity certainty, and thereby reduces the potentially disruptive effect of male-male copulatory competition within the pride.

In a few carnivore species—for example, wild dogs and wolves—it is common for a cooperative hunting pack to be made up of both males and females. After a kill they may carry bits of a carcass back to a home den, or devour it on the spot and regurgitate it later to feed cubs, the nursing mother, and nonhunting den mates. Young pups will induce the hunters to regurgitate by nipping at their mouths and mobbing them. The mother may also feed on regurgitated food. The crucial feature of these cooperative

hunting arrangements is that there is typically only a single reproducing female, who is often also the mother of many of the young hunters. Other females are kept from attaining sexual receptivity by a combination of social
behaviors and (probably) pheromones from the breeding female. Even infanticide of "illegitimate" pups may be used by dominant females to maintain their monopoly on the pack's resources. Only when the mother dies or
becomes too old to prevent another fully mature female from becoming sexually active will she be replaced. There is also typically only a single reproducing male in a pack.[5] He also uses threat of attack, copulatory exclusion
(copulatory "locking" together of male and female genitals for a period
after mating to minimize sperm competition from potential rivals), and
(possibly) pheromonal signals to suppress sexual behaviors of the other less
dominant males.

Unlike lions, male wild dogs and wolves play a major role in provisioning the young even when they are not the father of any pups. There is probably reasonable certainty of caring for others' offspring in these
circumstances. This might appear to be an unstable reproductive strategy
if looked at only in a single slice of time. However, as the mated pairs age
and are replaced by younger ones, most individuals will get an opportunity
to reproduce, and when they do, they will enjoy the benefits that group support confers by being able to raise litters that are much larger than a single
female or mated pair alone could raise. Both males and females in these
species employ a reproductive "strategy" that has one feature in common
with that employed by males in highly polygynous, "tournament" species.
Because of the intense male-male competition, males in highly polygynous
species don't gain access to females except for a very short period late in
their lives when they have reached their peak fighting ability, but during
this brief period they may be immensely successful reproducers. This "waiting" strategy can also lead to very prolific reproduction in social carnivores
that rise in status to become reproducers.

The sociality of wild dogs and wolves may also gain additional evolutionary support from kin selection, in the form of "helper-at-the-nest" behavior. This works best when the provisioning individuals are siblings of the
infants they provision. If the chances of successful reproduction are slim
on your own, it pays to contribute to your parents' sure thing. Offspring only
leave to reproduce on their own when they are mature enough to have the
possibility of procuring their own breeding territory and social group, when
the advantages of sticking around are providing diminishing returns, or
when the chances of inheriting the home territory are slim.

Wild dogs are far more efficient hunters in large groups, and eventually as the dominant reproducing males and females age they are replaced by younger individuals who at last can reap the benefits of being the core of a successful pack. Though they may not have reproduced for many years of their adult life, once they are reproducing they can be exceptionally successful, producing large litters in which the chance of survival to maturity will be quite good, given the support from a whole pack. They can make up in a short time for their "patience" earlier. If considerable excess meat were being produced, it would be advantageous for individuals to split off from the main group and start a separate pack rather than continue to provision another's offspring, relative or not. Otherwise, contributing to the reproduction of other pack members, biding their time, is the most promising route to successful reproduction for nondominant males and females.

These interesting exceptions to the rule that "male provisioning of young is mutually exclusive of large group size and social cooperation" serve to clarify the underlying principles of such a relationship. Group living and male provisioning can occur together only in instances where reproductive access is completely limited and unambiguous, as in the case of social carnivores. There can be no stable compromise pattern unless this principle is somehow maintained. And even in these cases, we often find that the provisioning comes more consistently from siblings or related females than from potential fathers.

These special adaptations to this problem are critical clues to understanding the ancestral human situation because almost all human societies, beginning probably with *Homo habilis* and *Homo erectus,* are exceptions to the general rule. Identification of these hominid species with the shift from polygyny to mostly pair bonding is supported by evidence that the great difference in male and female body sizes among the australopithecines—sexual dimorphism—seems to reduce to modern proportions by the time of *Homo erectus.*[6] This also coincides with the first development of stone tool technologies and the first increase in relative brain size above ape proportions (see Chapter 11). In the few other mammal species that appear to get around this constraint, it is accomplished by virtue of very distinctive social arrangements where specialized social communication plays a key role. In many of these species, we suspect that a major contributor may also be chemical communication—pheromones that may directly regulate the hormonal systems of group members. Unfortunately, as primates with diminutive smell organs, we humans are not well equipped for smell-governed social behavior. It should not be surprising, then, to discover that these intrinsically unstable reproductive arrangements in human societies are stabilized by a unique form of social communication, which can be as powerful and reliable as a social hormone.

A Symbolic Solution

The evolution of hunting among our distant hominid ancestors must be understood in this same social-reproductive context. Other primate species are known to eat meat. Both baboons and chimpanzees periodically capture young ungulates or monkeys and eat them, and chimpanzees have been regularly documented engaging in cooperative hunts. These hunts involve a group of males who intentionally stalk and surround a hapless monkey, cutting off all routes of escape, and then close in until one can grab the animal and kill or immobilize it long enough for others to join in for the grisly feast. But chimpanzee hunting behavior is considerably different from that of human hunter-gatherers in two major respects. First, all human hunters kill their prey and butcher the carcass with the aid of tools. Chimpanzees simply use their overpowering strength to immobilize their prey and dismember it, and use their longer, sharper canines to mortally wound and "butcher" it as they eat. Second, the meat that human hunters take from a felled animal is seldom all devoured on the spot, but rather is cut from the carcass and carried back to be shared with mothers, their young, and others who have not participated directly in the hunt. Chimpanzees engage in what might be called "trickle-down" provisioning of meat to females and their offspring. Only those females and their young who are in the vicinity of the kill site when the meat is being devoured will have a chance of gaining access to it. Typically, a male will allow a familiar female to beg away a remnant that is not prized.

Although we might consider meat as a source of provisioned food for chimpanzees, it is essentially a no-cost provisioning from the males' perspective, and it is, at best, a sporadic and unreliable resource for females and their young. This may be more evident when we consider how chimpanzees' hunting and sharing of meat differ in different conditions. Hunting probably increases during times of other food shortages, but the degree of sharing may also decrease during these times as well.

A kind of adventitious hunting and scavenging probably characterized the precondition for the evolution of stone-tool-assisted hunting, which appears to have begun about 2.5 million years ago. The transition to stone tool technology is evidence of a major change in the way meat was incorporated as a food resource. The manufacture and use of stone tools probably indicate that the use of meat took on a much more regular and necessary role in the diet of Pleistocene hominids, including a role as a regular food source for mothers with infants and young children.

Stone tools clearly reflect an important anatomical difference between early hominids and modern chimpanzees: early hominids did not have large canines. This is an unusual adaptation for a primate. Most primates have prominent canines that are used for threat displays, fighting, and to some extent for predator defense; this is true even for pair-bonding species such as gibbons. So why are canines almost completely reduced in all hominid ancestors, including australopithecines who appear to have been as sexually dimorphic as the most highly polygynous and feisty primates alive today? There is little agreement on this question. Some have suggested that it represents a reduction in male-male competition and fighting, or perhaps a shift to the use of hands and fists and wooden weapons. A reduction in male-male competition seems unlikely in the light of the extreme differences in body size of male as compared to female australopithecines, since this is an invariable indication of sexual selection on male fighting capacity with respect to mate competition. The argument that canines were replaced by "better" fighting implements also seems unable to offer a sufficient explanation. Even if we were to pretend that fists and sticks were better weapons than flesh-piercing canines, on the basis of being useful at a safer distance, it would not select against this formidable last line of defense. What violent man would give up his knife for a club when he could have both? Canines didn't just fade away; they must have been actively reduced by natural selection.

It seems likely that the loss of large canines in our australopithecine ancestors had more to do with eating than fighting. This is suggested by a look at the rest of the dentition in these species. The other atypical feature of early hominids' teeth is their incredibly robust and heavily enameled molars. These large, flat grinding surfaces, supported in relatively massive jaws—particularly in the bigger australopithecine species—almost certainly reflect an adaptation for chewing fibrous or hard foods that require considerable masticatory preparation before they can be digested. This suggests that they were eating things like tubers, grain, or even bark and pith. But if these were the dietary mainstay, then maintaining large self-sharpening canines, which must be honed against their upper and lower counterparts as the jaw is opened and closed, would pose a serious problem. It would significantly reduce the lateral mobility of the jaw, which is essential for effective grinding. Constant chewing would also significantly wear away at large canines, and possibly even increase the chances for self-inflicted damage or mouth sores where the canines were repeatedly rubbed. No matter how you look at it, big canines and big grinding teeth do not often go together. The few nonprimate species that are able to manage both large canines and a grinding-feeding strategy grow their canines outward and forward (as in elephants and boars), but this seems not to have been an option for our ancestors, thank goodness. In summary, canines were reduced *despite* their benefits for aggression and defense.

Almost certainly, given their upright posture, there was a shift from the use of the teeth to a use of the hands and objects in male-male aggression and defense during the evolution of australopithecines, but this was likely a response to the reduction of canines and not its cause. Nevertheless, an increased use of the hands for aggressive displays and physical combat, especially if it involved objects (something also observed among chimpanzees), may have increased the likelihood that eventually some would come to use tools for scavenging and hunting, as well as for male-male competition. In fact, lacking large canines, australopithecines would have likely been quite poor carnivores, even compared to chimpanzees. Stone tools would have been the only means by which they could enter a niche where meat was an essential food source.

The early stone tools indicate a shift to a diet that contains more meat, whether hunted or partly scavenged (stolen from other predators). But why a shift to include more meat? What did it offer? Again, chimpanzee foraging and hunting may provide some hints. As we have noted, there is some evidence that chimpanzees' hunting behaviors increase when other foods become more scarce. Meat is one alternative food source that can substitute for their preferred food, fruits. It is one component of a complex and variable feeding strategy. Chimpanzees are omnivorous: they eat leaves, insects, the new shoots of plants, bark, and even the pith from the centers of reedlike plants at various times of the year in different areas. Bark and pith are probably lowest on their preference hierarchy and are poor sources of nutrition. They are only turned to when all other foods become scarce, and they may serve more to provide bulk to stave off hunger than to provide nutrients. In fact, it is not clear how calorically expensive it is to extract these nutrients, in terms of processing the energy involved in finding, extracting, chewing, and digesting plant foods. Almost certainly there are diminishing returns if this strategy must be maintained for more than a month or two.

Chimpanzee hunting behavior also seems to increase during lean seasons, but unlike access to these other low-preference subsistence foods, meat is preferentially available to males. As a result, females and their dependent infants are probably the most affected during these periods. From an evolutionary point of view, intentionally directed provisioning of this now limited resource to a particular female and her offspring would carry few advantages to males, given their uncertainty of paternity. Nevertheless, in chimpanzees, there is some evidence of preferential sharing with certain females, and a suggestion that this may increase the chances of future matings between the sharing male and that female. This could offer some route for provisioning, but it would be indirect, irrespective of paternity, and secondary to the competition among males for female access. Such a selfish strategy has costs in terms of infant mortality rates, but the individual genetic/reproductive costs of adopting a nonselfish strategy pose a barrier to achieving a more efficient reproductive pattern overall. This equation could change, however, if scarcity were more extreme and more costly, or if a way could be found to gain more reliable access to a high-value subsistence food like meat.

For chimpanzees, hunting is not a major strategy for maintaining reproductive efficiency in the face of significant food shortages during seasonal droughts, primarily because it is not equally available to adult males, females, juveniles, and infants. Hunting or scavenging for meat can only provide a viable strategy for surviving seasonal drought if (1) success is highly predictable and regular, and (2) meat supplements the diets of nursing females and their young offspring. A supplementary food source that is mostly available to adult males or nonreproducing females would offer little or no reproductive benefit to its consumers, if it meant that reproducing individuals suffered an increased mortality rate.

The acquisition and provisioning of meat clearly would be a better strategy for surviving seasonal shortages of more typical foods than shifting to nutrient-poor diets of pith, bark, and poor-quality leaves, as do modern chimpanzees. But this is only possible if there is a way to overcome the sexual competition associated with paternity uncertainty. The dilemma can be summarized as follows: males must hunt cooperatively to be successful hunters; females cannot hunt because of their ongoing reproductive burdens; and yet hunted meat must get to those females least able to gain access to it directly (those with young), if it is to be a critical subsistence food. It must come from males, but it will not be provided in any reliable way unless there is significant assurance that the provisioning is likely to be of reproductive value to the provider. Females must have some guarantee of access to meat for their offspring. For this to evolve, males must maintain constant pair-bonded relationships, and yet for this to evolve, males must have some guarantee that they are provisioning their own progeny. So the socio-ecological problem posed by the transition to a meat-supplemented subsistence strategy is that it cannot be utilized without a social structure which guarantees unambiguous and exclusive mating and is sufficiently egalitarian to sustain cooperation via shared or parallel reproductive interests. This problem can be solved symbolically.

We should not underestimate what can be represented by nonsymbolic means. Almost any objects or events or even particular qualities of objects or events can be signified without symbolic reference, using iconic or indexical means. The more carefully we look at the social behaviors of primates, the more examples we find of calls and gestures that serve to refer to specific types of objects and activities. These include alarm calls that distinguish between types of predators, food calls that distinguish between types of foods, and many less dramatic grunts and gestures that determine identity, social spacing, and behavioral intentions. Almost certainly, individuals who have been long associated with one another in the same social group also develop proficiency at interpreting each other's subtle gestures and movements or characteristics of vocalization. Reports of the abilities of domesticated species to discern their master's emotional states have been around for a long time, and are probably only a superficial glimpse into such species' abilities to interpret the emotions of their own kind. Though there appear to be significant limitations on the numbers of innately preprogrammed calls and gestural signals that can evolve, the ability to supplement this by learning to anticipate the predispositions of specific individuals, and by acquiring certain habits of behavior conditioned by the context of others' responses, has the power greatly to amplify this set. Even coordinated social behaviors like group hunting appear to be organized quite easily on the basis of animals' abilities to learn to anticipate one another's habits of behavior.

But certain things cannot be represented without symbols. Indexical communication can only refer to something else by virtue of a concrete part-whole link with it, even if this has no more basis than just habitual coincidence. Although there is a vast universe of objects and relationships susceptible to nonsymbolic representation, indeed, anything that can be present to the senses, this does not include abstract or otherwise intangible objects of reference. This categorical limitation is the link between the anomalous form of communication that evolved in humans and the anomalous context of human social behavior.

Ultimately, the stability of human social groups is dependent on reciprocity. A relationship in which individuals at times sacrifice reproductive opportunities to others so long as there is a high probability that others will do the same in return at a later date is called "reciprocal altruism." Reciprocal altruism is found in other species, though it does not appear to be generally widespread. This is because a number of critical conditions must be met for it to persist, all having to do with guaranteeing the reliability of reciprocation. First, animals must be able to recognize individuals and remember their past behaviors. Second, they must be able to detect "cheaters" who do not reciprocate. Third, they must be able to exert control over the cheaters by withholding altruistic acts or ostracizing them. Ideally, there should be some means of preventing cheating from occurring in the first place. Fourth, the advantages gained by cheating must be significantly outweighed by the disadvantageous consequences of being "caught," and the costs of detecting and "punishing" cheaters must not be too high. One feature of social organization that tends to reduce the threshold for the development of reciprocal altruism is social familiarity. Individuals that spend all their lives in each other's company can come to recognize each other easily, and develop reliable expectations about each other's future probability of altruism on the basis of past experiences. Other features that are conducive to the evolution of regular reciprocal altruism include a high degree of relatedness, so that inequities have relatively smaller genetic consequences, and a low cost of helping, so that little is risked at any one time. All in all, these are still rather restrictive conditions, and make reliance on reciprocal altruism as a critical reproductive component a rather fragile evolutionary strategy.

How could these requirements have been met in early hominine social groups in order to maintain a reciprocally exclusive pattern of reproductive and food-sharing relationships? From what we know of other primate social patterns, it can be assumed that these groups were long-term, relatively stable ones, where individuals could come to identify each other, and that their functioning probably depended on the cooperation of either related females and/or related males. But what about the identification of who is following the rule and who is not? Where food sharing is involved, the relationships are all symmetrical and immediate. Food is either given or not. But in the case of sexual relationships it may not be so clear what makes one individual available and another unavailable, if it is not simple physical availability and receptivity or threat of harm.

How can there be reciprocal altruistic access to mates? Essentially, each individual has to give up potential access to most possible mates so that others may have access to them, for a similar sacrifice in return. A reproductive balance must be struck, so that most males and females have a roughly equal probability of access to reproduction or provisioning (respectively), over the course of a lifetime, in order for cooperative provisioning to be a stable strategy. But if there is no unequivocal marker of allowable and un-allowable sexual access, or appropriate and inappropriate sharing and provision of resources, how can anyone tell who is cheating and who is not? How does one tell who is obligated to whom?

The first requirement, then, is that there must be a means for marking exclusive sexual relationships in a way that all members of the group recognize. Sexual access and a corresponding obligation to provide resources are not just habits of behavior; they cannot be more or less predictable patterns, or just predictions of probable future behaviors. Sexual access is a *prescription* for future behaviors. No index or memory of past behaviors can represent this. Nor can any index of present social status or reproductive state mark it. Even the refusal or avoidance of sexual activity only indicates a current state and is not necessarily predictive. Sexual or mating displays are incapable of referring to what might be, or should be. This information can only be given expression symbolically. The pair-bonding relationship in the human lineage is essentially a promise, or rather a set of promises that must be made public. These not only determine what behaviors are probable in the future, but more important, they implicitly determine which future behaviors are allowed and not allowed; that is, which are defined as cheating and may result in retaliation.

The second problem is how to verify and guarantee the assent of the other individuals that could conceivably be involved, both as possible cheaters and as support against cheating. For a male to determine he has exclusive sexual access, and therefore paternity certainty, requires that other males also provide some assurance of their future sexual conduct. Similarly, for a female to be able to give up soliciting provisioning from multiple males, she needs to be sure that she can rely on at least one individual male who is not obligated to other females to the extent that he cannot provide her with sufficient resources. Unlike a pair bond in a species where the male and female remain isolated from other potential sexual competitors, establishing an exclusive sexual bond in a social setting is not just a relationship between two individuals. In the case of gibbon pairs in neighboring territories or pair-bonding birds that nest nearby other pairs, one aspect of maintaining the exclusive mating relationship may require that the female actively fends off other females and the male actively fends off other males that could be sexual competitors. This is only reliable when the mated pair tend to remain in the same vicinity. In the hominid case, however, where males are engaged in an activity that requires regular separation from females in order to hunt, mate guarding of this sort is not sufficient. Males and females must be able to rely both on the promise of a mate and, probably more importantly, on the support and threats of other males and females who are party to the social arrangement and have something to lose if one individual takes advantage of an uncondoned sexual opportunity.

In human societies, the breakdown of sexual exclusivity or even the threat of it often precipitates violent reprisals. Though this parallels interactions that are common in the social behaviors of other polygynous mammal species, the comparison is incomplete. Sexual jealousy may have the same roots in humans and other species, but in humans it involves something more abstract than just a threatening behavior. Though philandery, cuckoldry, and desertion are common consequences of reproductive competition in other species, adultery is more than this. It involves betrayal, and there can be no betrayal without prior explicit or tacit agreements. In nearly all societies, there are not only personal reprisals associated with sexual infidelity but also consequences imposed by the community. Even when explicit codes and punishments are not present in a society, those considered the victims of sexual infidelity are often allowed leeway to commit violent acts that might otherwise be proscribed.

The prevention of cuckoldry is partially supported by the potential of punishment from the entire social group. In no other species is there such direct involvement by the larger community in the maintenance of sexual exclusivity between individuals.[7] The problem is how to organize group behavior around something as intangible as a desired future habit of behavior. Each individual must share knowledge of these expectations; but more important, each must also be able to rely on the support of other group members to prevent violations of these patterns of conduct.

What I am essentially describing is, of course, the skeleton of what we recognize as a marriage agreement. As anthropologists have recognized for generations, marriage is not the same as mating, and not the same as a pair bond. Unlike what is found in the animal world, it is a symbolic relationship. But it is also not just a reciprocal set of promises between two individuals regarding sexual access and economics. As the French anthropologist Claude Lévi-Strauss and many others have emphasized, it is also the establishment of alliances: promises and obligations that link a reproductive pair to the social groups of which they are a part, and often a set of promises and obligations between the kin groups from which they arise. Marriage contracts establish both vertical lineal symbolic relationships and horizontal affinal symbolic relationships. Marriage, in all its incredible variety, is the regulation of reproductive relationships by symbolic means, and it is essentially universal in human societies. It is preeminently a symbolic relationship, and owing to the lack of symbolic abilities, it is totally absent in the rest of the animal kingdom. What I am suggesting here is that a related form of regulation of reproductive relationships by symbolic means was essential for early hominids to take advantage of a hunting-provisioning subsistence strategy.

Establishing such social-sexual relationships cannot be accomplished by indexical communication alone, that is, by systems of animal calls, postures, and display behaviors, no matter how sophisticated and complex. And yet, even extremely crude symbolic communication can serve this need. Only a few types of symbols and only a few classes of combinatorial relationships between them are necessary. But without symbols that refer publicly and unambiguously to certain abstract social relationships and their future extension, including reciprocal obligations and prohibitions, hominids could not have taken advantage of the critical resource available to habitual hunters. The need to mark these reciprocally altruistic (and reciprocally selfish) relationships arose as an adaptation to the extreme evolutionary instability of the combination of group hunting/scavenging and male provisioning of mates and offspring. This was the question for which symbolization was the only viable answer. Symbolic culture was a response to a reproductive problem that only symbols could solve: the imperative of representing a social contract.

Ritual Beginnings

The near synchrony in human prehistory of the first increase in brain size, the first appearance of stone tools for hunting and butchery, and a considerable reduction in sexual dimorphism is not a coincidence. These changes are interdependent. All are symptoms of a fundamental restructuring of the hominid adaptation, which resulted in a significant change in feeding ecology, a radical change in social structure, and an unprecedented (indeed, revolutionary) change in representational abilities. The very first symbols ever thought, or acted out, or uttered on the face of the earth grew out of this socio-ecological dilemma, and so they may not have been very much like speech. They also probably required considerable complexity of social organization to bring the unprepared brains of these apes to comprehend fully what they meant.

The success of Sherman, Austin, and Kanzi at acquiring the ability to use symbolic reference in a limited fashion demonstrates that a modern human brain is not an essential precondition for symbolic communication. Chimpanzees who have learned a modicum of symbols in the laboratory have the benefit of dedicated researchers who can construct an elaborate training context. The first hominids to use symbolic communication were entirely on their own, with very little in the way of external supports. How, then, could they have succeeded with their chimpanzeelike brains in achieving this difficult result? How could a social environment have arisen spontaneously, which possessed the necessary supports for overcoming the immensely difficult and complicated task of teaching symbolic relationships to individuals whose brains were not only unprepared but resistant to learning them? The transition to nascent symbolic culture probably began in fits and starts, with innumerable evolutionary trials and errors, before some semblance of stability was achieved. Some intense social evolution must have been responsible for creating such a context. But what sort of a context was it? Are the requirements for support of symbol transmission in an ape society so unusual that their spontaneous evolution will stretch credulity?

The ape symbol-training experiments provide an indication of what these requirements might be. First, they demonstrate that solving the problem requires a way of maintaining attention on many related indexical associations simultaneously, a way to shift attention to token-token relationships, and an ability selectively to suppress attention on immediate token-object associations. Like the chimps, early hominids were forced to learn a set of associations between signs and objects, repeat them over and over, and eventually unlearn the concrete association in favor of a more abstract one. This process had to be kept up until the complete system of combinatorial relationships between the symbols was discovered. What could have possibly provided comparable support for these needs in the first symbol-learning societies?

In a word, the answer is ritual. Indeed, ritual is still a central component of symbolic "education" in modern human societies, though we are seldom aware of its modern role because of the subtle way it is woven into the fabric of society. The problem for symbol discovery is to shift attention from the concrete to the abstract; from separate indexical links between signs and objects to an organized set of relations between signs. In order to bring the logic of token-token relationships to the fore, a high degree of redundancy is important. This was demonstrated in the experiments with the chimpanzees Sherman and Austin. It was found that getting them to repeat by rote a large number of errorless trials in combining lexigrams enabled them to make the transition from explicit and concrete sign-object associations to implicit sign-sign associations. Repetition of the same set of actions with the same set of objects over and over again in a ritual performance is often used for a similar purpose in modern human societies. Repetition can render the individual details of some performance automatic and minimally conscious, while at the same time the emotional intensity induced by group participation can help focus attention on other aspects of the objects and actions involved. In a ritual frenzy, one can be induced to see everyday activities and objects in a very different light.

This aspect of many ritual activities is often explicitly recognized as a means to help participants discover the "higher meaning" of the otherwise mundane, while at the same time promoting group solidarity. Thus, many ritual activities, from repeated prayer to pubertal initiation ceremonies, quite explicitly take the form of an ideal symbol discovery process. Of course, in all societies of modern humans, most ritual activities are in service of complex symbolic ideas and institutions. They are self-consciously employed for their ability to help define abstract social relationships and inculcate certain habits of thought and action. They are not, in this regard, good models of the earliest symbolic rituals. But they demonstrate one of the major ways that societies establish some of their most difficult symbolic precepts. In general, it is probably fair to conclude that the more difficult the social symbolic problem—either because of intrinsic conceptual difficulties or because of the intensity of the countervailing social forces that must be mediated—the more likely that highly ritualized means will be required to establish appropriate symbolic responses.

One of the most difficult social symbolic relationships to mediate is peace. This is not so much because of a conceptual difficulty, but rather because of the high potential and high cost of deception. The problem of establishing peace after a period of war also demonstrates the importance of the indexical substructure of a symbolic social relationship. Agreements and contracts concerning future behaviors and obligations are intrinsically symbolic. Because the referential link between symbols and their objects is indirect, the very same features that make symbolic reference the only means for definitively representing something that is as abstract and virtual as a promise or contract also open the door to misrepresentation and falsehood. The problem lies in determining whether a symbolic gesture for peace is made in earnest, particularly when it comes from a former enemy who in other circumstances may have employed misrepresentation and misdirection as a ploy.

An interesting example of this comes from the Yanomamö Indians of the rain forests of Venezuela and northern Brazil.[8] These slash-and-burn agriculturists live in small, lightly fortified villages with nearby gardens. Such villages are almost constantly at war with each other, and skirmishes are often initiated by surprise attacks. But there are times when it is necessary to cooperate with one's neighboring villages: for example, when a garden is becoming overgrown and unproductive, and it will soon be necessary to abandon one village and create a new one. At times like these a Yanomamö group might be at high risk of attack. How do you make peace in order to gain allies at this crucial time and decrease the chances of attack? Peacemaking is a difficult problem no matter what the conditions. It is a situation where the major communicative problems are not due to a lack of symbolic ability, but rather because no one is sure that others are using the symbols honestly. The situation is similar to not having reliable symbolic reference at all. What is necessary in order to reestablish reference—to ground it in the real world so to speak—is a reconstruction of the symbolic relationship from its component indexical relationships.

Demonstrating true symbolic reference is analogous to establishing symbolic reference in the first place. Symbols refer to relationships among indices, and are learned by first establishing these indexical associations. Regrounding questionable symbolic reference similarly requires a return to the indices on which it is based. So the question is: What system of indices does peace represent? Indications that members of the groups are not hostile. Indications that they would not engage in violent behavior even if the opportunity were to present itself. And possibly indications that they are disposed to cooperate with one another. Unlike symbols, indices are part of what they refer to, and this makes them reliable in ways that symbols are not.

In the case of the Yanomamö peacemaking process, an elaborate ritual provides the indices that are required. It is known as a "Feast." First, the hosts who wish to make peace prepare a meal. When their guests are due to arrive, dressed as for war and carrying their weapons, the hosts put their weapons away and the men recline on their hammocks waiting for the guests to enter their village. The guests enter, dancing and chanting, and circle around the camp stopping in front of each host. There they ritually threaten them, raising an ax or drawing a bow and arrow. The hosts must remain unmoved, trying to show no fear and no offense at provocative remarks. After this has been repeated for a while (and latent hostilities have not erupted in violence), the roles are reversed. The guests recline in hammocks, their weapons hidden away, while the hosts circle around the camp dancing and ritually threatening their guests. Finally, when it is clear that nothing untoward is likely to happen, they break off and the guests are offered food. Later they may chant together, barter and exchange goods, or even arrange a marriage.

Notice the similarity to the process that helped the chimps Sherman and

Austin over their own symbolic threshold. In a Feast, the participants establish the symbolic relationship of "being at peace" by posing the opportunity of its opposite. Assume the opposite and show that it is not true, or discover that it is. Like the trick used to push the chimpanzee to symbol discovery, the Feast constructs an abstract symbolic relationship using a sort of *reductio ad absurdum* strategy. Peace and war are *negations* of one another (not intending to kill and intending to kill); but being unable to trust the symbolic invocation of these abstract relationships, it is necessary to reconstitute the underlying system of negations in a controlled and complete manner. Peace is a generalization over each of the symptoms of being in a state of war (a sort of induction from them), and is only as firm as the indices that implicitly support it. This is a problem that does not go away with better symbolic systems. Establishing peace only gets more complex as societies get more complex. And the indices that are required to produce it only get more difficult to demonstrate fully, as the present state of the world amply shows.

This is an instructive example of how social contracts are implicitly symbolic. As a result, they depend on the same logic of symbol construction that is required at all levels of symbol learning. In either case, extra effort is required to provide the crucial indices in context with each other in such a way that the categorical negation relationships between them are explicitly juxtaposed. This parallel between symbols that are difficult to believe and those that are difficult to grasp also brings us back to ritual. The ritual context allows highly redundant, therefore reliable, associations between the markers of these relationships to become established. Because of this, it is possible to abandon reliance on the indexical associations themselves and come to rely on the symbolic (but virtual) reference to hold. Similarly, symbolic relationships that are difficult to grasp (as in the training of Sherman and Austin) require the highly organized and repetitive presentation of relevant patterns of indexical associations in order to help with the process of discovering the implicit higher-order logic of which they are instances. Not surprisingly, some of the most complex rituals in all societies grow out of efforts to convey some of the most intangible and ineffable symbolic notions: the meaning of existence, the nature of God, what it means to be a member of a particular ethnic tradition, and so on. They are different faces of the very same process, pushed to extremes either by learning difficulties or by mistrust of the virtual reference that symbols provide.

For a hominid species with a fairly typical ape brain, such as the first symbol users must have possessed, any symbolic learning would have been conceptually difficult. Moreover, the social dilemma in need of a stable solution is the sexual equivalent of warfare. In evolutionary terms, mistrust should be built into our sexual emotional responses, as indeed every daytime television soap opera suggests. Around the world, infidelity or the suspicion of cuckoldry is a major cause of violence and homicide. It is what comes of living in this otherwise unstable reproductive context. Thus, both in cognitive terms and in terms of the stakes involved, the otherwise simple symbols which establish the reproductive relationships that are at the core of human societies demand highly ritualized construction processes. Ritualized support is also essential to ensure that all members of a group understand the newly established contract and will behave accordingly. As in peacemaking, demonstrating that these relationships exist and providing some way of marking them for future reference so that they can be invoked and enforced demand the explicit presentation of supportive indices, not just from reproductive partners but from all significant kin and group members. Given both the cognitive and the social difficulties, it wouldn't be surprising to find that the earliest social symbols for establishing reproductive roles were created in a ritual *reductio ad absurdum.*

Marriage and puberty rituals serve this function in most human societies (though modern societies replace many of these concrete social symbolic rituals with more abstract religious and legalistic ones). The symbol construction that occurs in these ceremonies is not just a matter of demonstrating certain symbolic relationships, but actually involves the use of individuals and actions as symbol tokens. Social roles are redefined and individuals are explicitly assigned to them. A wife, a husband, a warrior, a father-in-law, an elder—all are symbolic roles, not reproductive roles, and as such are defined with respect to a complete system of alternative or complementary symbolic roles. Unlike social status in other species, which is a more-or-less relationship in potential flux, symbolic status is categorical. As with all symbolic relationships, social roles are defined in the context of a logically complete system of potential transformations; and because of this, all members of a social group (as well as any potential others from the outside) are assigned an implicit symbolic relationship when any one member changes status.

Out of these ritual processes for constructing social symbolic relationships, symptoms of the process itself (exchanged objects, body markings, etc.) can be invested with symbolic reference. Tokens that served an indexical purpose within the ritual symbol construction become symbolic because of it. Rings, ceremonial clothing, and ritual scarring are indices of having participated in a symbolic transition, and thereby can become symbols of the same relationship. Given the universality of human marriage and its embedding ritual and ceremony, it seems reasonable to imagine that hints of the original logical structure of the first symbolic rituals still echo in the marriages men and women make today.

The earliest forms of symbolic communication were therefore likely not speechlike or manual sign languages. They almost certainly included vocalizations along with conventional/ritual gestures, activities, and objects, all of which together formed a heterogeneous melange of indices transformed to symbols, each systematically dependent on the other, and defining a closed set of possible types of relationships. Vocal symbolizing in the earliest stages would probably have played a minor role, owing to the lack of descending motor control until brain size increased. Probably not until *Homo erectus* were the equivalents of words available. Since these individuals had brains that begin to approach the modern range, it is almost certain that they had some degree of vocal skill, and used it symbolically. What drove this shift to speech was not just constraints on the ease of use of objects and performances and manual gestures, but also the reciprocal effect of cortical enlargement in response to the demands of symbol learning and the consequences for manual and vocal flexibility. Physical object symbols are constrained to highly specific uses and contexts—e.g., use as physical icons or markers—and tend to keep complex performance symbols confined to use in specialized ritual contexts. The manipulation of vocalizations and hand gestures would have been far less constrained. The social evolution of symbolic communication would have selected against symbol systems that required the use of tokens (individual symbol objects or acts) that are difficult to produce or combine in a wide range of circumstances. Speech is by far the least constrained in these ways, and would have come to replace other token systems by social evolution alone as the neurological constraints loosened. The complicated co-evolution of social selection favoring vocalization, along with manipulative-cognitive demands of tool use and symbol learning, would have guaranteed that vocal symbols would not have waited to burst on the evolutionary scene late in the process. More likely a continual drift toward increasingly self-sufficient vocal symbol systems—languages—has always required using our articulatory abilities to their limits.

Nevertheless, we should not lose sight of the fact that symbols are still extensively tied to ritual-like cultural practices and paraphernalia. Though speech is capable of conveying many forms of information independent of any objective supports, in practice there are often extensive physical and social contextual supports that affect what is communicated. Language acquisition still relies on an extensive gamelike ritualization and regimentation of the symbol acquisition context, although the child's uniquely human computational supports enable this process to take place without explicit *reductio ad absurdum* grounding of all symbols and possible combinations in the system.

In conclusion, then, the theory of symbolic origins I have outlined is not just a new twist on Rousseau's "social contract" theory. It is not a theory of the origins of social behavior, but of the translation of social behavior into symbolic form. More important, it is not a scenario for how our intelligence triumphed over our reproductive competition, but rather how unique demands of reproductive competition and cooperation created the conditions that led to our unique form of intelligence. By answering the evolutionary question of how to take advantage of a new foraging trick, our ancestors unwittingly turned the tables of natural selection so that social evolution could reshape the brain in its own image. We reflect on this from the other end of an extensive co-evolutionary process, where the indispensable uses of symbolic communication as a social organizing tool were long ago relegated to being only one among a multitude of selection pressures mutually converging on making this communication more and more efficient. Two and a half million years of sustained selection in an unprecedented socio-ecological niche, maintained by an unprecedented communicational and cognitive trick have taken us far from these beginnings in both the physical changes in the brain that resulted and in the mental and cultural world that co-evolved with them. Figure 12.1 summarizes the parallel threads of adaptation that have ratcheted each other onward in this co-evolutionary process.

This brings us full circle to reconsider the other side of the paradox that began this book: the lack of simple languages in animals and humans. Human languages are immensely complicated and yet easily learned and used. The evolution of language was not the result of simply surmounting the symbolic problem. Solving this problem opened up a whole new realm of adaptive problems, all dependent on efficient and successful symbolic communication. This long co-evolutionary heritage has not just resulted in symbolic communication becoming easier and more efficient, but these simplifications of the task have upped the ante so that ever more efficient acquisition and ever more powerful employment of symbolic communication were an imperative. Although the problem of language origins cannot be answered in terms of a transition from simple to complex, or from less intelligent to more, it is clear that what has resulted is both an incredible enhancement of intellectual abilities and a competence to use a very complex mode of communication.

The origins of the first symbolic communication have nothing intrinsi-

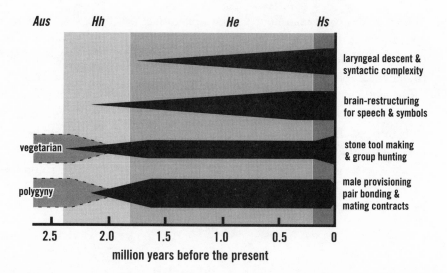

Figure 12.1 *Timelines of the correlates of brain-language co-evolution in hominid evolution from* Australopithecus *through* Homo habilis *and* Homo erectus *to* Homo sapiens. *The approximate epochs of each major recent hominid group are indicated by different gray backgrounds; the width of the bars indicates increasing importance/development of the respective features.*

cally to do with language per se. Most of the details that characterize modern languages have other later evolutionary causes. The argument I have presented is only an argument for the conditions which required symbolic reference in the first place, and which selected for it despite the great difficulties and costs of collectively producing and maintaining it. Much of the story of this intermediate evolutionary history, extending for over 2 million years from language origins to the present, has yet to be even imagined in any clarity. But putting evolutionary causes and effects in appropriate order and precisely identifying the anatomical correlates of this transition are a prerequisite for providing anything beyond "just so" versions of the process. The key to this is the co-evolutionary perspective which recognizes that the evolution of language took place neither inside nor outside brains, but at the interface where cultural evolutionary processes affect biological evolutionary processes.

The evolution of symbolic communication is special in this regard. It created a mode of extrabiological inheritance with a particularly powerful and complex character, and with a sort of autonomous life of its own. It is for this reason that the co-evolutionary process has played such a major role in shaping human brains and minds. It is simply not possible to understand human anatomy, human neurobiology, or human psychology without recognizing that they have all been shaped by something that could best be described as an idea: the idea of symbolic reference. Though symbolic thinking can be entirely personal and private, symbolic reference itself is intrinsically social. Not only do we individually gain access to this powerful mode of representation through interactions with other members of the society into which we are born, but symbols themselves can be traced to a social origin. Our uniquely human minds are, in a very concrete sense, the products of an unusual reproductive challenge that only symbolic reference was able to address—a concrete internalization of an ancient and persisting social evolutionary predicament that is uniquely human.

A Serendipitous Mind

The real danger is not that computers will begin to think like men,
but that men will begin to think like computers.

—Sydney J. Harris

The Fail-Safe Computer

Consider how remarkable it is that almost no human being fails to acquire at least some degree of rudimentary symbolic-linguistic ability—even in the face of considerable perinatal brain damage. Severely retarded people sometimes possess only crude grammatical abilities and a small vocabulary or exhibit difficulty articulating words and sentences, and yet they generally still understand the symbolic content of words and simple sentences. Contrast this with the near unlearnability of symbols by other species, even apparently very intelligent species. This book began by considering the curious lack of natural symbolic systems in all nonhuman species, the limited capacity to gain symbolic understanding in most, and the failure of domesticated animals—immersed in the dense web of human symbolic interactions—to discover more than a few rote associations of words and phrases. What are the implications of this species difference and its associated neurological basis?

Evolution has widened the cognitive gap between the human species and all others into a yawning chasm. Taken together, the near-universal failure of nonhumans and the near-universal success of humans in acquiring symbolic abilities suggest that this shift corresponds to a major reassignment of cognitive resources to help overcome natural barriers to symbol learning. Other species' failures at symbol learning do not result from the lack of some *essential* structure present only in human brains. As we have seen, chimpanzees can, under special circumstances, be brought to understand symbolic communication, though at best on a comparatively modest scale. The difference between symbolic and nonsymbolic communication may be a categorical difference in semiotic terms, but the neurological basis of our symbolic advantage is not due to a categorical difference in brain structure, only to a quantitative rearrangement of existing parts. Nevertheless, this shift in proportions spans a critical learning threshold that stands between indexical associations and symbolic reference. Although it is possible for other species to cross this threshold by learning and unlearning sets of associations in just the right way, it is incredibly unlikely. Yet in humans, a restructuring of the brain has acted like a catalyst, making the immensely improbable nearly inevitable.

In evolutionary terms, it would be accurate to say that the genetic basis for symbol-learning abilities has been driven to "fixation." In other words, it has become a universal trait of the species. Though there may be variations in this ability among people, essentially all of this variability is above the threshold necessary for acquiring symbols. Whenever most variation of a trait is eliminated, we can usually assume that selection for it has been and still is intense. There must have been some very significant reproductive advantages to symbol acquisition, and severe reproductive costs in cases of failure to acquire symbols. An individual born into a symbolic culture with an ape's bias against acquiring symbolic associations would be deprived of access to most realms of know-how and social influence, and have little chance to reproduce successfully. The ancestral lineages that succeeded best and left the most progeny were those in which symbolic abilities were able to develop despite a wide range of interfering influences. Language acquisition had to become fail-safe. After 2 million years it has clearly reached this status.

The simplest way to make something fail-safe is to design it far beyond the basic requirements. To ensure that a physical structure is totally safe, for example, it must be designed to handle loads that are far greater than those it would normally be called upon to support. It must be built to exceed the requirements of its one most extreme load, not the average load. In a word, it must be overbuilt. I want to suggest that the neuroanatomical evidence of massively altered brain proportions and the anthropological and clinical evidence for universality of symbol learning across a wide spectrum of circumstances indicate that the human brain has been significantly overbuilt for learning symbolic associations. Human brain structure is an exaggerated reflection—a caricature almost—of the special demands imposed by symbol learning. We are not just adapted for symbol learning, but for *fail-safe symbol learning*.

This symbol-learning insurance policy is provided by a comparatively overdeveloped prefrontal cortex, whose connections have gained the upper hand in numerous synaptic competitions throughout the brain. The extraordinary extent of this disproportional feature reflects its overdesign. It is a not-so-subtle clue to the central learning problems that make symbolic associations so hard for other species: the learning of higher-order conditional associations. But overbuilding for this one function has other consequences. The shift in cognitive strategy that underlies the assurance of symbol learning surfaces in a wide range of otherwise unrelated learning and attentional processes. If this exaggerated adaptation were merely a specialized sensory template or innately preprogrammed motor output, its effects might have been more domain-specific; instead, this peculiar adaptation has a largely supramodal character. It extends its influence into many learning realms. Symbolic abilities were recruited to serve other social and pragmatic functions than those which selected for their initial appearance. These ultimately became parallel "amplifiers" of the symbol-learning adaptations of the brain. This also applies to the component neurological biases that have made symbol learning fail-safe. Once present, these biases have become powerful sources of social evolutionary selection, independently recruited for functions in novel domains unrelated to language.

Important insights into the peculiarly human cognitive style have recently been provided by studies of the cognitive abilities of captive chimpanzees and other primates. Though the more carefully other primates' mental abilities are investigated, the more they seem able to do, some curiously simple tasks elude them. One characteristic example has recently been described by Sally Boysen of Ohio State University.[1] She set up a problem that chimpanzees could not seem to solve. Given a choice between two different-sized piles of a desirable food (like candy), chimpanzees consistently choose the larger pile, just as human children do. Boysen complicated matters by giving the larger pile not to the chimp who chose it but to a second chimp. In effect, one chimp was asked to choose the pile another would get, and by default, which would be left for himself.

When human children are presented with similar choices, most of them quickly catch onto the trick and learn to choose the smaller pile to give away (though infants under two years also have difficulty with the problem). Chimps, however, have extraordinary difficulty discovering the winning strategy. They repeatedly choose the larger pile, only to watch in agitated dismay as it is given away. Of course, one could argue that the chimps are just very unselfish creatures, but their emotional response suggests that generosity is not what they have in mind. I suspect that the task poses a difficulty, not because the chimps are ambivalent about sharing or cannot assess what they want, and probably not because they fail to comprehend the likely outcome after repeated trials, but because the presence of such a salient reward undermines their ability to use the stimulus information against itself. Being completely focused on what they want, they seem unable to stand back from the situation, so to speak, and subjugate their desire to the pragmatic context, which requires them to do the opposite of what they would normally do to achieve the same end. This is a very counterintuitive association for chimps to learn, because the indirect solution is overshadowed by the very powerful influence of its mutually exclusive and otherwise obvious alternative. The highly rewarding nature of the stimulus also reinforces the power of the competing association.

The chimps in Boysen's study were also trained to associate Arabic numerals with differing quantities, and succeeded in transferring these token-quantity associations to tests in which they used the tokens to make choices between larger or smaller piles of candies.[2] These experiments showed that when numerals were used instead of actual piles of candies, the chimps could learn to choose the one associated with less, and thus get the larger pile. Were the chimps able to use symbolic numeric information to overcome this cognitive "conflict"? Possibly, but a simpler account is sufficient to explain the result. If the chimps learned that the numerals were indices associated with perceived differences of quantity, choosing the lowest numeral to get the highest number of candies need no longer be interpreted as a "choose the pile you don't want" problem. When this associated stimulus is used in place of the candy, the chimp merely needs to learn which numerals correlate with getting the larger reward to succeed. It could thus be treated as a transfer learning problem of a particularly difficult form, because it requires a reversal of the pattern of associations. Irrespective of whether a chimp (or child) uses an indexical or a symbolic solution to this problem, the experiment clearly demonstrates how the indirectness of in-dexical and symbolic reference helps to reduce the power of the stimulus (icon) to drive behavior. Ascending the representational hierarchy pro-gressively frees responses from stimulus-driven immediacy, thus creating space for the generation and consideration of alternatives.

Choosing the opposite of what you want has all the hallmarks of tests that tend to be sensitive measures of prefrontal cortex damage in humans (sim-ilar incapacities to resist impulsive choices in patients with prefrontal dam-age are described by Antonio Damasio in his book *Descartes' Error*). It demonstrates the importance of this brain structure in shifting from a more "natural" habitual response to one that is its opposite; but more important, it also demonstrates that this requires overcoming the emotional immedi-acy of powerful reinforcers. The more powerful the past reinforcement and the more salient the competing alternatives, the more difficult the shift and the greater the need for prefrontal bias. This emphasizes the complex in-terdependency of the emotional as well as cognitive aspects of learning, and the special role that prefrontal cortex plays in both realms.

Am I suggesting that chimpanzees are like human patients with brain damage? Only in a metaphoric sense. Though healthy chimpanzees have well-developed and intact prefrontal lobes, and can solve many similar perspective-reversal and response-inhibition problems that other species can't, they run into difficulty when the power and salience of the alterna-tives are great. It is only when the solution to a problem demands domi-nance of prefrontal activities over very strong alternative tendencies that the greater prefrontal dominance in humans becomes evident. Indeed, it is in predicaments where the most intense impulses for immediate personal reward or self-protection are overcome in service of larger aims that we rec-ognize some of the most exalted expressions of humanity. A bias may only be a matter of degree, but in certain contexts the degree of prefrontal bias can make the difference between success and failure. This appears to be the case with symbol learning.

Of course, nothing is entirely fail-safe, and there are certain examples of abnormal human mental development where the human prefrontal bias seems to have been compromised, with both cognitive and emotional con-sequences. One such example is autism. Autism was discussed in Chapter 9 as an example of symbol acquisition difficulties, but correlated with this there are often both islands of spared "special" abilities and characteristic features of emotional impulsiveness, asociality, and stereotypic ritualized be-havior. Islands of spared or enhanced ability occur most often in the realms of spatial cognition, artistic talents, numerical skills, mnemonic fetishes, and even music. Such people are so-called idiot savants or autistic savants, like the character portrayed in the popular film *Rain Man*. We see in these cases a distorted reflection of what we generally consider to be genius. It is interesting that the most commonly recognized child prodigies tend to be those gifted with mathematical, artistic, or musical abilities. What unites all of these domains is a formal combinatorial system in which it is possible to produce results without interpreting them. Sums can be computed automatically, music can be learned by rote. Interpretation may even get in the way. Both mathematical prodigies and idiot-savant "lightning calculators" seem to do in their heads what normal people can only do with considerable external notational support and much more time. One might characterize the theory I have proposed about language abilities in these terms: prefrontal overdevelopment has made us all savants of language and symbolic learning. While still immature and exhibiting very limited learning abilities in most realms, two- and three-year-old children spontaneously solve symbol-learning problems "in their heads" that other species find essentially impossible even with considerable outside human support.

We humans are all like autistic savants in one other sense. We tend to apply our one favored cognitive style to everything. Like the character portrayed in *Rain Man*, who sees the world in terms of numbers of objects, we cannot help but see the world in symbolic categorical terms, dividing it up according to opposed features, and organizing our lives according to themes and narratives. From anthropologists studying the mythologies and kinship systems of diverse cultures to developmental psychologists studying young children's fascination with negation and use of it to regularize an inconsistent language experience or test social rules, we see not just a receptivity to symbolic relationships but a propensity to employ the biases that make symbols possible. We find pleasure in manipulating the world so that it fits into a symbolic Procrustean bed, and when it does fit and seems to obey symbolic rules, we find the result comforting, even beautiful.

But also like the autistic savant, we find that this global shift in cognitive strategy has other serendipitous consequences as well, which extend far beyond the realm of language.

The Sphinx

If language evolution correlated with a restructuring of the human brain on a global level, then there should be important consequences for a theory of the human mind and human nature. On the one hand, if language evolved as the result of the addition to the brain of a modular device whose effect on cognition was confined to a specialized domain, such as grammar and syntax, then we would be justified in thinking of human nature as ape-like, except in this one context. We could consider ourselves as chimpanzees-with-grammar. On the other hand, if the changes in the human brain were no more than a consequence of increasing brain size and general intelligence, then we would be justified in considering ourselves genius chimpanzees and thinking of chimpanzees as though they were dull and inept humans. Both of these caricatures of our place in nature have found widespread appeal over the years in popular expositions and even in the scientific literature. But the problem of comprehending where we fit in the context of the evolution of mind is considerably more difficult than either one suggests. Nonhuman minds are not just human minds with some special ability subtracted, nor are they human minds that are just considerably more dull and uninformed. Both views minimize the comparison problem.

The human case is difficult because we can neither rely on the logic of language to explain what happened to the brain, nor can we rely on some incremental increase in a general feature found in other species. Thus, when we use hindsight to analyze our own cognitive evolution—subtracting specific abilities or reducing overall thinking power—the simpler minds that we imagine as a result turn out to be both distorted images of ourselves and poor images of our ancestors and ape relatives. We end up misrepresenting other species' minds and not recognizing the oddities of our own. The result is an imaginary chimeric creature like those drawn in medieval bestiaries, half human, half animal, with a human head and an animal body like the Sphinx of ancient Egypt.

Because the human brain is organized differently, so is the mind that it gives rise to. The ways we think are the result of a new way of using the brain's resources. Brain-language co-evolution has significantly restructured cognition from the top-down, so to speak, when compared to other species. The prominent enlargement of prefrontal cortex and the correlated shifts in connection patterns that occurred during human brain evolution introduced strong biases into the learning process and gave human prefrontal circuits a greater role in many neural processes unrelated to language. Though intense selection was directed toward this one aspect of mind and brain, its secondary effects have also ramified to influence the whole of human cognition. Human beings approach the world of sensory stimuli and motor demands differently from other species, particularly with respect to higher-order learning processes, and these differences are evident even when our symbolic-linguistic abilities are uninvolved.

Let's begin with one of the most direct counterparts to language: our nonlinguistic vocal communication. With the evolution of language, there would inevitably have been changes in selection pressures affecting the hominid call repertoire. These changes would probably have included reduced selection for referential specificity, since such functions would have been better supported by symbolic communication. There would also be selection for changes in call and display functions that would be complementary to language functions. The effects of these selection pressures would have been unique to human evolution. In addition, there were likely incidental effects on the hominid vocal repertoire as a result of changes in vocalization circuits at the midbrain and brain stem levels.

Most species of primates have a modest repertoire of innate stereotypic calls; estimates range from fifteen to forty by different researchers studying different primate species. Though a small repertoire compared to human vocabularies, it seems large when compared to the few innate human vocalizations. These include laughing, sobbing, screaming with fright, crying with pain, groaning, and sighing. It is difficult to think of others. The list is embarrassingly small compared to call types cited for other primates. Although these primate estimates may be inflated by analytical subdivisions, and this human list does not include any variant subtypes, even conservative estimates of primate calls probably exceed the human repertoire in size. Human calls are not just suppressed or replaced by language. And transiently expressed calls are not produced by infants or released in adults as a consequence of cortical aphasia, though curses and expletive phrases may be.

Speech prosody is essentially a mode of communication that provides a parallel channel to speech. As we noted earlier, there are a number of reasons to suspect that this has been recruited from ancestral call functions. Like these systems, prosodic features are primarily produced by the larynx and lungs, and not articulated by the mouth and tongue. Though tonal shifts can be used as phonemes, the changes in tonality, volume, and phrasing that constitute prosodic features are most often produced without conscious intention. There are also likely to be many universal patterns of prosodic expression, though I am aware of few that have been systematically investigated. Finally, the communicative content of prosodic signals parallels that of other stereotypic vocalizations: they are symptomatic of arousal level, emotional states, and attention. But unlike calls of other species, prosodic vocal modification is continuous and highly correlated with the speech process. It is as though the call circuits are being continuously stimulated by vocal output systems. This may literally be the case. Displacement of midbrain axonal connections by cortical efferents likely includes prefrontal projections to midbrain and brainstem nuclei that also receive limbic and diencephalic inputs and contribute to stereotypic vocal displays in other species. Prefrontal output would be particularly likely to reflect rapid shifts in arousal and attention because of its role in organizing and anticipating the ordering and associative processes in speech production.

Laughter and sobbing are two innate human calls that are quite distinctive from those produced by other species. They are characterized by many invariant features in all normal humans. Their motor programs, like those of most primate calls, appear to depend on localized midbrain and brain stem circuits, and they are principally activated by limbic structures of the forebrain. They are the first two social vocalizations that children make, and they induce responses in others that are highly predictive of emotional states. They are highly contagious in social contexts, whether among adults or infants in a nursery. All of these features suggest that both were shaped by significant selection pressures for the social functions they perform.

As the most elaborated human calls, independent of language functions, laughter and sobbing may offer important clues about the context of language evolution. Laughter is highly socially contagious, and feigned or forced "social" laughter is produced frequently in many social contexts. These features suggest that laughter played an important role in the maintenance of group cohesion and identity during a major phase of hominid evolution. Another important clue is that both crying and laughter play significant roles in the social communication of infants with their caretakers long before language develops later in childhood. I think that it is particularly informative that these two vocalizations involve inverse breathing patterns: spasmodic breathing on inhalation (sobbing) versus spasmodic breathing on exhalation (laughing). This indicates that their sound configurations were selected with respect to one another, as a result of disruptive selection against intermediate ambiguous forms. These two vocalizations must have played very important roles in social communication to have turned out so distinctive and so independent of speech systems. Both seem to be powerful influences for group cohesion and for promoting shared emotional experience. The fact that both are expressed in infancy, long before any use of language is possible, offers evidence that they may at least in part have played crucial roles in bridging the gap in social communication during the extended period before the brain is able to approach the difficult task of symbol acquisition. But their production also persists into adulthood, and serves to initiate some of the more intense social-affiliation responses humans engage in.

But correlated with the co-evolution of these human calls with language, and the changes this produced in the brain, are some unique features. Sobbing is a call that has many nonhuman counterparts, particularly in young animals separated from their mothers. Its social function and the powerful selection pressures that shape it apply equally to both humans and nonhumans. Curiously, however, human sobbing is also commonly associated with loss, not just fear and isolation. Sorrow is an emotion that almost certainly has been exaggerated and modified by symbolic cognition, due to the power of symbolization to aid the mental representation of what-if scenarios. How much more painful is the loss of loved ones, because we can imagine what life would have been like had they remained?

The same sort of mix of nonsymbolic and symbolic correlates characterizes laughter. The role of laughter and smiling as vehicles for social bonding in parent-infant interactions may be important, but the function of their ubiquitous inclusion in casual adult conversation is less obvious, though probably similar. The role of laughter as a play signal, especially in mock aggression (as in response to tickling), may also offer some hint as to an older evolutionary role. But the ability of jokes and other forms of humor to elicit this stereotypic call is the most curious feature of all. Rather than a function under any selection pressure, it too has probably evolved as an incidental correlate of these other social functions and a somewhat serendipitous connection to the peculiar symbolic pirouette that constitutes humor.

The enigmatic connection between laughter and humor may be a side effect of the adaptation for symbol learning. The hierarchical nature of the processes that determine symbolic reference is particularly well exemplified by a joke, because humor provides an additional layer to the hierarchy of interpretive processes, beyond the level we generally consider word or sentence meaning. This is evident in cases where a person understands the meaning of every word and sentence of a joke and yet still fails to "get" the joke. The interpretive process involves more than understanding what is said. To "get" a joke is to understand the way the logic of the punch line both fulfills and undermines some "expected" conclusion. This ability simultaneously to entertain inconsistent alternative perspectives extrapolated from the same initial context is something that only we humans have. The structure of puns, jokes, and other forms of humor involves conflicts of conditional and hierarchic relationships. The shift of representation from one system of associations to another parallel but previously unrecognized one is common to both interpreting jokes and reconstructing symbolic relationships. Both are discovery processes, or insights, wherein previously acquired information and expectations must be quickly recoded according to a previously obscure but implicit logic. It is probably not accidental that such discovery processes are associated with positive emotions and a kind of release of social tensions.

Consider the intensity with which contemporary humans pursue mysteries, scientific discoveries, puzzles, and humor, and the elation that a solution provides. The apocryphal story of Archimedes running naked through the street yelling "Eureka!" captures this experience well. The positive emotions associated with such insights implicate more than just a cognitive act. The reinforcement that is intrinsic to achieving such a recoding of the familiar may be an important part of the adaptation that biases our thinking to pursue this result. A call that may primarily have been selected for its role as a symptom of "recoding" potentially aggressive actions as friendly social play seems to have been "captured" by the similar recoding process implicit in humor and discovery. In both conditions, insight, surprise, and removal of uncertainty are critical components. These features of humor again implicate prefrontal functions that enable mutually exclusive associations to be juxtaposed. Perhaps the "release" of laughter, like the sudden disinhibition of a suppressed automatic response, reflects the disengaging of prefrontal control. The link between a type of cognitive operation and a stereotypic call is curious and probably reflects the increased cortical control over vocal call production.

It is often the case that jokes have an aggressive social undercurrent to them, and yet they trivialize these tendencies as well. This may be a clue to a presymbolic function. Laughter is probably most common not in humorous contexts but in uncomfortable social contexts, where it displays both a nonaggressive stance and a kind of group assent. But laughter is not just an expression of emotion. It is a public symptom of engaging in a kind of mental conflict resolution. For this reason, whereas continuous crying is taken as a symptom of a seriously depressed state, continuous laughter is seldom interpreted as a sign of continuous elation, but rather as a symptom of a disturbance of reason, such as confusional or delusional states, and the associated conflicted emotions.

The unusual size of the prefrontal cortex in comparison to other not so enlarged brain structures may have costs as well as incidental side effects. These may be realized in the form of susceptibilities to disorders centered on prefrontal dysfunction. Even in resting attentive states, human prefrontal cortex tends to be more metabolically active than most other cortical areas. Human brains may thus be far more sensitive to prefrontal disturbances than are other mammal brains. Schizophrenia, manic-depressive disorders, obsessive-compulsive disorders, panic syndromes, Parkinsonism, and Alzheimer's disease are all associated with major changes in prefrontal cortex metabolism, and by implication, activity in the prefrontal cortex. Manic states, panic states, and obsessive-compulsive disorders appear to be associated with hyperactivity in the prefrontal cortex, whereas schizophrenia, depression, Parkinsonism, and Alzheimer's disease appear to be associated with hypoactivity of the prefrontal cortex.

Many of these disorders are associated with neural regulatory systems that serve to rev up or slow down certain general neural processes in different regions during different modes of activity. They are in this sense analogous to hormone systems and autonomic nervous functions in the rest of the body, which selectively tune the levels of visceral processes, repair mechanisms, and metabolic processes to adapt to certain major modes of activity. In the brain, this selective tuning is under the control of a small number of groups of neurons with connections that span many regions of the brain. They also employ a related set of neurotransmitters. Three major groups use aromatic monoamine neurotransmitters (e.g., dopamine, serotonin, and noradrenaline) and have vast branching arbors of axons that extend from small midbrain populations of cells and are distributed over vast forebrain areas. These neurotransmitters play crucial roles in adjusting the gain in the motor system, the experience of reward and pleasure, the regulation of attention, the process of learning, induction and maintenance of sleep, dreaming, and many other global brain processes. They are also the major targets for a great many psychotropic drugs, including cocaine and LSD, used for their mood- and mind-altering effects; L-DOPA, used for Parkinsonism; many antipsychotic agents; and Prozac, which has gained wide appeal as an antidepressant.

It is notable that these populations of neurons are located in the midbrain and upper brain stem, and these are regions that remain unexpanded in the human brain. Though we do not know yet whether these groups of cells have escaped the cell-number-determination processes of the brain regions they are born into, if they do not, then human brains will possess far less of these regulatory cells per regions of the brain to be regulated than any other mammal. Moreover, one of the major cortical target zones for many of these projections is the prefrontal cortex. This may make this system far more sensitive to slight disturbances of regulation, and more easily driven to pathological states of operation.

From studies of these projection systems in animals, it has become clear that one way the brain compensates for loss of connections or downregulation of neurotransmitter release, as often occurs with the use of psychotropic drugs, is to upregulate the sensitivity of cells that receive inputs from them. Thus during aging, drug sensitization, or pathological degeneration of these systems, the brain can extend the limits of control by becoming hypersensitive to even low-level inputs from these systems. But being close to the edge, so to speak, increases the sensitivity to perturbations and biological mistakes. Because of their brain proportions, humans might be predicted to have prefrontal cortex neurons that are relatively hypersensitive to changes of monoamine inputs. Since other systems of the human brain are also likely to be far more dependent on prefrontal functions than in other species' brains, it probably makes this entire relationship highly vulnerable to radical malfunction in response to lesions, metabolic disturbances, or genetic abnormalities. Depending on how it is impacted, the effects produce social mood disturbances or disturbed symbolic thought processes. These seem to be part of a spectrum of prefrontally associated disturbances of both social and symbolic cognition that can undermine normal function after the development of symbolic abilities, as compared to Williams syndrome and to autism, which disturb the initial development of social and symbolic abilities.

No Mind Is an Island

Because of our symbolic abilities, we humans have access to a novel higher-order representation system that not only recodes experiences and guides the formation of skills and habits, but also provides a means of representing features of a world that no other creature experiences, the world of the abstract. We do not just live our lives in the physical world and our immediate social group, but also in a world of rules of conduct, beliefs about our histories, and hopes and fears about imagined futures. This world is governed by principles different from any that have selected for neural circuit design in the past eons of evolution. We possess no brain regions specially adapted for handling the immense flood of experiences from this world, only those adapted for life in a concrete world of percepts and actions. These unsuited neural systems have been forced into service, and do the best they can to accommodate to an alien world and recode its input in more familiar forms. The consequences are both marvelous and horrendous.

One of these alien realms is the realm of other minds. Philosophers have long struggled with the problem of how we know that we are in a world populated with other minds. The problem was brought into precise focus by René Descartes in a classic meditation on the problem of whether we can be certain that other people really exist. Though the question merits no practical consideration, it challenges both our conception of self and mind, and is directly relevant to the symbol/nonsymbol distinction. Indeed, it is not just coincidental that Descartes was convinced that only people have minds. Other animals he thought were mere mechanisms, mindless clockworks. The dualistic dichotomy between mind and mechanism, subjective experience and material causation, is implicit in common sense psychology and has been the major theoretical subject of scientific psychology since that time.

Descartes was interested in whether we can ever know beyond any doubt if the bodies of friends and neighbors we encounter daily also have their own subjective experiences. Could it be possible that we are surrounded by the illusions of other beings, as in our dreams? When someone tells me their thoughts, are the words I hear just sounds produced by a biological robot, a mere mechanism? In a serious modern parody of this question, artificial intelligence researchers have constructed programs that are capable of fooling people into thinking that a person rather than a program is communicating with them by computer keyboard. This is what was known as a Turing test, suggested by a thought experiment proposed by the English mathematician Alan Turing as a test of whether a mechanism can be considered intelligent (though Turing actually had a somewhat less ambitious formal problem in mind). Such exercises testify to human ingenuity and gullibility alike: many people are fooled.

The problem with other minds is that the glimpses we get of them are all indirect. We have a subjective experience of our own thought processes, but at best only an imagined representation of what goes on in others' subjective experience. When we speculate about others' "inner" states, the only data we have to go on are what they tell us and what we observe of their physical states. Like the subject in the Turing test, we are forced to make assessments on rather limited and indirect data. We can, it seems, have direct knowledge only of ourselves. In philosophy, this argument is aptly termed *solipsism* (from the Latin *solus*, alone, and *ipse*, self).

In the post-psychoanalytic age, we are now painfully aware of a troublesome extension of this problem: we often do not even know ourselves. Not only have we forgotten much about our pasts, but Sigmund Freud convinced us that we can often be wrong about our own memories and beliefs about ourselves (a view that even non-Freudians hold, though disagreeing on the cause or interpretation of the error). The idea that an unconscious process might "rewrite" our personal memories to cover up past trauma puts us in doubt of even our direct experience of self. In other words, if our mental experiences are mediated by representation all the way down, then there is no direct knowledge. In these terms, the problem of representing the subjective experience of another and the problem of representing one's own subjective experience both entirely depend on the nature of the representational processes involved. The problem is not whether some knowledge is representation and some is direct and unrepresented. The problem is, rather, what sort of representation is involved, and what knowledge this provides of our minds and the minds of others.

If thought and experience are information processes, then the problems of representing other minds and representing our own minds ultimately become the same problem. Both forms of knowledge depend on a person's or animal's interpretive abilities. For this reason, as Descartes saw, the human/nonhuman difference in representational abilities inevitably enters into the question of knowledge of minds. If symbolic referential abilities are essentially confined to humans, there should indeed be important implications for animal and human minds, beyond communication alone.

In our everyday dealings with one another, we constantly try to anticipate others' behavioral plans or decision-making processes, either just to get along with them or else the better to manipulate them. Of course, even human social behavior is at least 90 percent social habit, and most of our actions are probably performed with no more than a fleeting consideration of what "anyone" would tend to do in some situation. Learned social habits provide a sort of unanalyzed folk psychology that gets us by. It is usually only in circumstances where we are especially surprised by others, or when we are intensely motivated by love, anger, fear, or avarice, that we strive to understand the workings of someone else's thought processes. Our frequent poor guesses about what others are thinking or planning attest both to the mostly "theoretical" nature of the process and to how complicated minds actually are.

Human social institutions, etiquette, and morality are predicated on the assumption that we *are* capable of such thinking about other minds. Even if this is a very fallible exercise, we expect people to engage in such activity under a variety of circumstances and are rightfully indignant or angered when they don't. We do not, however, hold other species to these standards. When we observe chimpanzees hunting down, capturing, and dismembering colobus monkeys alive as they devour them, we often feel a mixture of horror and acceptance. We do not call the chimpanzees cruel, as we would a human torturer causing his victim horrific pain. We have the intuition that somehow these otherwise very intelligent creatures fail to empathize with their victims, not out of conscious effort or the intention to cause suffering, but because it just doesn't occur to them. Of course, we exercise the same restrained judgment with respect to young children's acts of unkindness. The ability to take another's perspective is not inborn. It takes time and some degree of enculturation both to develop the ability to think in these terms, and to learn under what circumstances others expect you to apply it.

The problem of other minds has recently found its way into studies of animal social behaviors—often phrased in terms of a "theory of mind." Basically, the question is whether animals know that there is some*one* behind their perceptual experience of others. Do animals act socially on the basis of what they think others are thinking? To have a "theory of mind," in this nonphilosophical sense, is to have a mental representation of the subjective experience of others. A number of ingenious experiments have attempted to test whether animals tend to behave in response to what they infer others know, feel, or desire. Not surprisingly, the results are often hard to interpret.

It is a difficult question to answer experimentally, because it is possible for an intelligent and perceptive animal to learn to anticipate another's behavior in response to different conditions, irrespective of any model of why they are responding the way they do, and so give a false impression of representing a mental cause as opposed to merely developing an unanalyzed, conditioned response. This experimental difficulty is analogous to the problem we encountered earlier of determining whether a given referential behavior is symbolic or merely indexical. The parallel is not just a superficial one. Both are questions about the mode of representation. Is the representation of the expected behavior of the other mediated simply by a pairing of past instances of others' responses to similar circumstances, or by some additional representations about how these responses are caused by (recoded as) the hidden experiences of others?

Apart from postulating an innate "theory of mind module," how could such a mental representation be produced? In other words, what is necessary in order for information from another's behavior to be interpreted as representing another mind? The nature of this representation will depend on whether the individual is only able to interpret the behavior patterns as indices of future tendencies to behave, or is able also to interpret them as indices of mental states. The lack of symbolic referential abilities in non-human species limits them to representation of associations between stimuli, including the behaviors of others. These relationships can be quite numerous and complex, providing the animal with extensive familiarity with the predispositions of others. Probably most of the intimate behavioral knowledge we gain about friends and family members is of this type.

But the additional step to building an independent mental representation of the subjective experience of another, represented as if one were able to trade places, requires a symbolic jump. Since indexical reference grows out of repeated experiences of explicit pairing of stimuli or events, there is no possibility that pairing with another's mental state could be learned by association. It is an abstract attribute, inferred from the systematic patterns of the indexical relationships between observed conditions and behaviors, and identified with and opposed to one's own experiences. Constructing another point of view requires the ability to maintain awareness of the indexical information, and to use this to produce a representation of what it would be like to experience it from a different perspective, a complicated double-negative referential relationship. This is a cognitive task that almost certainly requires prefrontal cortex involvement, and evidence of failure to be able to take the other's perspective is a common disability of patients with significant prefrontal damage. Some striking examples of related difficulties of taking another "perspective" are described by Antonio Damasio in *Descartes' Error.* In this regard the hypersociality of Williams syndrome patients, and their intense monitoring and solicitation of others' responses in social interactions, may also be understood not just as a function of modified affect but as a shift in cognitive style, in which an exaggerated prefrontal bias may lead to an exaggerated reliance on symbolic prediction of others' behaviors. This may be particularly important if indexical representation of the predispositions of others is impaired. The inverse argument may also apply to autistic individuals. Because of a difficulty with perspective shifting, they may only develop knowledge of another's predispositions to behave by virtue of attention on physical behaviors as indices. Thus, the ability to form a representation of another's mental experience, as opposed to just their predispositions, is both mediated by symbols and dependent on many of the same mental operations and neural substrates as are critical to symbolic abilities.

As novelists and poets amply demonstrate, the range of personas and experiences that can be conveyed through symbolic media is unbounded. In a very real sense, this gives us the ability to share a virtual common mind. Because symbolic representation maintains reference irrespective of indexical attachment to any particular experiences, when an idea or a narrative of someone's experience is reconstructed by another, it can be regrounded, so to speak, by interpreting it in terms of the iconic and indexical representations that constitute the listener's memory. Symbolic reference is interpreter-independent, because each interpreter independently supplies the nonsymbolic ground for it.

Without symbolic representations at their disposal, it seems unlikely that other species could behave according to a theory of others' minds, much less share representations of others' experiences. Sharing common intentions, interests, goals, and emotions is the most effective means for coordinating behavior, and being able to imagine and anticipate another's mental and emotional responses is a powerful tool for social manipulation. If I am correct about the social-reproductive dilemma that served as the initial impetus for symbol evolution, then the ability mentally to represent other minds is one of the primary functions of symbolization.

Animal societies are complex, and animals often cooperate and appear to share common intentions and emotional states. And many species have a well-developed sensitivity to the emotional states of others, even others that are different species. Without the ability to represent others' experiences, how is this possible? Is it an error of anthropomorphic attribution to see empathy in these cases? The problem of empathy is not necessarily a problem of representing another mind; rather it is a problem of arriving at an emotional state that is the same as, or parallel with, that of another. Empathy involves both communication and interpretation, in which one individual (animal or human) interprets signs produced by another in a way that induces a corresponding emotional state. Symbols are not necessary to this process. The question we must ask is not whether language is a good vehicle for conveying and inducing emotions in others, but rather whether the ability to produce a symbolic interpretation of signs of another's emotional state provides empathic capabilities that are inaccessible without it.

The problem of empathy is a problem of shared emotions, and although language may play a role in communicating about one's emotional state, it is far less effective as a means of conveying emotion than the numerous nonverbal forms of communication that evolved for precisely this purpose. Tone of voice, posture, facial expression, and such specialized vocalizations as laughter and crying are incomparably more powerful conveyers of emotional state. This sort of communication is at least as well developed in other species as in ourselves, and as many commentators have observed, we humans are often less in touch with the emotional states of others than are other species such as our own pets. We value the performances of skilled actors precisely for bringing to the words of a play the nonverbal signs that can release our own emotional engagement. It would therefore seem that language evolution could have contributed only modestly to empathic ability, and may even interfere with it. One might be tempted to argue that the development of language abilities goes hand in hand with a decline in emotional sensitivity and empathy. But this only considers language as a medium for communicating emotion. What is also relevant is that language, as the most sophisticated symbolic system, provides a medium for building complex symbolic representations of emotions.

Following the representational hierarchy, we can identify three ways in which the emotional states of individuals can be brought into concert with one another. First, the response can be mediated by spontaneous mimicry (iconic); second, it can be mediated by a reaction to a common extrinsic stimulus (indexical); and third, it can be mediated by a representation of the state of the other (symbolic).

As we have seen, two familiar examples of empathy produced through spontaneous mimicry are the contagiousness of laughing and sobbing. Even newborns in a nursery exhibit this mechanism for producing empathy. Laughter and sobbing are not just evolved stereotypic indices of emotional state, they are vehicles for coordinating the emotions of a social group. Evolution has built in the reference of these innate calls by selecting for a stereotypic response that is concordant with the state of the "caller." It is a sort of preestablished harmony. Though clearly there are times in which the receptive state of the interpreter is not appropriate to produce a concordant response, even then the tendency to produce it (overcome by conflicting emotions) provides an interpretation.

To put the communication of emotions in context, we need to realize that it is only one facet of a more general process whereby information about one animal's state influences another's. Often this is not cooperative. It should be remembered that a large fraction of a species' communicative repertoire may have evolved for manipulative purposes, and not to solicit another to assume the same state. Whether one responds with empathy, sympathy, or antipathy depends not on the communication but on the specific social and evolutionary context in which it is produced. Evolution has selected for mechanisms of social signaling that are generally reliable indices or symptoms of the state of arousal and mood of the signaler. Ineluctable production of a particular vocalization or display when in a given state of arousal is the best guarantee of reference, but it does not determine the interpretation.

Alarm calling provides an example of a more complex, indexically mediated means for producing concordant arousal and emotion. In addition to sharing a concordant emotional response to the same call, the object of the call becomes an independent determinant of common response. It is significant that in both the iconic aspect of laughter/sobbing and the indexical aspect of alarm calling the representation of a corresponding mental-emotional state is grounded by common physiology and common perception, and as a result need not be an intentional activity. But symbolic information offers no similar guarantee. Listening to other people describe their discomfort or reading a third-person account of someone else's hardships can induce empathetic or sympathetic responses. Though these responses are often spontaneous, they depend both on the sophistication of the symbolic interpretation and the willingness of the listener or reader to carry on the interpretation process. Not only does this involve symbolic interpretation, but a sort of virtual experience in parallel to one's own.

The ability to interpret a narrative as a sort of simulated experience often requires the generation of complex mental imagery. Powerful mental images can elicit a vicarious emotional charge that makes them capable of outcompeting current sensory stimuli and intrinsic drives for control of attention and emotion, resulting in a kind of virtual emotional experience. The power of mental images to displace arousal on sensorimotor signals doubly depends on the ability of prefrontal activity to predominate over other systems, because of the requirement to maintain linked but opposed mnemonic traces.

This suggests that our most social cognitive capabilities may serendipitously grow out of the learning and attentional biases of the prefrontal bias that made symbolization tractable in the first place. So, although symbolic production of a state of emotion is far less reliable and predictable, it can produce far more profound and complete empathy than any other means. Unfortunately, it also has a dark side that other forms of empathy do not. Being able to imagine the suffering of another, by extrapolation from one's own subjective experience, is also the source of the most despicable and heinous of human practices. Though many other species harm and threaten one another in service of individual goals, only human beings can conceive of torture and the threat of death as means to an end. Indeed, threatening to cause suffering to another's loved ones is the most powerful of all forms of coercion—a uniquely human collision of the most base with the most noble powers of symbolic consciousness.

The best and worst of what it means to be human arose with the dawn of symbolic abilities. The slow evolution of prefrontal predominance probably correlated with an equally slow growth of the ability to get inside of one another's minds. Along with the power to organize reproduction around collectively enforced agreements and expectations of reciprocity, these other abilities inevitably also contributed to selection for the restructuring of the human mind. With increasingly powerful symbolic abilities comes an increasingly sophisticated ability to model the world that symbols enable us to represent. We do not hold young children and animals responsible for their failures to take the feelings of others into account, because we intuitively recognize that they are still limited in their ability to perform the mental gymnastics this requires. We also legally recognize the possibility that under certain states of psychotic delusion, adults too may be unable to model the "other" perspective well enough to make informed moral judgments.

The ethical stance is not, however, intrinsic in human nature. It cannot be innate because of its reliance on symbolic representation. Nor do I think that it is directly rooted in the "simpler" social behaviors of other species. Sophisticated predispositions for cooperative behavior or for caring for other individuals have evolved in many social species, and need not depend on symbolic reflection to anticipate the social consequences of one's actions. Social manipulation based upon predictions of others' reactions, and restraint of actions that might lead to negative consequences, can be based on nonsymbolic learning and mental imagery abilities, and tens of millions of years of social evolution have produced a brain whose arousal system is highly sensitive to both anticipated and received social signals. Though there are likely to be significant species differences in the tuning of the social-affect systems which drive attachment, status-seeking, play, aggressive, and appeasement behaviors, it seems unlikely that the human repertoire has added or deleted elements from this collection.

Ethical considerations are something in addition to the complex set of social-emotional responses we have inherited. The symbolic constructions of others' plausible emotional states, and their likely emotional responses to our future actions, are analogous to a whole new sensory modality feeding into our ancient social-emotional response systems. This ability to let our emotions be activated by the virtual experiences constructed with the aid of symbols probably makes us the only species in which there can be a genuine conflict of simultaneous emotional states. No wonder we often come off as out of touch with our social surroundings and less perceptive of others' arousal state than our own pets, or that we often feel as though our lives have become unnatural and envy the simpler state of connection to the physical world that we imagine other animals to have. We can easily become overloaded and confused by our ability to generate and respond to multiple what-if worlds. Perhaps this further explains our susceptibility to a range of psychiatric conditions that only superficially correspond to what other species can experience.

The ability to generate models of others' emotional states, and to exercise restraint or determination with respect to them, must be discovered through the manipulation of symbols. This makes such interpretations conditional on the maturation, experience, and even cognitive capacities of people. We intuitively hold the most intelligent, most educated, and most experienced people to higher ethical standards, and are more willing to give children and even adolescents the benefit of the doubt. This is because we recognize that these symbolic insights require some of the most counterintuitive shifts of perspective and recoding processes of any symbolic activity. When these cognitive difficulties are considered in the context of choices that may be in conflict with immediate self-interest, it becomes obvious why ethically guided self-control is both uncommon and fragile.

Such Stuff as Dreams Are Made on

Thirty spokes share the wheel's hub,
but it is the hole in the center that provides its usefulness.

—Lao Tsu, from the *Tao Te Ching*

Ends

A s a species, we seem to be preoccupied with ends, in all senses of the word. We organize our actions around imagined extrapolations of the consequences they will produce. We struggle in vain to comprehend the implications of our own impending cessation of life. And we weave marvelously elaborate and beautifully obscure stories to fill our need to find purpose in the fabric of the universe. This fills no obvious adaptive need. Our evolution never included selection favoring anything like this intense and desperate drive. And yet it is so powerful as to be able to overcome some of the most irresistible predispositions that evolution has provided. If we are language savants compared to other species, then a preoccupation with ends is the special exaggerated compulsion that complements our unique gift.

Symbolic analysis is the basis for a remarkable new level of self-determination that human beings alone have stumbled upon. The ability to use virtual reference to build up elaborate internal models of possible futures, and to hold these complex visions in mind with the force of the mnemonic glue of symbolic inference and descriptive shorthands, gives us unprecedented capacity to generate independent adaptive behaviors. Remarkable abstraction from indexically bound experiences is further enhanced by the ability of symbolic reference to pick out tiny fragments of real world processes and arrange them as bouys to chart an inferential course that predicts physical and social events. The price we pay for this is that our symbolically mediated actions can often be in conflict with motivations to act that arise from more concrete and immediate biological sources. Arguments in support of the classic notion of free will frequently cite this capacity to use reason (that is, symbolic inference and model building) to overcome desire and compulsion. One might respond that calling some actions "free" and others not oversimplifies what is really only a matter of the degree of the strengths of competing compulsions to act, some compulsions arising from autonomic and hormonal sources and others from our imagined satisfaction at reaching a symbolized goal. But there is an important sense in which these competing compulsions are not equal.

Those that arise from purely physiological sources, or physiological sources mediated by conditioned associations, could be called bottom-up processes for producing action. They are much more tied to mechanism and thus exhibit few degrees of freedom and limited spontaneous variation. They are comparatively predictable, though any organismic process inevitably exhibits tangled paths of causality. But symbolically mediated compulsions to act are far more chaotic, in the technical sense of that word, far more susceptible to the influence of tiny initial differences in starting assumptions or ways of dividing up experiences and qualities symbolically. This is because symbolically mediated models of things—whether theories, stories, or just rationally argued predictions—exhibit complicated nonlinearity and recursive structure as well as nearly infinite flexibility and capacity for novelty due to their combinatorial nature. It is not so much that our actions arise from a totally unconstrained and compulsion-free center of intentions, but that the potential starting point, the intended purpose we have modeled, can be drawn from such a vast variety of alternatives with little initial difference in motive power.

Final causality, according to Aristotle, is exhibited when processes are driven not by antecedent physical conditions but by ends. In some ways this is like time reversed. In hindsight it is easy to infer that certain past conditions were necessitated by the way things turned out. Deductive inference is a lot like this sort of reflective inversion of temporal and physical order. The consequence is already implicitly included in the premises. In symbolic thinking, this results in what might be called a sort of symbolic compulsion. Certain statements compel certain others. Aristotle reserved another term for such compulsion—*formal causality*—but I think there is an important way that this links to the other, classic conception of cause in terms of the ways that symbols work. Little of our reasoning is so precise as to be called deductive, and yet the way that certain beliefs compel others can have nearly this force. Ideologies, religions, and just good explanations or stories thus exert a sort of inferential compulsion on us that is hard to resist because of their mutually reinforcing deductive and inductive links. Our end-directed behaviors are in this way often derived from such "compulsions" as are implicit in the form that underlies the flow of inferences. So one might say that thinking in symbols is a means whereby formal causes can determine final causes. The abstract nature of this source makes for a top-down causality, even if implemented on a bottom-up biological machine.

Though the evolution of brains has been about systems for modeling and predicting events in the world, the evolution of symbolic abilities has not just amplified this ability far beyond that in any other species, it has also introduced an insidiously inverted modeling tendency. The symbolic capacity seems to have brought with it a predisposition to project itself into what it models. The savant, instead of seeing a field of wildflowers, sees 247 flowers. Similarly, we don't just see a world of physical processes, accidents, reproducing organisms, and biological information processors churning out complex plans, desires, and needs. Instead, we see the handiwork of an infinite wisdom, the working out of a divine plan, the children of a creator, and a conflict between those on the side of good and those on the side of evil. We carry a nagging doubt about anything really being accidental. Coincidence isn't just coincidence, it's a sign, and bad luck and disease don't just happen, perhaps a sorcerer has wished harm on the village. Wherever we look, we expect to find purpose. All things can be seen as signs and symbols of an all-knowing consciousness at work, or the marks of mythical events that occur in a dreamtime, behind the scenes of the universe. We are not just applying symbolic interpretations to human words and events; all the universe has become a symbol.

This is the evidence that we have become symbolic savants in the deeper sense of that metaphor. We are not just a species that uses symbols. The symbolic universe has ensnared us in an inescapable web. Like a "mind virus," the symbolic adaptation has infected us, and now by virtue of the irresistible urge it has instilled in us to turn everything we encounter and everyone we meet into symbols, we have become the means by which it unceremoniously propagates itself throughout the world.

It is clear that we feel more comfortable in a world that is meaningful, living a life that has meaning. The alternative is somehow too frightening. But why? Why should the ability to acquire symbolic abilities and conceive of things symbolically also bring with it a powerful urge to see it in every conceivable context? It could be seen as part of the predisposition to acquire symbols in the first place, part of the overdesign of the mind to ensure that symbols get discovered. But I think it may be a more mundane feature of cognitive and sensorimotor biases in general. The autistic savant is in this way no different from the kitten that sees every small mobile object as a representative prey toy, or the baby who interacts with every holdable object as a thing to be put into the mouth—for reasons that probably flow ineluctably from the Darwinian-competitive structure of neural information processing. Brains are spontaneously active biological computers in which activity patterns incessantly compete for wider expression throughout each network. Under these conditions, the dominant operation simply runs on its own and assimilates whatever is available. In us, this appears to be the expression of what I have called front-heavy cognition, driven by an overactive, busybody prefrontal cortex. It gets expressed as a need to recode our experiences, to see everything as a representation, to expect there to be a deeper hidden logic. Even when we don't believe in it, we find ourselves captivated by the lure of numerology, astrology, or the global intrigue of conspiracy theories. This is the characteristic expression of a uniquely human cognitive style; the mark of a thoroughly symbolic species.

One of the essentially universal attributes of human culture is what might be called the mystical or religious inclination. There is no culture I know of that lacks a rich mythical, mystical, and religious tradition. And there is no culture that doesn't devote much of this intense interpretive enterprise to struggling with the very personal mystery of mortality. Knowledge of death, of the inconceivable possibility that the experiences of life will end, is a datum that only symbolic representation can impart. Other species may experience loss, and the pain of separation, and the difficulty of abandoning a dead companion; yet without the ability to represent this abstract coun-terfactual (at least for the moment) relationship, there can be no emotional connection to one's own future death. But this news, which all children even-tually discover as they develop their symbolic abilities, provides an unbid-den opportunity to turn the naturally evolved social instinct of loss and separation in on itself to create a foreboding sense of fear, sorrow, and im-pending loss with respect to our own lives, as if looking back from an im-possible future. No feature of the limbic system has evolved to handle this ubiquitous virtual sense of loss. Indeed, I wonder if this isn't one of the most maladaptive of the serendipitous consequences of the evolution of symbolic abilities. What great efforts we exert trying to forget our future fate by sub-merging the constant angst with innumerable distractions, or trying to con-vince ourselves that the end isn't really what it seems by weaving marvelous alternative interpretations of what will happen in "the undiscovered coun-try" on the other side of death.

In many ways this is the source both of what is most noble and most pathological in human behaviors. Supported by these interpretations, rea-son can recruit the strength to face the threat of emptiness in the service of shared values and aspirations. But the dark side of religious belief and powerful ideology is that they so often provide twisted justifications for ar-bitrarily sparing or destroying lives. Their symbolic power can trap us in a web of oppression, as we try through ritual action and obsessive devotion to a cause to maintain a psychic safety net that protects us from our fears of purposelessness. The interaction of symbolic cultural evolution and un-prepared biology has created some of the most influential and virulent sys-tems of symbols the world has ever known. Few if any societies have ever escaped the grip of powerful beliefs that cloak the impenetrable mystery of human life and death in a cocoon of symbolism and meaning. The his-tory of the twentieth century, like all those recorded before it, is sadly writ-ten in the blood that irreconcilable symbol systems have spilt between them. Perhaps this is because the savantlike compulsion to see symbols in everything reaches its most irresistible expression when it comes to the sym-bolization of our own lives' end. We inevitably imagine ourselves as sym-bols, as the tokens of a deeper discourse of the world. But symbols are subject to being rendered meaningless by contradiction, and this makes al-ternative models of the world direct threats to existence.

Almost certainly this is one of the other defining features of the human mentality: an ever present virtual experience of our own loss. And yet we know so little about what it is that we fear to lose. Perhaps if we understood this symbolic compulsion, and the consciousness it brings with it, we might find this emptiness at the center a bit less disturbing.

To Be or Not to Be: What Is the Difference?

Throughout the history of philosophy one question above all others has constantly occupied center stage. What is the nature of consciousness? Unfortunately, terms like *consciousness, mind, thought,* and *idea* have many conflicting meanings, and this question is often confounded with a number of other related philosophical questions about thought, reason, agency, and the existence of an immortal soul. Is consciousness one thing? Is it epiphenomenal? Does consciousness require agency (or free will)? Are nonhuman species conscious? Are there different types or levels of consciousness? Is self-consciousness an essential ingredient of all consciousness? Is reason an essential ingredient in consciousness? Can consciousness only be experienced by living organisms? The absence of answers to such questions attests to the deep confusions that still surround the nature of consciousness. Though the subjective perspective from which each of us views the world is the mystery we ultimately hope to address by these questions, it seems difficult to determine where to begin the investigation because we are not even entirely sure what we mean by the questions.

A number of well-reasoned and elegant arguments about the nature of consciousness have recently brought these questions back into the spotlight of legitimate scientific and philosophical debate. These efforts to rethink this classic mystery are motivated by the growing body of new information from the neurosciences and the computing sciences. The more that is known about how the brain works, and how this might be similar or dissimilar to the operation of the information-processing devices we build, the closer we feel we are coming to the Rosetta Stone for translating subjective accounts of experience into empirical accounts of physical causes and effects.

The whole problem of consciousness is far too complicated to consider in the closing pages of a book on human brain evolution, nor am I prepared to deal with the many incredibly complicated issues it poses. Many of the subquestions that need to be answered first will not be approachable until we have a better understanding of the global information-handling principles around which brains are designed. But it is a question that can be approached in pieces. In contemporary discussions of the relationship between brain process and consciousness, three general problems are often treated separately. The first is often called "the binding problem" and refers to the problem of how the separate activities of millions of brain structures produce a unified subjective experience of self. The second is less often given an explicit name, but is sometimes referred to as "the grounding problem." It is the problem of explaining how our thoughts and words are guaranteed a correspondence with reality. Much of this book has been concerned with explaining this second problem. The third is often confused with the first two, but also has its own special difficulties. This is the problem of "agency"—explaining the experience we have of originating and controlling our own thoughts and behaviors. It is closely related to the more general concept of sentience. In older philosophical contexts, agency might have been part of the question of whether we have free will.

Many aspects of each question are beyond the scope of this book and my expertise. But the grounding problem is particularly relevant to the language origins problem. Indeed, I have approached the origin of language as essentially a comparative and evolutionary version of the grounding problem, in the sense that it requires us to provide both a mental and a physical account of the crossing from more concrete to more abstract forms of representation. The approach I have taken to the evolutionary question thus suggests a parallel approach to the grounding problem: analyzing the grounding of intentional consciousness in terms of levels of representational relationships.

The distinction that is often made between different *levels* or *types* of consciousness is closely related to the classic Cartesian claim that animals are mere mechanisms, whereas human beings have minds and are capable of reason. Few animal behaviorists these days would fully agree with Descartes' claim. Chimpanzees, dogs, and many other highly social species appear to consider alternative modes of action, anticipate consequences, and give the impression of sharing a similar range of emotional experiences. The minor differences in brain structure between these animals and human beings—which even backed Descartes into a corner (specifically into the pineal gland) in trying to find an essential human difference—provide no support for such a dichotomous view. And yet few would be willing to say that the consciousness of a dog or cat is of the same sort that we ascribe to humans.

Even a very much loosened notion of consciousness as extended to these relatively intelligent mammal species runs into trouble when we begin to probe more distant branches of the phylogenetic tree. Do mice exhibit the same sort of consciousness as apes? Are worms and insects conscious at all? Do lobsters suffer when plunged into a pot of boiling water? Is it more immoral to harm a cat than a fish? And so on. Arguments about comparative intelligence and comparative consciousness appear to overlap in these questions, and almost certainly the intuition we have about the level of consciousness experienced by other brains has influenced our notions about the evolution of intelligence. What we mean by "consciousness" is tied to issues of complexity. We intuitively believe that simple information-processing devices could not be *as conscious* as more complicated ones.

This brings us back to an argument that has reappeared a number of times in this book: to what extent is the evolution of the human mind adequately explained in terms of an increase in the *quantity* of information processing that can be supported? I think that the answer is the same as for language. The information-handling capacity of the brain is important, because there are certain minimal numbers of operations that must be performed and minimal memory requirements for symbolic processing. There is, therefore, probably a threshold of neural complexity below which symbolic processes are not possible. It is not, however, clear that our access into symbolic communication was just determined by surmounting this threshold, because there are other critical requirements as well, in the form of a very specifically structured learning and memory organization. In this book I have argued that the restructuring of information processes has been a much more critical determinant of the evolution of symbols. Similarly, I think that the human difference in consciousness is not merely a quantitative difference.

The idea that there must be grades of consciousness in the animal world suggests that the question of what *is* consciousness might be subject to a separate analysis from the question of whether there can be different extents or types of consciousness. Many species' brains may be constructed in such a way that they produce some limited conscious experience, and similarities in the neural architectures of human and nonhuman brains offer evidence that this is likely. Those theories of brain evolution that cast intelligence in quantitative terms suggest that the difference is not in some special consciousness-producing neural architecture, but simply in how much of it is present. On this view, it would seem to make better sense to forget the philosopher's predisposition to use human experience as the starting point, and instead solve the consciousness problem by work with frogs, flies, or leeches—whichever brains are deemed simplest but still capable of some basic level of consciousness. Clearly, the brains of different species can support very different amounts of information processing and storage. This recognition is paralleled by common sense ideas about how much we are *conscious of* under different circumstances. When asleep, I am not conscious of what is happening elsewhere in the room. When I avoid reading newspapers or watching television news, I am not conscious of what is happening "in the world." When I inadvertently hurt someone's feelings, I am not conscious of their needs or expectations. Though this might better be described in terms of "awareness," the two ideas are obviously entangled with one another in our thinking about the problem. Are animals which are more aware of their surroundings the species that we would call more conscious? Or is awareness only one necessary contributor to consciousness? What a complicated mess!

The question of human origins inevitably finds its way into discussions of the nature of consciousness because of the presumption that there is something special about human minds. The classic dichotomy between conscious and unconscious has been linked with the dichotomy between human and animal minds from the beginning. Religious traditions have historically played a significant role in guiding people's intuitions on the special nature of human consciousness. In Judeo-Christian tradition, only humans are supposed to have immortal souls, which presumably have something to do with consciousness and "personhood," but in innumerable tribal traditions and in many Eastern religious traditions, such as Hinduism, which include beliefs about the transmigration of souls, it is imagined that one's spirit could inhabit the body of any number of "lesser" creatures. The belief that only humans could have an immortal soul has been a major influence in the Western scientific and philosophical traditions, which tend to draw categorical distinctions between animal and human minds, and has therefore contributed to the historic reactions to Cartesian dualism or other claims of human mental uniqueness. But these polar responses—*either* that only humans have ("true") consciousness and other species lack it, *or* that all brainy species have consciousness and humans are not qualitatively different in this regard—are mostly the result of not carefully separating these different dimensions of the consciousness problem.

But are the alternatives posed by contemporary philosophical categories and cognitive science models really able to frame the full range of possibilities, or are we, as Dewey warns, faced with implicit alternatives that don't divide the world in a useful way? I think that there is a middle ground, but I doubt that any compromise framed in terms of the dichotomous alternatives of mind and mechanism can succeed in discovering it. Without trying to address the many larger philosophical issues that lurk behind these phenomena, I think that a reinterpretation of symbolic cognition can at least help to reframe one important assumption implicit in all these discussions. These arguments are all based on the notion that the different aspects of the mind-brain phenomena are distinguished by requiring *opposed* and *mu-*

tually exclusive forms of explanation. The clue to be drawn from the evolutionary transition to symbols is that it must be understood as a difference of *levels* of representational process.

This leads me to two proposals. First, I suggest that the oppositional form of most mind-brain theories reflects a failure to recognize the intrinsically hierarchical and emergent nature of the referential processes that underlie mental processes. Second, I suggest that the three contrasts implicit in these arguments—mind/body, intentional/mechanical, human/animal—do not all address the same categorical difference. Specifically, I think that the concepts of mind and intention need to be decomposed with respect to level of representation, requiring us separately to consider iconic, indexical, and symbolic aspects with respect to their mechanistic correlates. This allows us to recognize that the human/animal distinction is different from the others, in that it only involves a difference in symbolic ability, and does not address mind/body issues at all. Animals can have conscious minds, without sharing all the attributes of human consciousness. So, in the end, I do not suggest that explaining this transition gets to the heart of the consciousness problem, but it may help pare off some related problems that have been long confused with it. My point can be illustrated by reexamining some current exemplary arguments about the nature of cognition in the context of these representational distinctions.

The dominant alternative to the Cartesian perspective is exemplified by the theoretical claim that the mind is like the sort of "computation" that takes place in electronic computers. In simpler terms, minds are software (programs) run on the hardware (neural circuits) of the brain. The strong version of this materialistic reductionism (i.e., that mind is nothing more than mechanism) has recently been given its clearest expression in a theory called "eliminative materialism." The claim is that notions such as mind, intention, belief, thought, representation, and so on will eventually be eliminated in discussions of cognitive processes in favor of more mechanistic synonyms that refer to chemical-electrical signaling processes of the brain. Mentalistic terms, it is suggested, are merely glosses for more complex brain processes that we at present do not understand.

An analogy is often made to the relationship between a "high-level" computer programming language like C or Basic and the sets of machine operations that they specify. In the sorts of digital computers that we are most familiar with, the words and phrases of these programming languages are first translated into strings of machine codes, which bear only a vague resemblance to the higher-level language elements, and these are further translated into machine "addresses," to which a signal can be sent to activate a corresponding processor operation, store or retrieve a certain processing result, and so on. The languagelike structure of high-level code words is merely a convenience, a helpful shorthand. They are completely eliminable because they are nothing more than alternative ways of marking and activating specific machine operations. But is this a good analogy to the way "high-level" cognitive processes map onto brain "operations"?

Cognitive scientists have long argued over whether the ultimate "language of thought" is more like a computer algorithm, or more like collections of images. Our personal experiences of what goes on in the mind during our brief "down times," when walking between offices or washing dishes, suggest that both are right, that the mind is filled with many sorts of mental objects, and moreover that they overlap and blend into one another almost seamlessly. Arguments that pose mental images and symbolic "computations" as alternative or complementary "codes" for mental processes also miss the point that these are not representational processes at the same level. They are simultaneous facets of representational processes at different levels that are dependent on one another in specific ways.

Part of the danger in current computer metaphors comes from our tendency to call typographical characters "symbols," as though their referential power was intrinsic, and to call the deterministic switching of signals in an electronic device a "computation," because it simulates operations we might perform to derive an output string of numbers from an input string according to the laws of mathematics. We fall into the trap of imagining that the sets of electronic tokens (data) that are automatically substituted for one another in a computer according to patterns specified by other sets of tokens (programs or algorithms) are self-sufficient symbols, because of their autonomous parallelism to the surface features of corresponding human activities. This brackets out of the description the critical fact that the "computation" is only a computation to the extent that someone is able to interpret the inputs and outputs as related in this way. As the philosopher John Searle has suggested,[1] we might be able to discover a way of mapping numerals and operations of some mathematical schema onto the molecular events occurring in a container of chemicals, and in so doing treat the chemical reaction as a minimal computer as well, but this interpretation applied to what was already disposed to occur in nature would not invest it with any additional intentional properties. It is still just a simple chemical reaction. All the representational properties are vested in the interpreter. Applying this fundamentally incomplete analogy to brain function yields a model of thought as a program, and has suggested to some cognitive scientists that interpretive processes performed by brains are just the results of running neural programs. Though there is likely some sense in which the *physical* isomorphism between brains and certain kinds of electronic devices holds, this version of a *representational* isomorphism is empty, because it includes no account of the determination of reference in either part of the analogy.

In a now famous thought experiment that John Searle calls the Chinese Room experiment, he asks us to imagine a man locked in a room with an instruction book containing Chinese character strings in which input strings are paired with output strings.[2] The man in the room knows no Chinese, but through a slot in the wall he is given pieces of paper with Chinese characters written on them, and he is expected to respond by looking up that sequence in his book, copying the characters specified for the response onto another piece of notepaper, and handing it back out. If the book provides an exhaustive pairing of inputs to sensible outputs (and the look-up procedure could be complicated in a number of ways to make it more sophisticated), then to those on the outside, whoever or whatever they imagine to be in the box will likely be assumed to be reading, thinking, and responding in Chinese. In short, communicating symbolically. Searle's point is simple. We can imagine ourselves in this man's position, and we know that under those circumstances we would not be doing the conscious processes that are attributed to us. We would just be performing a "mindless" procedure that even a machine could do (and indeed we program computers to do similar things all the time). This model of cognition as a set of rule-governed procedures, an algorithm or program, reduces it to mere mechanism. Such a model does not explain cognition or consciousness. It merely simulates input-output relationships that could be produced by actually reading and understanding the messages, and consciously, intentionally answering.

A number of philosophers and cognitive scientists, including Daniel Dennett, have criticized this argument by pointing out that it does not really demonstrate that the man could not himself also be something equivalent to a Chinese Room.[3] Inside the room, consciousness and intention are irrelevant to the token-substitution task. So, can we be sure that we too are doing something different when we interpret and respond to similar strings of symbols? Are "consciousness" and "intention" just other labels for such a process? The man inside could be replaced by a computer running a similar kind of look-up translation program. So, just as Searle suggests that such an algorithmic account is insufficient to provide a constructive demonstration of an intention-based cognitive process—consciousness—it can also serve as the basis for computer modelers to retort, "So what! If an algorithmic account can produce language input-output relationships that are indistinguishable from those that people produce, what more is necessary? Isn't it possible that a similar rule-governed process is all there is to consciousness, irrespective of how we experience it?" Dennett calls the Chinese Room thought experiment an "intuition pump," because it offers neither a demonstration nor a proof. It doesn't prove that something is missing from the algorithmic account of mind, it instead challenges us to figure out why it is so unsatisfying. It emphatically begs the question: What's wrong with this picture? But does it help us to identify what additional kinds of processes underlie the way that we interpret words and sentences?

I suspect that what is missing and what is present in this analogy do not correspond to the difference between mind and mechanism. Minds emerge from the actions of mechanisms, and neither side in this contemporary debate is prepared to deny a link between them, nor is either side willing to adopt a dualistic interpretation. Instead, without its explicitly being described as such, the difference that is being highlighted by this thought experiment is a difference between indexical and symbolic representation processes, irrespective of the underlying mechanism. The man in the Chinese Room is working with indexical relationships, inputs that point to outputs, but he is oblivious to the fact that to those outside, the tokens he shuffles in and out are interpreted symbolically. I want to reinterpret this analogy, then, not as directed at the problem of consciousness in a general sense, but specifically at the problem of those aspects of human consciousness that have a symbolic basis: rational intention, meaning, belief, and so on.

To see the index/symbol distinction that is implicit in this problem more clearly, we can change the story a bit. Instead of following a list of English look-up rules by rote, which muddies the distinction, the man in the Chinese Room can be imagined to have learned these patterns of behavior by trial and error as a result of receiving a shock for wrong answers and food for correct answers. I don't believe this modification changes Searle's intent, but it makes clear just how superficial the algorithmic account of symbolic processing actually is. We do not even need to allow the man in the room to have rules or instructions to guide his actions. He could be replaced by a bevy of trained pigeons. It is the assumption of the eliminative materialists that, so long as there are enough pigeons to account for all the possibilities and they have each learned correctly to complete their operation on the Chinese character substitutions, then what they are *collectively* doing is language processing, even though individually no pigeon knows even a bit of Chinese. Each pigeon's response characteristics could be precisely specified by an algorithm. Indeed, each pigeon could be replaced by a computer with the appropriate input and output mechanisms running that algorithm.

What I find wrong with this metaphor of the mind as an algorithmic process is that it confuses the map with the territory, the indexical markers of points in a process with the process itself. The tokens and their manipulations are only fragments of the symbolic process. They have been abstracted from their iconic and indexical bases, and these have been discarded due to the isolation of the actor in the room. Symbolic reference is grounded by the relationships between a system of token-token relationships and a system of token-object relationships, but the walls of the Chinese Room make symbolic interpretation impossible because they make it impossible for the man to discover any relationships between the token-token system in the book he has access to and the systems of token-object and object-object relationships that are unavailable to him. The symbolic is reduced to the indexical (this is not unlike an archeologist's pondering ancient writing with no clue for translation). The only relationships available are the correlations among the typographical stimuli themselves.

It is important to point out that these minimal relationships would only be indexical to a human or animal actor in this thought experiment, because the token-token relationships are not in any way intrinsically indexical either. If the person's role in this room were instead implemented by some bit of software, I would argue that even indexical reference would be absent. Putting a human being in the room makes the stronger statement that even if we grant the representational capacity of a person in this circumstance, we can still see that the resultant capacity cannot be symbolic. Searle apparently intends this as a model for the more encompassing conscious/nonconscious (or mind/mechanism) distinction, which I would rather describe as a difference between referential and nonreferential processes in general. Maintaining the distinction between these two problems—the symbol problem and the reference problem in general—is important. One thing that makes the symbol problem troublesome is the tendency to confuse these two. This is what Descartes does when describing animal behavior as mere mechanism. And although I think we may be close to explaining the symbol problem, we are further from an answer to the reference problem and an account of the conscious/nonconscious distinction. So, for the moment let's just consider the simpler problem and see if we can discover how solving it might add insight to the more fundamental problem as a result.

What Searle's man in the Chinese Room is denied access to are two other levels of systematic indexical relationships that would be necessary to construct the symbolic reference relationships (and abandon the indexical ones). He lacks access to the system of indices implicit in the relationships between Chinese characters and outside objects and events, and he lacks the set of indices that links these outside objects and events to one another. The indexical relationships that are implicit in the combinations, sequences, and substitutions of characters that he receives and prints out cannot therefore be any more than this. A system of symbolic references emerges from the recognition of how all three systems of indices are linked, each iconic (isomorphic) of the other, each indexical of (correlated with) the other. The missing information that would be necessary in order to transform this indexical interpretation process into a symbolic one is between inside and outside. This is why the Chinese Room analogy undermines the argument that a program or algorithm for making token-token associations is the sufficient basis for an understanding of what they mean. Even an algorithmic system that completely captures all possible syntactic relationships between tokens does not in itself provide any representation—only the potential for internally circular indexical association. This is why a self-sufficient mental language or mentalese is an impossibility. No set of preprogrammed algorithms can smuggle symbolic reference into Searle's Chinese Room, because symbolic reference cannot be solely inside.

Symbolic reference therefore cannot be an intrinsic quality. This is why Searle can claim that there can be no "eliminative" strategy that can reduce intentional (read "symbolic"') processes to neural programs. The source of symbolic reference is not in the brain at all. This is why it is pointless to look for the basis for symbolic consciousness in a lower-level essence that is only associated with brains, or to invoke special physical laws that undermine the deterministic character of neural mechanisms in order to explain intentional consciousness. Symbolic reference does not derive from anything particularly special about the brain, but from a special sort of relationship that can be constructed by it.

Though the Chinese Room thought experiment implicitly exemplifies the distinction between indexical and symbolic reference, and the differences in consciousness associated with each, it is not a model of the animal mind/human mind distinction, but rather the computer/human mind distinction. These are very different. Though it is common to find researchers referring to simple brains, such as insect brains, as computers, the representational architectures of computers on the one hand and simple brains on the other are essentially inverted. The computer and the Chinese Room are "windowless monads." The collection of indexical relationships embodied in the program and the set of instructions, respectively, are only internally and circularly referential. Characters point to other characters, strings point to other strings. In contrast, an animal mind, even if it is of minimal computational capacity, constructs and processes internally generated indices with respect to an external world to which it is partially adapted. Its capacity to generate indexical responses may be limited, but the scope of reference is open-ended. Small brains may be capable of only a limited scope of iconic and indexical reference, but it is a mode of representation none the less. Algorithms that are not adaptive in a deep sense have none of this quality. Not all indexical processes are conscious, however. When reference is circular it too is mere mechanism, because there is nothing else for the indices to refer to but themselves. Where there is nothing other to represent, there is nothing to be conscious of. A program—whether in a computer or in a brain—could be an object of consciousness but not its source. These arguments do not really contribute any new explanation of consciousness; they merely illustrate certain minimal conditions for symbolic interpretation. But it is significant that this indexical/symbolic distinction seems central to the problem of consciousness.

No matter what else various theorists might claim about the nature of consciousness, most begin by recognizing that to be conscious of something is to experience a representation of it. The subjective experience of consciousness is always a consciousness of something. This is not to suggest that consciousness is something separate from the representation process itself (a view that Daniel Dennett caricatures as the "Cartesian Theater" perspective).[4] It is simply the realization that experiences arise as the brain progressively transforms neural signals which have been modulated by external physical events into correlated patterns of neural activity in other parts of the brain, which themselves modulate other patterns of neural activity, and so on, each re-presenting some formal aspect of the initial interaction in an additional neural context. In very general terms, both the neurophysiological and subjective information processes can be described as generating representations and interpreting them in terms of others.

For example, the pattern of electromagnetic waves reflecting off an object and entering the retina and the pattern of neural signals ramifying through circuits of the visual areas of the brain are both part of the causal chain on which the experiences of color are based. The color does not inhere in the object alone, nor is it merely a mental phantom. Something intrinsic to the object is re-presented in the pattern of light waves and again re-presented in the pattern of neural signals. But it is also re-presented in the subjective experience of color. There is no jump from material stuff to mental stuff in this process. The material and cognitive perspectives both recount relationships in form between successive and simultaneous points in a process, and these formal relationships, I submit, can be either iconic, indexical, or symbolic at various stages and levels of the process.

If consciousness is inevitably representational, then it follows that a change in the nature of the way information gets represented inevitably constitutes a change in consciousness. Consciousness of iconic representations should differ from consciousness of indexical representations, and this in turn should differ from consciousness of symbolic representations. Moreover, since these modes of representation are not alternatives at the same level but hierarchically and componentially related to one another, this must also be true of these modes of consciousness as well. They form a nested hierarchy, where certain conditions in the lower levels of consciousness are prerequisite to the emergence of consciousness at each higher level.

All nervous systems support iconic and indexical representational processes, irrespective of their size and complexity. They are the basic ingredients for adaptation. To some extent, I suspect that every living nervous system exhibits consciousness with respect to the iconic and indexical representations it can support. It's just that for some, this is a very limited realm. Their interpretive capacity will determine their capacity for consciousness. The differences between species in this regard are not qualitative, but quantitative. In species with more complex brains, representational states will be more numerous, more diverse, have a greater range of arousal amplitude, and will integrate across signals that cover a greater scope in both space and time. Statistics of large numbers and immense differences in magnitude have a way of making quantitative differences appear to be qualitative differences. It is therefore easy to imagine that the human difference is a difference of this kind, a significant quantum increase in the capacity. And to some extent it is. It's just not the only or even the major difference.

To appreciate why human beings are able to experience conscious states unprecedented in evolution, we do not need to have solved the mystery of consciousness itself. We do not need to understand the mechanism underlying conscious states in order to recognize that since they are based on representations, any difference in representational ability between species will translate into a difference in the ability to be conscious of different sorts of things. The formal characteristics of the interpretation process, whether iconic, indexical, or symbolic, will define the elements of a creature's conscious universe. So the development of an unprecedented form of representation—symbolic representation—while not the origin of consciousness, has produced an unprecedented *medium* for consciousness. This doesn't deny generic consciousness to other species; it only denies a particular aspect of consciousness that is based on symbolic abilities. Our brains share a common design logic with other vertebrate brains, and so we also share all those aspects of consciousness that are mediated by the iconic and indexical representation that these other species experience. Since iconic and indexical referential relationships are implicit and essential components of symbolic reference, the modes of consciousness that other species experience are an essential ground for consciousness of the symbolic world. We live most of our concrete lives in the subjective realm that is also shared with other species, but our experience of this world is embedded in the vastly more extensive symbolic world.

The evolution of symbolic communication has not just changed the range of possible objects of consciousness, it has also changed the nature of consciousness itself. Common sense psychology suggests that a lot of thinking gets done in the form of talking to oneself, editing and re-editing imaginary future or reconsidered past conversations, even when this also involves writing or typing out these thoughts to see how well the shorthand of imagined monologues translates into a coherent argument. Of course, these sorts of internal conversations must be unique to human brains while the majority of the other modes of thought are not. And given that our brains have only recently been "made over" to aid language processing, it is likely that the proportion of neural space and time dedicated to these various mental activities strongly favors the nonlinguistic. This does not necessarily imply that other species do not "replay" troubling past experiences over and over again, or that they are incapable of actively imagining possible experiences in some immanent future. They simply do not do so with the aid of symbolic reference or linguistic mnemonics. It also does not imply that imagistic thinking in humans lacks symbolic character and symbolic logic, though these forms of cognition are capable of following chains of association that are also uninfluenced by language.

The Russian cognitive psychologist L. S. Vygotsky suggested in the 1930s that a significant number of normal human psychological processes could be understood as internalized versions of processes that are inherently social in nature.[5] He gave language a central role to play in this because its fundamentally social nature provides a mental tool for gaining a kind of subjective distance from the contents of thought, that is, from our own subjective experiences. By importing, as it were, an implicit speaker-listener relationship into cognition, we create a tool for self-reflection by a sort of virtual social distancing from our own thought process. Thus we can talk to ourselves as though talking to others. Vygotsky conceived of mental development as a process of condensing and streamlining this internalized social process.

Language functions as a sort of shared code for translating certain essential attributes of memories and images between individuals who otherwise have entirely idiosyncratic experiences. This is possible because symbolic reference strips away any necessary link to the personal experiences and musings that ultimately support it. The dissociation allows individuals to supply their own indexical and iconic mnemonics in order to reground these tokens in new iconic and indexical representations during the process of interpretation. My imagistic and emotional experience in response to the episodes described in a novel is distinct from that of anyone else, though all readers will share a common symbolic understanding of them. The "subjective distance" from what is represented confers a representational freedom to thought processes that is not afforded by the direct recall or imagining of experiences.

This is crucial for the development of self-consciousness, and for the sort of detachment from immediate arousal and compulsion that allows for self-control. Self-representation, in the context of representations of alternative pasts and futures, could not be attained without a means for symbolic representation. It is this representation of self that is held accountable in social agreements, that becomes engaged in the experience of empathy, and that is the source for rational, reflective intentions. According to Vygotsky, this sense of self emerges slowly as children mature. It becomes progressively more facile in perspective shifting and at the same time consolidates greater control over the other aspects of self that derive from nonsocial sources, such as the experience of pain and effort, the arousal of basic drives, or the physical boundaries of control over events. And, as studies of this process in various mentally and socially handicapped children suggest, the extent to which it is developed depends both on the extent of exposure to relevant social-symbolic experiences and on the symbol-processing capacity of the individual.

Unlike the interpretation of icons and indices (a process which is uniquely personal and insular within each brain), symbolic representations are in part externally interpreted—they are shared. For example, though each of us supplies the interpretation of the words and phrases we hear and use, on a moment-by-moment basis, the implicit injunctions and constraints that determine each individual interpretation are borrowed from the society of users, and the symbolic reference that results is only reliable insofar as each interpretation corresponds with those performed by others. Imagine that Washington Irving's character, Rip Van Winkle, had remained in his magical slumber for many centuries. Upon awakening, not only would he be out of touch culturally, but he would find himself constantly misinterpreting the meaning of many still familiar-sounding words and phrases being spoken by those around him. As languages evolve and meanings and patterns of use drift away from older patterns, reference is maintained by continuity but not fidelity to the past. Symbolic reference is at once a function of the whole web of referential relationships and of the whole network of users extended in space and time. It is as though the symbolic power of words is only on loan to its users. If symbols ultimately derive their representational power, not from the individual, but from a particular society at a particular time, then a person's symbolic experience of consciousness is to some extent society-dependent—it is borrowed. Its origin is not within the head. It is not implicit in the sum of our concrete experiences.

Consciousness of self in this way implicitly includes consciousness of other selves, and other consciousnesses can only be represented through the virtual reference created by symbols. The self that is the source of one's experience of intentionality, the self that is judged by itself as well as by others for its moral choices, the self that worries about its impending departure from the world, this self is a symbolic self. It is a final irony that it is the virtual, not actual, reference that symbols provide, which gives rise to this experience of self. This most undeniably real experience is a *virtual* reality.

In a curious way, this recapitulates an unshakable intuition that has been ubiquitously expressed throughout the ages. This is the belief in a disembodied spirit or immortal "pilgrim soul" that defines that part of a person that is not "of the body" and is not reducible to the stuff of the material world. My ability to appreciate symbolic reference is not "reducible" to the indexical or iconic reference I use to ground my interpretation, though it is dependent on these lower-level modes of reference. Symbolic reference is also independent of any particular interpretive process, and retains its referential invariance despite interpretation by very different iconic and indexical processes in different minds. Its virtual nature notwithstanding, it is the symbolic realm of consciousness that we most identify with and from which our sense of agency and self-control originate. This self is indeed not bounded within a mind or body, and derives its existence from outside—from other minds and other times. It is implicitly part of a larger whole, and to the extent that it too contributes to the formation of other virtual selves and worlds, it is virtually present independent of the existence of the particular brain and body that support it. This may seem a shallow sort of disembodiment that pales in compared to mystical images of "out of the body experience"—it is more similar to the legacy of self composers leave in their music or great teachers bequeath to their students—but this symbolic aspect of self is nonetheless the source of our internal experience of free will and agency.

Abstract symbolic objects, like the Pythagorean theorem, guide the design and construction of innumerable human artifacts every day. Imagining counterfactual conditions, like what I might have done if I were the one who had stumbled on a reported accident scene, can cause me to enroll in medical first-aid training and perhaps some day aid an accident victim. Even imagined worlds—Olympus, Valhalla, Heaven, Hell, the "Other Side"—influence people's behavior in *this* world. Indeed, assumptions about the "will" of an ineffable God have been among the most powerful tools for shaping historical changes. These abstract representations have physical efficacy. They can and do change the world. They are as real and concrete as the force of gravity or the impact of a projectile.

On the other hand, the self that persists to influence others and continue shaping the world independent of the brain and body that originally animated it is detached from the specific iconic and indexical experiences that once grounded it in a personal subjective experience. This is precisely what makes it available for regrounding in the subjective experience of others, of becoming part of the self that controls, and feels, and connects with yet other selves from the locus of a different body and brain. In this regard, this part of personal identity is intersubjective in the most thoroughgoing sense of the term, and is capable of true transmigration, though not necessarily as a unified whole.

By this twist of logic, or rather its untangling, we also again return to Descartes' elaboration of the religious insight that only humans have a soul, and that this core of the self derives from a realm that is of the nature of language, pure mathematics, or geometry. Descartes' insight, currently maligned as archaic and antiscientific, seems to have more than a passing similarity to the notion I have developed here. But his rationalist assumptions—like those represented in theories of innate knowledge of language, computational theories of mind, or claims that human brains have a special essence that imbues them with intentional ability—reflected an implicit analytic or top-down perspective on the nature of symbolic reference. By failing to appreciate the constitutive role of lower forms of reference, iconic and indexical reference, this perspective kicks the ladder away after climbing up to the symbolic realm and then imagines that there never was a ladder in the first place. This leaves symbolic reference ungrounded and forces us to introduce additional top-down causal hypotheses, such as the existence

of an ephemeral soul or the assumption that there can be forms of computation or mental language (mentalese) that are intrinsically meaningful, in
order to fill in for this missing causal role in the explanation.

But unlike the eliminative materialist alternative, the perspective I have
outlined does not suggest that this top-down experience of self is all epiphenomenal nor that some of the claims about the nature of the mind which
derive from it are based on mystical notions. The symbolic representation
of self is solidly grounded in simpler representations of self, derived from
simpler forms of representation, and yet the arrow of cognitive processes
points neither from lower to higher nor from higher to lower forms of reference. As symbolic reference and symbol minds co-evolved from the nonsymbolic, each level of process drawing adaptive novelty from the other, so
do the levels of self-representation that constitute our experience bring
themselves into being in a moment-by-moment coevolutionary process. As
the symbolic process can be the co-author of our unanticipated brains, so
can the symbolic self be the co-author of the component neural processes
that support it. We live in a world that is both entirely physical and virtual
at the same time. Remarkably, this virtual facet of the world came into existence relatively recently, as evolutionary time is measured, and it has provided human selves with an unprecedented sort of autonomy or freedom
to wander from the constraints of concrete reference, and a unique power
for self-determination that derives from this increasingly indirect linkage
between symbolic mental representation and its grounds of reference. With
it has come a more indirect linkage between mind and body, as well. So this
provides a somewhat different perspective on that curious human intuition
that our minds are somehow independent of our bodies; an intuition which
is often translated into beliefs about disembodied spirit and souls that persist beyond death. The experience we have of ourselves as symbols is in at
least a minimal sense an experience of just this sort of virtual independence—it's just not an independence from corporeal embodiment altogether. Though this might seem like a weak consolation in comparison to
the freely transmigrating homunculus of mythical tradition, we should not
underestimate the miraculous power of symbols to break down even vast
barriers of space, time, and idiosyncratic experience that would otherwise
separate us impenetrably.

As we have seen, the symbolic threshold is not intrinsic to the
human–nonhuman difference. It is probably crossable to some extent in
many different ways by many species. This means that we are not the only
species that could possess such a "pilgrim soul," to use William Butler Yeats's
elegantly descriptive phrase. It was a Darwinian accident, or miracle, of na-

ture that this ability arose once and persisted for so long; but it has provided each of us with the opportunity to participate in bringing new "souls" into the world, not by procreation, but by allowing our own symbolic selves to be shared by other human beings, and perhaps by other animals, or perhaps eventually even by artifacts of our own creation.

Reinventing the Mind

Will we someday build devices that understand symbols, that have symbolic consciousness? Yes, I think so. And it probably will happen in the not too distant future. Will they look like human brains? No, probably not. Silicon would do. But the logical design, the way they acquire and "compute" symbolic relationships will of necessity be the same. This probably will not significantly constrain their physical design, however, or whether they are built from artificial neurons or transistors. Are present-day computing devices anything like this? No, mostly not. What's missing is not just symbolic interpretation, but something that is a prerequisite to all representational processes: sentience. I think this property of mental processes is analytically separable from the mode of representation that is used to achieve it. Though we commonly fail to distinguish sentience from consciousness, I think it makes sense to use the term *sentience* to refer to a more generic property of organisms with brains—their spontaneously adaptive, self-versus-nonself organization—and to use the term *consciousness* to refer to the way they represent aspects of the world to themselves.

The computer on my desk that is assimilating these notes as I type them in is not sentient, much less capable of comprehending symbolic relationships. It does not comprehend icons or indices either, because its informational architecture is essentially passive and closed. But this is not because of what it is made of. I suspect that even my 80-megaherz, slow electronic substitute for paper and pencil could run software able to exhibit some inkling of sentience; could generate some modicum of iconic and indexical representations, though it may be too simple to run software capable of symbolic processes. The key to this trick is not in the machinery itself but in the flow of patterns through it.

The first requirement in building a device capable of iconic and indexical processes is that it must be actively and spontaneously adaptive. It must continuously evolve new means of fitting with and anticipating its environment, even if this environment is constrained by whatever very limited input (keyboard?) and output (video screen?) devices can assess and respond to. And I really mean evolve, not in a genetic or phylogenetic sense, but through a moment-to-moment natural selection of patterned information. It must be capable of generating new patterned information where none previously existed. This is very different from the sort of data architecture comprising a massive list of templates to which an input is compared and matched. The latter is the way most computing is done these days, and many people have envisioned brains as doing just this: matching an input to a set of models in memory—some learned, some innately prespecified. Indeed, there is an approximation to this sort of process going on in most of what brains do. Most cognitive processes aim to achieve just this: completely automated, unconscious, mechanical, input-output matching. It is the goal of most cognitive processes to make information processing unconscious and automatic—as quick, easy, and efficient as possible—because these sorts of processes take comparatively little in the way of neural representation and energy to manage, compared to the active adaptational processes we experience as consciousness.

Like any evolutionary process, consciousness is messy, anything but streamlined and efficient. Vast numbers of alternative patterns, corrupted and modified by the incessant noisiness of the molecular processes underlying neural activity, must be generated, set into competition with one another for recruiting neural substrates, and mostly culled by selective processes due to their lack of correlation with sensory and mnemonic information. Out of this "bloomin', buzzing confusion," the temporary winners become the recognition processes and responses of each moment. This neural microevolution (microgenesis, it is sometimes called) is the generator of representations: that process to which sensory input patterns are represented and with respect to which they are re-cognized as icons. Such iconicity is, as we have seen, the bottom step of the hierarchy of representational processes in which symbolic representations are a distant possibility.

A microevolutionary basis for mind follows inevitably from the way brains are constructed in the first place. As we have seen, at every step the design logic of brains is a Darwinian logic: overproduction, variation, competition, selection. And ultimately, these biological processes are mediated by the information that flows through the nervous system. Brains are built using their own information processing—raised by their own bootstraps, so to speak. This design logic is the key to understanding how functional capabilities shifted in our own evolution, and also how symbolic and linguistic functions distribute themselves in our mature brains. It should not come as a surprise that this same logic is also the basis for the normal millisecond-by-millisecond information processing that continues to adapt neural software to the world.

There have been many recent insightful speculations on this notion of mind as a "Darwin machine," as many authors refer to the process. It has been around in some form for generations: it was obliquely suggested in some of the holistic psychologies of the early twentieth century, and was vaguely implicit in Donald Hebb's conception of how synapses become strengthened or weakened. However, with the maturation of evolutionary theory since the discovery of molecular genetics, and especially after the growth of interest in nongenetic evolutionary processes made popular by Richard Dawkins's coining of the term *meme* to refer to units of cultural evolution, many have recognized that Darwinian processes could also account for the evolution of ideas within the brain as well as without. A number of book-length discussions of Darwinian models of neural/cognitive processes are available that demonstrate the range of possibilities. Unfortunately, though it would touch on many of the problems I have addressed in this book, a detailed discussion of the Darwinian nature of neural information processing would take us too far from the theme of the book—symbolic processes. Rather than review these diverse approaches to what the Nobel Prize-winning biologist Gerald Edelman has called Neural Darwinism, and what others have called Darwin machines, I will focus on two aspects of this process that offer particularly useful insights about the relationship between Darwinian neural-processing models and the symbol acquisition problem.

The first insight, discussed at some length in chapters 9 and 10, is the recognition that different neuronal computations may be in competition with one another for representation in the brain both during development and within each cognitive activity. But if brains are designed by competitive processes, then the resulting architectures reflect a sort of stable microevolutionary balance between competing signal processing pathways. The differentiation of functions from one another and the segregation of competing computations is never quite complete. They are always in flux at some level. This probably results in considerable functional redundancy between nearby regions, even in adult brains, and probably also dictates that many functions are multiply and fractionally represented in diverse distributed networks. This should be especially true of computational processes that are somewhat abstracted from direct sensorimotor experiences, as symbolic processes are. It also predicts that representation of specific symbolic operations will dynamically shift depending on cognitive demand, recruiting additional computational support from regions more commonly recruited for other functions. All of these features seem to be reflected by data from brain-imaging studies (see Chapter 10). Thus, the cognitive experience of internal conflicts, interferences, resonances, and ambiguities among ideas and mental images has a direct correspondence to underlying patterns of neural signal processing.

But there is a second aspect of the subjective experience of consciousness that is well modeled by the Darwinian model of mind. Unlike the mechanistic view of mind that is suggested by digital computer metaphors, mind as evolution provides a way of understanding that aspect of our experience that is least like clockwork: our experience of being the originators of our own thoughts, perceptions, and actions. We do not experience ourselves as merely devices on which prespecified operations are "run," like programs in a computer—of being vehicles of cognition only. Some have suggested that these subjective introspections of self-determination and intention are unreliable epiphenomena. But a Darwinian interpretation of neural information processing offers two general reasons to suspect that we genuinely are self-determining and intentional creatures: one, because the operations are not prespecified; and two, because there is no clear dividing line between neural signal processing and neural architecture in a system where the circuits are created by patterns of signal processing. Evolution is the one kind of process able to produce something out of nothing, or, more accurately, able to create adaptive structural information where none previously existed. And the raw materials are the ubiquitous noise and context of the process. So an evolutionary process is an origination process—perhaps, as Richard Dawkins once remarked, the only known mechanism capable of being one. Evolution is the author of its spontaneous creations. In this regard, we do not need to explain away the subjective experience. We are what we experience ourselves to be. Our self-experience of intentions and "will" are not epiphenomenal illusions. They are what we should expect an evolutionlike process to feel like!

Symbolic representation processes have added a new level to this evolutionlike process. Their power of condensing representational relationships, and supporting extended flights of virtual reference, creates a whole new landscape in which the evolutionary process of mind can wander. By internalizing much of the physical trial and error, and even internalizing abstract models of physical processes that can be extrapolated to their possible and impossible extremes, we are capable of what genetic evolution is not: forethought. Representational processes are the substrate for "final causality," that everyday use of imagined ends to guide the selection of present means. Symbolic processes have freed this process from the bounds of the immediate present and possible.

But building a sentient symbolic processing device need not recapitulate the tortured path that humans took to this capability. The evolution of symbolic abilities in our species was a freak rarity only because of the way natural selection had biased brain evolution against its ever occurring. Most of what determines the structure of human brains was baggage accumulated on this meandering course prior to the evolution of symbolic communication. In designing an artificial symbolic device, it may be possible to take many shortcuts that avoid much of the presymbolic and visceral regulatory baggage that human brains carry to this process. Indeed, symbolic processing in a very simplified domain—adapting to a limited "environment" of potential signals—may not require anywhere near the massive information storage and retrieval capacity that human brains utilize. Neither mnemonic capacity nor learning rate blocked the evolution of this ability in other lineages; indeed, these capacities were to some extent a source of impediments. Symbolic capacities might be within the grasp of far less sophisticated computing devices than those that even the simplest mammals possess, so long as they were strongly biased toward the peculiarities of symbolic analysis and the symbol system was of only modest complexity.

This sort of mind is something with which we do not yet have any familiarity: a simple symbol processor. It might seem quite alien and odd. However, Williams syndrome may offer a human analogue to this sort of shortcut to symbolic abilities. Recall that this genetic abnormality produces individuals who are severely mentally handicapped in most ways but are relatively spared in certain language abilities. Their vocabularies are not nearly so severely reduced as their low IQ would suggest, and they can produce remarkably sophisticated lexical associations and grammatical analyses. But their lexical (symbol-token to symbol-token) understanding is supported on a very minimalistic iconic and indexical base as demonstrated by a poor ability to understand the pragmatic correlates of words and sentences. This, in a far more exaggerated form, is what we should expect from a man-made device capable of symbolic learning: not highly concrete rote (indexical) associations between tokens and their referents, as is typical of most animals, but in some ways the opposite. Symbolic representation with a minimum of iconic and indexical support could be creative, productive, and complex, but mostly vacuuous and circularly referential—almost pure language games. Nevertheless, there may be many contexts where this might be a useful tool.

How much further do we need to progress in our knowledge of brains and computing architectures to reach this stage? Probably not as far as one might think. In laboratories throughout the world, molecular biologists are capable of shuttling selected genes from organism to organism in order to study their effects or take advantage of their expression in a novel context. They do this with only a basic understanding of the way the complex molecular machinery of life operates, but ignorance of the global logic of the process does not preclude an ability to manipulate some of the details to see what happens. In a similar way, "neural network" designers are able to copy the organization of isolated bits of neural circuits and simulate their logic in software or in electronic circuits without really understanding what they are copying. Even if informed by only a dim understanding of the function of these circuit patterns in living brains, scientists are not precluded from studying the pieces to see how they work. Although a simulation must simplify many attributes of what it models, it is quite possible that those attributes of neural circuits that are essential to many high-level mental operations do not require any "deep" isomorphism with neural cellular properties. The isomorphism necessary for mental-like functions may only depend critically on aspects of the overall architecture and basic synaptic response properties, and not on the details. Evidence from human evolution has demonstrated that the achievement of symbolic referential abilities did not spontaneously emerge simply because of a threshold increase in computing power. So we should not expect symbolic capabilities simply to jump out as bigger, faster, and more sophisticated computing devices are built. It is not the size of the network that is crucial to symbolic processes, but the special logic of the relationships between learning processes. This is a global network feature, not a function of microarchitecture, and the design strategy may only need attend to rather large-scale network structures to succeed.

I have no doubt that nonbiological devices with nonbiological minds can, and probably will, be built. Nature itself produced minds by blind trial and error. So we know that a theoretical understanding of the nature of minds is not an essential ingredient for producing them. The science of "mental" engineering could proceed in a parallel trial-and-error fashion. Even without ever sorting out the theoretical issues, we may eventually achieve the physical result simply by copying nature. Mind is a physical process, and physical processes can be copied whether we understand what we are copying or not.

So, does it matter whether or not we have a complete account of consciousness, if it can be achieved without it? Is there more than an aesthetic value, a satisfaction of a deep curiosity, to be derived from pursuing a deep theoretical understanding of the relationship between mechanism and mind? Finding a way out of the philosophical maze presented by the problem of mind could, of course, greatly aid in the design and engineering of intelligent machines, but I do not believe this is the most pressing reason for pursuing the issue. Eventually, we will be forced to make judgments that distinguish between a simulation and the real thing—ethical judgments. Someday, these may be relevant to objects designed and built in factories. Will we know when we have crossed this line?

This leads us back once more to the mystery of subjective experience, and the extent to which we can know of it in another. The question of whether there could be independent, nonsubjective means for determining the presence of consciousness has long plagued philosophers. As we have seen, this mystery led to one of the most troubling doubts that confronted Descartes in his skeptical meditations on the nature of mind: how can we be sure that there are other minds in the world? A more mundane way to put this question is to ask whether we can ever truly share subjective experiences, as opposed to simply imagining that we do. In more recent times, the Turing test has addressed the parallel question of whether a machine can be determined to have intelligence. The Cartesian version of the Turing test might be phrased: how would I know if my computer program were conscious? Many computer scientists continue to debate this in a framework that is in many ways only a modern rephrasing of Descartes' problem. In the Turing test the question is whether a computational system can produce answers to queries that cannot be distinguished from those a person would provide under the same conditions. If a sophisticated questioner with extensive resouces can be entirely fooled into mistaking a device for a person, then the test purportedly indicates that the device possesses intelligence analogous to that of a person. In the Cartesian version of the Turing test, we might ask the examiner to judge whether or not the respondent is conscious and responding intentionally.

As Turing test competitions for programmers have demonstrated, people can be fooled in this regard, though generally they can still outwit most programs. This begs Descartes' question. Is it possible for a program to fool everyone all the time and still not be either intelligent or conscious? And likewise, is a person just a naturally-evolved program suffering from delusions of intentional consciousness in itself and other persons? Because the information available to assess this is circumstantial and indirect, in theory it should be possible for a program to fool everyone, including itself! The simple thought experiment of collating millions and millions of interviews between people and implementing them in a supercomputer as a look-up table of possible questions and relevant responses (as also might be imagined in a Chinese Room set up with access to such a supercomputer data base) suggests a trivial answer to the question. In principle, given enough data from prior interviews between real people and enough computing power to store and search through them, it will always be possible to fool even the most sophisticated questioner and yet do so without any semblance of consciousness or sentience. This is the gist of Searle's critique.

If we expand the nature of the "test" a bit, perhaps we can find a way beyond this impasse. Searle's test for the "merely algorithmic" is a clue about which information handling procedures are insufficient. An understanding of the way that symbolic interpretive processes are structured differently from this might provide a test for what is sufficient, irrespective of the answers produced (this is important, because human respondents in Turing tests can be mistaken for mindless computers!). In daily life, we sidestep the Cartesian dilemma simply by recognizing that the people we live among are physically just like us, and assuming that this means they have thought processes like our own, that they too are persons with conscious, intelligent minds. Descartes and Turing have disallowed this shortcut: it is in principle no more reliable a guide than the information provided in a Turing test. The questioner in a Turing test is denied knowledge of whether the answers come from a human subject or an electronic gadget, but if we are asking whether the output is the result of conscious intelligence it does not matter. However, I think there is a way that information about the architecture of the device (like knowing about sharing a common architecture with another human subject) can be independently informative.

Ultimately, the structure of the Chinese Room situation was the key to assessing the presence or absence of symbolic representation. This is because symbolic processes have a particular structure that defines and determines them. Similarly, I would argue that sentient processes in general have an identifiable structure—a Darwinian structure—the absence of which would mean the absence of subjective experience, and the presence of which would attest to the presence of at least a minimal form of subjective experience. If the features of this structure can be discovered by investigating the logic of underlying signal processing relationships in a device or organism, then it should be possible to discover something about the sentience and level of mental representation of that device or organism. In other words, an indirect inductive inference—a guess—would not be the only way to assess the possible subjective experience of another being or mechanism. I think that there can be an empirical basis for making that judgement, if one has access to the right information about the underlying information-processing architecture. Another way to put this is that, in principle, we can hope to determine the nature of the consciousness of another more reliably from bottom-up information than we ever could from top-down information, which Descartes and Turing showed can at best only provide a plausible guess. The process of symbolic interpretation will have distinctive physical attributes as well as produce appropriately interpretable signs. If we know how to recognize its logical architecture, we can know if it is there or not. The determination of subjectivity may not be merely a subjective matter.

We don't just need to know this in order to assess whether our computers are thinking or not. Many contemporary ethical dilemmas are unresolvable due to our inability to think clearly about the distinctions between mind and non-mind, conscious and nonconscious states, and the boundaries of personhood. Abortion and euthanasia, animal rights, the insanity plea in criminal justice, and care for the severely mentally retarded are just a few of the circumstances that force us to make ethical decisions without sufficient knowledge of this most relevant fact. These cases raise troubling questions about when a person begins and ends existence, which aspects of brain function are essential to personhood, what constitutes human rationality, and to what extent other animals share the experience of subjective consciousness. It is not clear that understanding these distinctions will make any of our choices easier, but at least we will know the consequences.

Albert Einstein once remarked that what is most miraculous about the universe is its understandability. This is also one of the most unsettling things about science: its ability to lift the veil of mystery, only to replace what was once magical and miraculous with dry, cold clockwork. An understanding of the basis of human consciousness will signal the final surrender in a long-fought Copernican revolution that has progressively usurped our privileged place in the universe. What are the ethical and aesthetic consequences for us, if consciousness and the capacity for the subjective and intersubjective experience of self are features that can be built into machines? Would this know-how somehow cheapen the value of our subjective experience of self? Would it diminish the intrinsic value we ascribe to conscious beings to know that they could be produced from silicon and wire?

On the one hand, like an audience unwilling to hear an explanation of a joke that they didn't get, we fear that this prospective fact almost certainly will diminish the sense of specialness we feel about our sentience. Something of the sanctity of personhood may be violated, if we ever succeed in explaining how it works and how it came about. However, it should be remarked that without this knowledge we currently mass-produce new consciousnesses anyway in a free-for-all of out-of-control population growth. Our cherished belief in the specialness of human consciousness has not prevented us from thoughtlessly treating people as throw-away tools. On the other hand, how much less thought will we give to the mistreatment of conscious devices, mass-produced in factories? The question before us is whether we will begin to treat people like unconscious computers, or come to treat conscious computers like people.

There is also a possibility that in a more accurate knowledge of the basis and nature of consciousness we will discover the cure for the angst that has been growing in Western societies ever since we began to doubt the Cartesian dichotomy of mind and body. If we conceive of human consciousness as an alien addition to an otherwise dead world filled with clockwork mechanisms, discovering that we ourselves are mechanisms appears to imply that we don't really exist, at least not as the intentional, self-determining persons we thought we were, and that there is no one else out there either. But discovering how such mechanisms work may be what is necessary to shatter this persistent belief. Unmasking the source of the subjective experience behind human consciousness is less likely to demonstrate how mental processes can be eliminated from material explanations than to demonstrate how they are implicit in them. And this may help us to recognize that the universe isn't, after all, the soulless, blindly spinning clockwork we fear we are a part of, but is, instead, nascent heart and mind.

Notes

Part One

CHAPTER 1

1. This cult classic science fiction movie was based on the notion that in the earth's future, other apes would evolve upright posture and speech, and would recapitulate a sort of ape version of human societies.
2. For an excellent recent review of studies of animal communication in the wild, see Marc Hauser (1996), *The Evolution of Communication*.
3. Some linguists, however, argue that language is so human-specific and dependent on innately built-in language knowledge that translation to and from a truly alien language would be impossible. We will return later to the question of how radically different symbolic languages could be from one another; probably the

most extreme examples are between spoken and manual signed languages like American Sign Language. Though manual languages such as this have to some extent evolved with respect to local spoken languages, borrowing some words in translated form, most of these are not fragmentary, partial, or parasitic languages, but complete and unique forms that often differ significantly in structure from those spoken around them. See Bellugi and Klima (1982).

4. One recent sophisticated addition to this genre, extrapolating from animal behavior to language, is presented in a recent book by Robin Dunbar (1997), *Grooming, Gossip and the Evolution of Language.* See also Dunbar (1992 a, b).

5. The hopeful monster approach to major evolutionary innovations was championed by the evolutionary theorist Richard Goldschmidt. See his 1952 review in *American Scientist.*

6. Moliére's play *The Imaginary Invalid (Le Malade Imaginaire)* can be found in *The Misanthrope and Other Plays,* translated by John Wood (Baltimore: Penguin Books; 1953).

CHAPTER 2

1. See, for example, Bickerton (1990).

2. Seyfarth, Cheney, and Marler (1980).

3. For example, a number of different species of birds use similar-sounding alarm calls to distinguish between such predators as hawks and owls. Peter Marler (1959) suggested many years ago that the localizability of these calls may have been a critical factor shaping their evolution. Calls that are used for "mobbing" a raptor need to be localizable, but those for warning of a stealthy, nearby predator need to be difficult to localize. Consequently, it is argued that the sound structures of alarm calls in different species probably resemble one another due to convergent evolution under these influences (and possibly also to the advantages of being able to recruit support from more than one species). See a recent review of other examples in Hauser (1996).

4. Such a scenario has been explicitly outlined by a number of researchers, including the linguist Derek Bickerton, in his book *Language and Species* (1990); and with a very different emphasis by the cognitive psychologist Philip Lieberman, in his book *Uniquely Human* (1991).

5. Cheney and Seyfarth (1990).

6. See Grice (1969). I would argue that this is incidental to linguistic communication, though almost ubiquitously present. In the next chapter and again in the final chapter, the role of symbolic reference for the development of knowledge about other minds and its relevance to intentionality will come up again.

7. Frege (1879).

8. This example provides an interesting exception that demonstrates a general rule about the assignment of word meanings in a language. Seldom do two words in

the same language share the same exact scope of reference. This particular exception snuck past the general tendency in English and in many other languages precisely because the common reference is not generally known. So these terms appear on the surface to have different references when in reality they do not. The debatable question is whether this tendency derives from a predisposition to avoid redundant reference or redundant sense.

9. Most notably these views are argued by the philosophers Saul Kripke (1972) and Hilary Putnam (1975).

10. An excellent short selection of Peirce's writings has been edited by Buchler (1955), including a number of his semiotics writings.

11. See Herrnstein, et al. (1980).

CHAPTER 3

1. de Saussure (1916). I have simplified this account considerably from what de Saussure described. He recognized an orthogonal "plane" to that of reference, associated with the combinatorial and diachronic (time-dependent) processes of language.

2. Though I will avoid most of Peirce's complex terminology and only obliquely suggest the unique philosophical context in which he embedded his semiotic theory, I believe the following analysis will stay close to his original insights, which focused on representation as process, not as a static relationship, and recognized these sign types as hierarchical levels of representation, not as opposed categorical alternatives. For a brief set of excerpts of some of Peirce's core writings on the topic, I recommend Buchler, ed. (1955). Many of Peirce's major writings on the semiotic basis of logic were compiled in the Hartshorne and Weiss, eds., collection of Peirce's papers (1978), especially Vol. II.

3. The symbolic nature of humor will be discussed in some more detail in Chapter 13.

4. See Savage-Rumbaugh, et al. (1978; 1980); and Savage-Rumbaugh (1986). In the stripped-down account presented here, I focus only on the two of the four chimps who were most successful (Sherman and Austin) and who continued to participate in subsequent language experiments.

5. Described in Savage-Rumbaugh and Lewin (1994).

6. Rumbaugh (1977).

7. I have simplified the paradigm a bit by omitting mention of an additional key that signaled the start of a trial (sometimes glossed in a sort of tongue-in-cheek fashion as "Please") and one to signify its end (sometimes glossed as a period). These were essentially irrelevant to the problem except as added distractors, since they did not vary from trial to trial, and such glosses could be quite misleading.

8. Savage-Rumbaugh and Lewin (1994).

9. Köhler (1927).

10. Piaget (1952).

CHAPTER 4

1. Chomsky (1972; 1980; 1988).
2. See an early formalization of this argument by Chomsky and Miller (1963).
3. Christiansen (in press). The analogy of language to an organism has been used as a heuristic principle for understanding language change by comparative linguists since early in the last century. One of the most prominent was the linguist August Schleicher, who in the middle of the nineteenth century proposed that languages should be analyzed as organisms with their own family trees, which could be traced like phylogenetic trees, using common inherited traits.
4. Berlin and Kay (1969).
5. Rosch (1978); and see other discussions in the same volume for more details and examples.
6. I have borrowed this term from Nelson Goodman's (1955) discussion of a similar point about the problems of accurately "projecting" color reference to future uses, where he argues that there is insufficient basis in past exemplars to guarantee consistency of reference into the future. This is part of a more general critique on the nature and weakness of the argument from induction. Goodman suggests that only social "entrenchment" (essentially habit) is responsible for stability of word reference projection to future uses, and that this is far too fragile to provide a basis for referential certainty. The co-evolutionary argument suggests an alternative source of entrenchment that can provide an incredibly robust projectability. Since Goodman's intention was also to question the projectability of scientific induction, a similar argument can be made about the evolution of scientific terms and theories.
7. Jackendoff (1992; 1994).
8. See Savage-Rumbaugh and Lewin (1994).
9. Greenfield and Savage-Rumbaugh (1990; 1991).
10. Gold (1967).
11. A long history of debate in the philosophy of science has pursued this theoretical problem of projecting general rules from a finite set of instances. There is general agreement that rules which we apply to describe natural events (e.g., the laws of physics) are inevitably underdetermined by the inductive inferences that we use to justify them, and that in general extrinsic constraints or special conditions are implicit in general rules that predict real-world regularities. So real-world inductions invariably rest on a very infirm foundation of assumptions. See also note 6 on Goodman's critique of induction.
12. Ramsey and Stich (1991).
13. Newport (1990; 1991).
14. For example, a fragment from the lower-left corner of a hologram of a face would need to be viewed obliquely from its lower left, and only a view of the face as seen from below and to the left would be possible. Unfortunately, the sort of holograms used on credit cards or sold as pictures (known as white light

holograms) provide a far more limited range of perspectives than true laser-illuminated holograms, and can be viewed at only a few angles. In these partial holograms, full holographic recording of information must to be sacrificed in order for the hologram to be viewable in the mix of frequencies and phases found in normal light.

15. Elman (1991; 1993). See also Elman, et al. (1996), *Rethinking Innateness*, for a more detailed discussion of how neural network models can help explain the role of learning biases in language development.

16. Consequently it would not be expected to avoid grammatically correct nonsense sentences, as exemplified by Noam Chomsky's well-known example: "Colorless green ideas sleep furiously." However, if the input strings avoided such syntactically correct nonsense, the statistical likelihood of the network predicting it would be minimal. This further exemplifies the fact that the network has not embodied abstract rules that can be applied arbitrarily and generatively to any possible novel input, but rather has internalized the positional statistics of patterns in the input strings. It is a little like having a vast list of possible word combinations to choose from, each ranked in order of their statistical likelihood.

17. A kernel sentence is a completely simplified (i.e., not compound) sentence such as "A dog chases the ball," "The dog is playful," "The ball rolls under the chair," etc., that can be combined with other tense and phrase markers to form a compound sentence such as "The ball that rolled under the chair was chased by a playful dog."

18. For a more detailed review of these differences between learning problems, the tricks used to overcome them, and the relevance this has for all sorts of normal biological learning problems, I refer the reader to a recent paper by Clark and Thornton (in press). They distinguish two types of learning problems, designated type 1 and type 2, where type-1 problems are solvable directly using statistical regularities present in input patterns, but type-2 problems have highly distributed and therefore marginally presented regularities in any part of the input corpus. In general, they demonstrate that type-2 problems cannot be solved by any sort of "uninformed" learning paradigm, no matter how powerful, but that specific information (as suggested by strong nativism) need not be required if there are ways to constrain learning parallel to the ways regularities are "buried" in the input. Incremental learning (as in Elman's simulations), and recoding strategies, essentially accomplish the same reduction of type-2 problems to type-1 problems and render them solvable.

19. The term refers to "the spandrels of San Marco," a metaphor from Gould and Lewontin's (1979) paper on the dangers of pan-selectionist arguments in evolution. Spandrels are the symmetric, triangular arch structures that support the upper domed roof of the cathedral of San Marco in Venice, and are elaborately decorated in ways that attempt to integrate their number, shape, and symmetries into the content of the religious icons decorating the cathedral. The point of this example is that the integration of these structures into the artistic design

may make them appear as necessary thematic elements, when instead they just happened to be unavoidably part of the background and so were incorporated into the artwork after the fact.

20. Bickerton (1981; 1984; 1990). For a summary, see also Bickerton (1983).

21. For a detailed survey, see Todd (1974). A brief overview of some the major arguments (not completely up-to-date) and numerous examples are also provided by David Crystal in the *Cambridge Encyclopedia of Language* (1989).

Part Two

CHAPTER 5

1. Holloway (1995).

2. There have been studies which suggest that under supernormal levels of stimulation, often referred to as "enrichment," brain size and other measures of brain structure such as cortical thickness can be induced to increase. In general, however, this difference has been determined with respect to "control" animals that have been housed in stimulus-impoverished environments (typical laboratory cages), and so I think it likely that what has been demonstrated might more accurately be interpreted as the effects of sensorimotor deprivation. I think it is a good bet that lab animals housed in "enriched" environments are also somewhat deprived of stimulation and environmental challenges when compared to their feral cousins.

3. The idea of assessing species intelligence by brain/body ratio rather than total brain size was first promoted by the eighteenth-century physiologist Albrecht von Haller and his contemporary, the great comparative anatomist Georges Cuvier. By the latter half of the nineteenth century, biologists had heated debates over the validity of absolute brain size versus brain/body ratio in studies of human intelligence differences and the ways they were linked with sexual, ethnic, or racial differences. Because different analyses produced differing trends, it was nearly always possible to find an analysis that fit one's intuitions (or ideological biases). Stephen Jay Gould provides a cautionary treatment of the effect of these biases on the brain science of the last two centuries in his book *The Mismeasure of Man* (1996).

4. It is unclear whether Snell (1891) actually knew about Brandt's (1867) analysis.

5. Harry Jerison, *The Evolution of the Brain and Intelligence* (1973).

6. Related anatomical interpretations had previously been offered in slightly different terms by two of Jerison's famous predecessors in the brain size/intelligence field, E. Dubois (1913) and G. von Bonin (1937).

7. I generally think that the relative metabolic cost of possessing a large brain tends to get overemphasized in evolutionary arguments. It has been cited as a major cause of heat production and heat dissipation problems in the "radiator" theory of hominid brain size increase (Falk, 1990), as the major constraint affect-

ing foraging requirements (Allielo and Wheeler, 1995), and as limited in size by the amount of energy that a body can supply (Armstrong, 1990). These assumptions are extrapolated from the order of magnitude of greater brain metabolism per mass at rest compared to that of other tissues. But the key phrase is "at rest." Resting or basal metabolism may not be the most useful measure in evolutionary terms. Just what fraction of the energy of overall metabolism is demanded by brains under normal active conditions is difficult to answer, but compared to total energy consumption it is periodically dwarfed by what is required when we are not at rest and need to make extensive use of muscles, even in daily activities, such as foraging. Unlike muscles, which can fluctuate from minimal metabolism to vastly more than the brain during peak exertion, the brain's highs and lows of energy use exhibit a much narrower range from sleep to active cognition. It's easy to work up a sweat just climbing a few flights of stairs, but few people work up a sweat solving the *New York Times* crossword puzzle. Indeed, neural activity alone is most likely to generate perspiration only during times of intense emotional arousal when the fight or flight response triggers the sympathetic nervous system to prepare the body for imminent intense *physical* exertion. Nevertheless, the approximately 10% of resting metabolism that is accounted for by the brain's energy demand cannot be ignored in the overall metabolic economy of the body, especially when whole life history estimates are considered.

8. D'Arcy Thompson (1917).

9. Ewen Macphail (1982).

10. See Bitterman (1975; 1988) on vertebrate/nonvertebrate comparisons of learning.

11. This work is reviewed by Duane Rumbaugh and colleagues in a 1996 paper in *Japanese Psychological Research*. The theory and experiments behind the work are also discussed in Rumbaugh and Pate, 1996.

CHAPTER 6

1. However, see the recent book by Stanley Coren, *The Intelligence of Dogs* (1995), for a lively discussion and some comparative tests of canine IQ.

2. The slope for brain size/body size scaling within a species is only about 0.1 to 0.2, in contrast to the 0.6–0.8 slopes observed within whole orders of mammals. This was first demonstrated at the beginning of this century by Lapicque (1907a, b), in a study of dogs.

3. Coren (1995).

4. Kruska (1988).

5. The first of these genes were discovered by investigating the causes of fruit-fly mutations that produced duplicated additional body segments, such as one that produces an additional thorax segment with a duplicated set of wings, called *Ultrabithorax*.

6. It should be noted, however, that this vertebrate/invertebrate comparison is

complicated by two curious and still poorly understood differences: the HOM genes in flies form a linked group or family and are arranged in serial order along the same chromosome. This order corresponds to the front-to-back order of their expression along the body axis. But in vertebrates there appear to be duplicate families of corresponding Hox genes (four in mice and humans), and although their expression is segmental, it is also overlapping, whereas in flies it is more strictly segmental.

7. General overviews of the significance of these homeotic genes is provided by Holland (1992), and Finkelstein and Boncinelli (1994). Reports of Emx and Otx gene expression in the developing brain are from Boncinelli, et al. (1993a, b). Here I describe the expression of these genes in the embryonic brain, where they are most prominent, but some of these genes are also expressed in other regions of the body and change their expression patterns over developmental time. This is characteristic of many homeotic genes, and demonstrates a kind of diverse multifunctionality, which belies any claim that they are like brain blueprint genes. They are probably better thought of as surveyor genes, whose function is to establish borders and property rights that constrain which other genes can "settle" where.

8. These headless Lim1 knockouts actually develop in the womb but die after birth apparently due to a failure to breathe normally. See the discussion of Williams syndrome in Chapter 9 for discussion of a related human gene "knockout" (of the gene Lim1-kinase) that occurs naturally and affects language and cognition.

9. One continually recurring error is a failure to control for the effect of confounding a part with the whole when the part in question is a significant fraction of the whole. Though this is the only way to track what fraction of the whole is made up by the part, assessing the scaling relationship in such terms can lead to very misleading claims about underlying causes. The figure below graphically depicts the artifactual consequences of including the part in the whole in allometric analyses. One particularly troublesome example of this artifact involves predictions about the scaling of the cerebral cortex with brain size. The cerebral cortex comprises as much as 70% of the whole brain in a large primate, when considered with underlying white matter. A number of recent analyses of primate and nonprimate brains fail to correct for this, however, with the result that they greatly underestimate the variance among species, underestimate the human divergence from the trend, and bias the assessment of the trend to appear close to isometry (Deacon, 1990a). This error can be avoided by restricting comparisons to nonoverlapping structures, or else only by comparing part-whole relationships where the part is a small fraction of the whole (and where the non-independent variables contribute only insignificant biases). The graphs I have included in the text, which compare brain structure scaling relationships, have all been calculated from nonoverlapping measurements.

10. These problems have been dealt with in more detail, along with other allometric difficulties of comparing brains, in Deacon, 1989 and 1990.

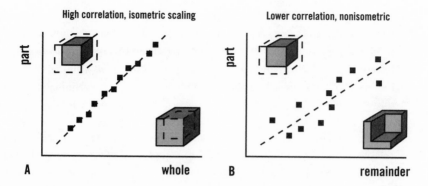

High correlation, isometric scaling

Lower correlation, nonisometric

part

part

A   whole   B   remainder

Figure i. *Analytical problem of confusing part and whole in statistical analyses of brain structure sizes.*

A: Part and whole are confounded, artificially shifting the trend and reducing the apparent the variances from it.

B: Segregating units of analysis avoids this problem and often produces very different results.

---

11. In this process, called *neurulation,* the outer cell layer (ectoderm) is induced to fold inward to form a tube, by the signals produced by contact with inducer cells from the middle layer (mesoderm).

12. Work by Balaban, et al., 1988.

13. Deacon, et al., 1994.

CHAPTER 7

1. Deacon, et al. (1994); Isacson, et al. (1995); Isacson and Deacon (1996).

2. This view shares many points in common with Gerald Edelman's (1987) theory of "neural Darwinism," which attempts to frame a global theory of brain functions on a Darwinian-like theory of neuronal group interactions. The Darwinian nature of neural developmental processes plays a critical role in this theory, since Edelman hypothesizes that the initial nonspecificity of connections provides the total initial variety upon which selection processes (in the form of learning) can later act. Though my approach also models neural function on Darwinian-like competitive neural processes during development, I make no assumptions about the sources of variation available to the selective processes. Specifically, whereas Edelman begins with the assumption that the entire source of informational variety available during cognition derives from the early nonspecificity of axonal and dendritic growth, I suspect that initial nonspecific connectivity accounts for only a small initial fraction of the variety of information available to selection/learning processes during mature function. The rest is supplied "on line" by continuous, spontaneous, nonspecific sources, including the effects of metabolic noise on random neural activity and continuous axoden-

dritic growth processes. More important, I argue that the initial connectional nonspecificity and axonal competition principally influences cognitive processes by introducing global biases in neural computations that result from global patterns of regional parcellation. Large-scale quantitative relationships determined by this developmental Darwinism thus play a major role in producing species differences.

3. Hebb (1949) originally envisioned his theory as a way to explain basic reinforced learning processes in the brain: how connections could become altered in their relative signal transduction "strength" in response to the patterning of signals carried on these connections. Much of his original insight has survived the test of time. Many neural net computing paradigms (see the discussion in Chapter 4) employ Hebbian learning rules and demonstrate their effectiveness as a substrate for learning. The general theory Hebb proposed applies not only to processes thought to play a major role in modifying synaptic strengths during learning, but also to those that determine which synapses are preserved and which are lost as connections are formed during early development. The continuity of mechanisms at these two levels of information generation in the brain lends support for a more global Darwinian theory of the logic of neural information processing, in which early wiring and later learning processes are considered two ends of the same spectrum.

4. The neonatal cortical translant experiments reported by Stanfield and O'Leary (1985) offered a powerful demonstration of the absence of any strong connectional predispositions distinguishing the different areas of the neocortex. Many related studies have followed and are reviewed in O'Leary (1992).

5. This developmental displacement approach to brain evolution is presented in an extensive review of problems and theories of mammalian brain evolution in a paper entitled *Rethinking Mammalian Brain Evolution* (Deacon, 1990b).

6. O'Leary (1992).

7. See the studies by Doron and Wollberg (1994) and Heil, et al. (1991), on the connections between the visual and auditory centers of the brain of the blind mole rat. Though this is one of the most easily recognizable examples of evolutionary displacement of neural circuits, more subtle variants on this process are likely very widespread. Other likely natural examples are offered by Taylor, et al. (1995), on frog species in arboreal, aquatic, and subterranean habitats by; and by Roth, et al. (1993; 1995), on miniaturization effects on the brain structure in amphibians.

8. Among the many reasons for not taking this story too seriously is the fact that it would require a number of only partially successful transplant experiments to explain the numerous intermediate changes that paleontologists observe in hominid brain size evolution.

9. Holloway (1979) showed that human visual areas were disproportionately smaller than predictions extrapolated from the trend observed in other primates.

This was noted previously by Passingham and Ettlinger (1974), though not emphasized as a divergence. The original data were collected by Heinz Stephan and his colleagues some years earlier (1969; see also Stephan, et al., 1981, for an updated data set, used in the analyses in this book) but had not been subjected to statistical analyses that could specifically discern the divergence of human from primate brain structure proportions. Nevertheless, Holloway's insight was largely ignored at the time and was not noticed by me until I again independently derived the same result a few years later.

10. A posterior shift of the lunate sulcus (marking the anterior boundary of primary visual cortex) in australopithecine endocasts was interpreted by Ralph Holloway (see Holloway, 1985, for an extensive review, and Holloway and Shapiro, 1992, for recent evidence) to mean that australopithecines had a proportional reduction of visual areas, similar to that he had noted in modern human brains (see also note above). Thus, he concluded that this species exhibited reorganization toward a "humanlike" pattern, even prior to evolutionary increases in brain size. Another paleoneurologist, Dean Falk (see her review of this same controversy in Falk, 1989), has disputed his identification of this feature and argues that australopithecine brains retain the morphology characteristic of ape brains in this regard.

Though I believe that Holloway's identification of this endocast feature is probably correct, developmental analysis leads me to doubt his interpretation of its cause and significance. Since australopithecine brain/body proportions were far more typical of modern apes (and they had ape-size eyes), neurodevelopmental mechanisms should not produce a reduction in the neural proportions of this area. This does not rule out *morphological* reorganization of the brain, however, which I suspect may have been produced by the shift to upright posture in australopithecines. Vertical orientation of the skull with respect to the spinal cord would have caused the brain stem and cerebellum to become oriented more directly below the cerebrum than in living apes. Since the cerebellum fills a considerable part of the space below and between the cerebral hemispheres posteriorly, a shift of this structure downward and forward in australopithecine brains would vacate some of this interhemispheric space and allow the posterior cortical areas, mostly visual cortex, to become more folded into the posterior midline. The morphological consequence would be a shift of the lunate sulcus to a more posterior position. Additionally relevant in this case is the fact that most of the debate is based on assessment of the natural endocast from the Taung specimen, which was a young child at the time of death. Since the cerebellum is the last major brain structure to complete its growth and grows extensively during postnatal development, at this stage it probably occupied even less interhemispheric space than in an adult australopithecine brain, thus exaggerating the relative posteriorization of the lunate sulcus even further. So, I conclude that both antagonists in this debate are half right. The morphology of the brain does share posterior features that are humanlike rather than apelike, but the actual neural proportions and internal organization were probably more apelike than human (see also Deacon, 1992c). For views of endocast data used to discuss the evolution of language, see also Holloway (1983), Tobias (1981; 1987), and Falk (1983).

11. An interesting complication of the evolution of the human motor system is that the thalamic motor nuclei (the ventro-anterior and ventro-lateral nuclei) are not as reduced (data from Armstrong, 1980b; see also Deacon, 1988b, for a review of other data as well). Although these thalamic nuclei provide the principal projections to the motor and premotor cortical areas, they receive their afferent inputs from a number of structures of varying sizes. One major source of inputs to this nuclear complex is the cerebellum, one of the most enlarged structures of the human brain. Thus, the neurons of motor and premotor cortex are at the connectional juncture between expanded input projections and reduced output targets. Though the expanse of motor cortex appears to be constrained by its targets (depicted in Fig. 7.7), it seems likely that the projection fields of these thalamic nuclei extend beyond their ancestral motor targets, probably into prefrontal and somatic fields on either side. The functional consequences of this are not obvious.

12. Thus, efforts to correlate the size of any given brain structure with some behavior run the risk of the anatomical equivalent of quoting something out of context. The sizes of brain structures are not independently determined, but are competitively parceled by afferent axons with respect to other brain areas during development. Ignoring the systemic interdependence of the sizes of these many brain structures creates the false impression that there has been independent size determination of this one brain region to subserve some highly modular localized function.

13. Not all quantitative analyses of human prefrontal proportions have concluded that they are disproportionately enlarged above typical primate proportions on the basis of MRI reconstructions, and mammal trends. For example, Semendeferi, et al. (1997), have suggested that frontal cortex is of equal proportions in ape and human brains, and Uylings, et al. (1990), suggests that its proportion changes little from rats to humans. However, the identification of a disproportion is entirely dependent on (a) how the analysis controls for species variance (unless the nonhuman species included in the analysis contained considerably more varaince than between the human and the average nonhuman value, including the human value would strongly bias the assessment); (b) the identification of homology of architectonic boundaries across diverse species (it is not clear, for example, that the relevant architectonic distinctions are actually homologous in primates and rodents and the analysis by Semendeferi, et al., lumped motor, premotor, and prefrontal areas because of MRI limitations); and (c) whether the confounding effects of part/whole errors are controlled for (the prefrontal area is roughly 25% of the cortical surface in large apes). When these

sources of analytical artifact are controlled for, these data sets do not produce conflicting results and human prefrontal cortex is found to be divergent from nonhuman trends.

My own assessments (see also Deacon, 1984; 1988) are based on primate data sets collected decades ago by careful architectonic neuroanatomists (see Brodmann, 1912; Blinkov and Glezer, 1968). Data sets were analyzed separately, controlling for part/whole and extrapolation effects (trend lines are produced separately, including and excluding the human data to test for a bias effect). Both analyses converge on a value that is roughly two times greater than predicted by nonhuman primate trends. An assessment of the degree of gyrification in human and other primate brains by Armstrong, Zilles, and colleagues (1989, 1995a, 1995b) also found that the prefrontal region in the human brain is a significant outlier with respect to primate trends, because of its increased folding. This would be expected from the typical allometry of increasing cortical folding with brain size. Finally, these findings are in agreement with many qualitative assessments of neuroanatomists for over a century, and are consistent with the predictions of developmental parcellation processes discussed above. So I believe that a twofold increase above the trend is a reliable estimate. Nevertheless, a more current and complete data set would be valuable for settling lingering questions and also could provide a more detailed assessment of the differences between primate and nonprimate prefrontal areas.

14. The fibers in question are called mossy fibers and they extend from a structure called the dentate gyrus to the part of the hippocampus called the CA fields. Some of these learning trade-offs are described in Schenk, et al. (1995), and in terms of strain differences in learning in Lipp, et al. (1995).

CHAPTER 8

1. For reviews, see Jurgens (1979), Ploog (1981), and see Deacon (1992) for a theoretical discussion of the relationships of midbrain call systems to language. Jurgens and Pratt (1979) provided some of the most detailed evidence for central gray vocalization control.

2. For example, Newman and MacLean (1982) showed that lesions of the tegmental regions adjacent to the central gray area could modify the form of certain calls.

3. Balaban, et al. (1988), used transplants of midbrain from very young quail embryos to chick embryos whose midbrain regions had been previously removed (also discussed in Chapter 6). The quail-chick chimeras produced a number of typically quail fixed-action patterns after hatching, including quail vocalizations. Not only does this demonstrate a localized representation of these motor programs, but also that the remainder of the brain, even from another species, can interact with it as a sort of module—a feature that is unlikely to be true of brain regions supporting language (see discussion in Chapter 10).

4. Studies of circuits controlling tongue and facial musculature are from Sokoloff and Deacon (1989; 1990; 1992). Related information from a broad comparative study of corticospinal connections is provided by Nudo, et al. (1995).

5. For detailed examples, see Jane Goodall's book *The Chimpanzees of Gombe: Patterns of Behavior* (1986).

6. Jurgens and Pratt (1979).

7. Jurgens, et al. (1982).

8. Much of this work on the psychology and physiology of laughter comes from the research of Robert Provine, University of Maryland. See his 1996 review of this work.

CHAPTER 9

1. Goldman-Rakic (1987).

2. Barbas and Mesulam (1981).

3. Barbas and Mesulam (1981); Pandya and Barnes (1987); Deacon (1992).

4. For in-depth reviews of prefrontal deficits and theories about prefrontal function, see Fuster (1980); Perecman (1987); Stuss and Benson (1986).

5. Jacobsen (1936).

6. This is very similar to the hidden object problems demonstrated in young children by Jean Piaget (1952).

7. Passingham (1985).

8. Reviews in Luria (1980); Kolb and Whishaw (1990); Stuss and Benson (1986); and Perecman (1987). Particularly relevant studies are reported by Grossman (1980; 1982), who demonstrates human Broca's area defects in hierarchical and reversal-learning problems, and by Petrides (1982; 1985; 1986), who reports localized primate prefrontal lesion defects on a number of conditional learning paradigms that involve dependent combinatorial analysis.

9. See Guilford (1967).

10. An explanation of some of these techniques and more detailed examples are provided later in this chapter and in Posner and Raichle's book *The Illuminated Brain* (1994).

11. Ojemann (1979).

12. The case for a neuroanatomical dichotomy was most clearly articulated by Myers (1976), and a more detailed description of how these systems are anatomically linked in language function is provided in Deacon (1992d).

13. Hydrocephaly can result from an occlusion of the canal linking the forebrain ventricles with their outlet to the brain stem and spinal cord. This blocks cerebrospinal fluid drainage, which can cause the ventricles to expand and compress the temporal and parietal cortex (hydrocephaly). The condition has been associated with a number of cases of hyperlexia with relatively poor comprehension.

14. The LIM1 kinase gene deletion in Williams syndrome is reported in Frangiskakis, et al. (1996).

15. There are many excellent books discussing autism, including autobiographical texts by Temple Grandin (1986, 1995). See also Frith (1989, 1991) for detailed accounts, and Trefert (1989) for an excellent nontechnical description.

16. Prefrontal abnormalities associated with schizophrenia, depression, and obsessive compulsive disorders are reviewed in Kolb and Whishaw (1990); Posner and Raichle (1994); Stuss and Bensen (1986). For an insightful account of frontal lobe disturbances of emotion, emotional assessment, and social cognition, see Antonio Damasio's book *Descartes' Error* (1994).

17. Cerebellar contributions to nonmotor cognitive processes have been demonstrated by PET studies of word retrieval (Petersen, et al., 1988) and in response to cerebellar damage (Leiner, et al., 1989 and 1993; Barinaga, 1996; also see discussion in the next chapter).

CHAPTER 10

1. The classic studies for which these aphasic syndromes are named are Paul Broca's *Sur la faculté du langage articule* (1865) and Carl Wernicke's *Der aphasische Symptomencomplex.* (1874). An excellent historical review of aphasia research is provided in Lecours, et al. (1983).

2. For the original discussion of Broca's aphasia agrammatism, see Goodglass (1968; 1972; 1973), and for alternative explanations see Kean (1977); for current *in vivo* imaging of Broca's aphasic brain impairment, see Tramo, et al. (1988).

3. Regional cerebral blood flow studies are from Larsen, et al. (1978); Lassen, et al. (1979; 1980); and Roland (1985). PET studies use a radioactively labeled glucose analogue, 2-deoxy-glucose, which is taken up by cells but does not get readily broken down in the normal metabolic machinery of the cell. As a result, it accumulates in metabolically active tissues. Though all tissues take it up, more highly active tissues take up more of it over time (usually activity is assessed over many minutes), and this allows researchers to measure relative metabolism reflected in differences in radioactive output (original studies reported by Posner, et al., 1988, and Petersen, et al., 1988; reviewed with additional examples and explanations in Posner and Raichle, 1994).

4. Wilder Penfield first demonstrated that stimulating the supplementary motor area could interrupt speech. He provides a summary of his research in his book, with L. Roberts, *Speech and Brain Mechanisms* (1959).

5. Broca's original demarcation of the frontal language region included the entire inferior third of the frontal and prefrontal cortex on the left side; but over the years researchers have progressively restricted the designation of the effective site to include only the premotor region, just in front of the motor mouth region.

6. As does functional magnetic resonance imaging (MRI), which is also based on blood flow changes but allows data to be collected during much shorter intervals, thus making it possible to avoid repetitive tasks necessary for cumulative effects in other techniques. The general agreement with rCBF and PET data

is quite good and so these will not be described separately. See the brief comparative description of methodology in Posner and Raichle (1994).

7. These are summary schematizations of data that are a bit more complex.

8. This brief account simplifies and summarizes neuroanatomical studies by the author (Deacon, 1984; 1992).

9. The syntagmatic/paradigmatic distinction was introduced to linguistics by the theories of Roman Jakobson (1956), who also predicted that frontal and prefrontal brain damage should disturb the former and posterior brain damage should disturb the latter. Numerous brain damage studies have shown that at least in general terms, this dichotomy is accurate.

10. Recent reviews of research on lateralization of language can be found in Corballis (1992), MacNeilage (1995), and see Kolb and Whishaw (1990) for a review of evidence of the role of cerebral asymmetries. Gainotti (1972) and Heilman, et al. (1983), provide reviews of asymmetric affective functions. See also Kinsbourne (1978) for an evolutionary perspective on the origins of the crossed pattern of connections.

11. For an interesting example of right hemisphere language comprehension, see Gardner, et al. (1983); also see Larsen, et al. (1978; 1980), for evidence of right hemispheric functions contributing to speech.

12. See Fabbro (1992) for a review of this bilingual brain research.

Part Three

CHAPTER 11

1. Baldwin (1895; 1896a–d; 1902). See also the excellent historical account of Baldwin's work in Robert Richards's (1987) book on late nineteenth-century theories of the evolution of mind.

2. Durham (1994).

3. Many are also discussed by Durham (1994).

4. Pinker (1994). See also his short (1990) summary.

5. These adaptations for speech are discussed in more detail below.

6. Casts of the inner surface of the skull show it to be subtly shaped by the pressures of the developing brain. Bumps and depressions on the bone record the slight differential in pressure of underlying gyri and sulci (bumps and fissures, respectively). Fossil skulls retain this impression, though their interpretation is often not without ambiguity.

7. See Philip Tobias's (1981) and Dean Falk's (1983) assessment of the gyral and sulcal landmarks on endocasts in the position of the modern Broca's area.

8. The most suggestive evidence was provided by a fossil designated *KMNER* 1470, one of the exemplar specimens of *Homo habilis* (though recent proposals suggest placing this individual in a separate species designated *Homo rudolfensis*

because of its many other australopithecinelike features), who had a brain size of approximately 750 cc.

9. Though many have tried to draw the debate into posing these as opposed alternatives, I leave it to paleontologists to quibble over the number and size of the "increments."

10. Baldwinian evolutionary processes are also involved in the process of language evolution discussed in Chapter 4 (though I focused on change without a Baldwinian effect) and so have also played a role in the evolution of children's learning biases as well.

11. For example, the archeologist Nick Toth taught Kanzi (the bonobo chimpanzee who spontaneously learned lexigram communication) to make and use stone flake tools in order to cut a rope and open a box hiding food.

12. A wonderful compendium of such examples is provided by the historian James Burke in his television series and book, *Connections* (1978).

13. Gould and Vrba (1982).

14. Philip Lieberman (1984; 1991).

15. See the excellent historical review by Gordon Hewes in Rumbaugh (1977). A recent version of the hypothesis that gestural language preceded vocal language is outlined by Corbalis (1992).

16. An excellent survey of the evolution of the larynx is provided by Wind (1970).

17. Initial strong claims about Neanderthal's inability to speak were presented by Lieberman and Crelin (1971). Later, more cautious claims about the correlation with speech evolution can be found in Lieberman (1984; 1991), and evidence for incremental changes in this morphology throughout the last 2 million years of hominine evolution is presented by Laitman, et al. (1992).

18. See Liberman, et al. (1967), Liberman and Mattingly (1985; 1989), and Liberman (1984) for extensive examples. (Note: Liberman and Lieberman are not the same.)

19. See Lisker and Abramson (1971).

20. Chinchillas were the first animals to show a voice-onset time categorical perception like that shown for humans (Kuhl and Miller, 1975).

21. This follows for the same reason that a number encoded in binary notation (i.e., using only the numerals 0 and 1), used for computer memory addresses, is on average composed of many more digits than its representation as a decimal (i.e., base 10) or hexadecimal (i.e., base 16 code, also used for computing and made up of the numerals, 0, 1, 2, 3, 4, 5, 6, 7, 8, 9, A, B, C, D, E, F) number. For example, 11,111,110 in binary = 254 in decimal = FE in hexadecimal.

22. See the insightful and critical reflection on assuptions influencing paleontological theories of human origins by Misia Landau (1991).

23. Hublin, et al. (1996), recently verified that these complex cultural artifacts from the late middle Paleolithic were associated with Neanderthal skeletal materials. This is one more piece of evidence which undermines the tradi-

tional view that Neanderthal culture was lacking in tool variety and artistic expression. It does not support the view that Neanderthal cultural communication was distinctly inferior to that of contemporary anatomically modern populations.

24. See also Holloway (1981; 1983).

25. Arensberg and Tillier (1991) review the evolutionary significance of the discovery of a Neanderthal hyoid bone that demonstrates modern anatomical structure (first described in Arensberg, et al., 1989), and specifically conclude that Neanderthals would not have exhibited any significant difficulties producing speech.

CHAPTER 12

1. Jones (1993) and Caird (1994).

2. Statistically speaking, most human marriage relationships are monogamous, even though most societies allow or encourage polygynous marriages. Even within a highly polygynous society, most relationships are monogamous for the simple reason that polygyny requires significant accumulation of resources and power by particular males at the expense of many others, leaving the rest with insufficient resources to become polygynous. Introductory summaries of these patterns can be found in Daly and Wilson (1978) and Durham (1994).

3. The special status of prostitution and the rarity of polyandry in human societies are two exceptions that prove the rule. In the few societies where polyandry is practiced, husbands of the same wife are often close relatives, such as brothers (for details, see Durham, 1994).

4. This is evident in the special case of large rookeries, where hundreds of pair-bonded bird "couples" are forced to roost together because of limited nest sites. Despite the absolutely critical need for biparental care, the degree of cuckoldry increases as group size increases; in addition, infanticide of neighbors' chicks, nest destruction, and nest site stealing are common noncooperative behaviors in these conditions.

5. Some social carnivore species with large litters can produce littermates from multiple fathers. In these cases, the pattern is more complicated. So long as the dominant male is able to sire more of the offspring than his genetic competitors, the cooperative social arrangement is evolutionarily stable.

6. High degrees of sexual dimorphism (difference in male versus female body size) are highly correlated with degree of polygyny in a species. This is favored by sexual selection mediated by threat and fighting behaviors, where being larger confers an advantage for gaining access to mates.

7. Unfortunately, such negative consequences are seldom applied equitably to men and women or to rich and poor.

8. A Yanomamo Feast and the warfare context in which this method of peacemaking occurs is decribed in Chagnon (1983).

1. This experiment is reported in Boysen, et al. (1995 and 1996), and was part of a study investigating numerical abilities in chimps. See also Boysen, et al. (1993), for the first report of ordinality and transitivity learning by chimps.

2. A parallel study by Hauser, et al., 1995, showed ordinality learning ability in wild rhesus monkeys. The ability to learn transitive associations has also been demonstrated in rats (Eichenbaum, et al., 1996) in an ABC paradigm where choosing A over B and B over C can yield knowledge that A over C is also correct. This can be expanded to a series of at least 5 ordered stimuli (ABCDE), and rats can still generalize to the novel choice of BC on probe trials (Eichenbaum, personal communication). Though ordinality and transitivity can each be acquired as indexical associations, they are (respectively) the token-object and token-token associative components of numerical symbols, and so can under the right conditions serve as the basis for generating the concept of number. Whether any of these or the chimp experiments accomplish this step is uncertain, but it could be demonstrated if the animals could generalize from known ordinal associations to novel transitive relationships. However, not only has this not been shown, as far as I know, but all experimental results to date seem to approach a ceiling in abilities when quantity exceeds about 5. This suggests to me that there is no symbolic generalization in these cases and we should not too quickly jump to the conclusion that ordinal discrimination of quantity necessary implies symbol ration.

CHAPTER 14

1. Searle (1980) notes that computational interpretations could be mapped onto manipulations of many objects, for example, manipulations of collections of bottles and cans, and yet such a mapping would not demonstrate this to be a cognitive process. A mapping of symbolic representations onto semiconductor currents and potentials in a computer is no different from this and therefore is not intrinsically more mindlike. I believe the logic of my example is essentially parallel. In fact, many chemical reactions can occur at thresholds of relative concentration, heat, or time, and so can produce essentially digital responses to initial conditions, analogous to those in a computer. Indeed, a number of molecular biology labs are currently experimenting with the use of biological macromolecular systems as simple computing elements for exactly these reasons.

2. The Chinese Room analogy proposed by John Searle in a 1980 paper has been argued and reargued many times by philosophers and cognitive scientists. Searle provides a retrospective interpretation in his more recent (1992) book.

3. Dennett (1991).

4. Dennett (1991).

5. Vygotsky (1978).

Additional Readings

CHAPTER 1. For some recent alternative scenarios for language origins, see: *Language and Species* (1990) and *Language and Human Behavior* (1995) by Derek Bickerton; *The Lopsided Ape* (1991) by Michael Corbalis; *Origins of the Modern Mind* (1991) by Merlin Donald; *Gossip, and the Evolution of Language* (1997) by Robin Dunbar; *The Biology and Evolution of Language* (1984) and *Uniquely Human* (1991) by Philip Lieberman; *Grooming, The Ape That Spoke* (1991) by John McCrone; and *The Language Instinct* (1994) by Steven Pinker. Also, see a marvelous historical survey of language origins theories by Gordon Hewes in *Language Learning by a Chimpanzee: The Lana Project* (1974), edited by Duane Rumbaugh.

CHAPTER 2. For an extensive and up-to-date survey of animal communication research that takes a quite different approach to human language evolution, see *The*

Evolution of Communication (1996) by Marc Hauser; and for an introduction to topics in animal cognition, see *The Animal Mind* (1994) by James Gould and Carol Grant Gould. For background and historical discussions about the nature of word reference and language function in general, I recommend *The Cambridge Encyclopedia of Language* (1992) edited by David Crystal and *The Science of Words* (1991) by George Miller.

CHAPTER 3. Besides a major influence from C. S. Peirce, I owe some of my thoughts about the hierarchy of communicative processes to some now classic ideas presented in *Steps to an Ecology of Mind* (1972) and *Mind and Nature* (1979) by Gregory Bateson. Recalling his analogy of play communication in animals to a *reductio ad absurdum* proof helped me rethink the chimpanzee symbol-learning problem. I have also been influenced by a classic text on *Symbol Formation* by Heinz Werner and Bernard Kaplan (1963). Finally, I recommend reading Helen Keller's incredible personal recollection of first discovering the symbolic nature of finger spelling in *The Story of My Life* (1903). For books by some of the researchers involved in ape language-training experiments, see: *Koko's Story* (1987) by Francine Patterson; *The Mind of an Ape* (1983) by David Premack and A. Premack; *Gavagai! Or the Future History of the Animal Language Controversy* (1986) by David Premack; *Language Learning by a Chimpanzee: The Lana Project* (1974), edited by Duane Rumbaugh; *Ape Language: From Conditioned Response to Symbol* (1986) by Sue Savage-Rumbaugh; *Kanzi: The Ape at the Brink of the Human Mind* (1994) by Sue Savage-Rumbaugh and Roger Lewin; and *Nim: A Chimpanzee Who Learned Sign Language* (1979) by Herbert Terrace.

CHAPTER 4. For discussions of modular theories of mind, see: *The Adapted Mind* (1992), edited by J. H. Barkow, L. Cosmides, and J. Tooby; *Modular Approaches to the Study of the Mind* (1984) by Noam Chomsky; and *The Modularity of Mind* (1983) by Jerry Fodor. For critiques of modularity theories, see: *Rethinking Innateness* (1996) by Jeff Elman, Elizabeth Bates, and colleagues, and *Beyond Modularity* (1992) by Annette Karmiloff-Smith. Besides Richard Dawkins's introduction of the concept of meme in *The Selfish Gene* (1980), there are some interesting recent treatments presented in *Virus of the Mind* (1996) by Richard Brodie and *Darwin's Dangerous Idea* (1995) by Daniel Dennett.

CHAPTER 5. For the view that the evolution of relative brain size produced increased intelligence, see *The Evolution of Intelligence* (1973) by Harry Jerison and *The Runaway Brain* (1993) by Christopher Wills. And for some cautionary views, see *The Mismeasure of Man* (1996) by Stephen J. Gould; *Frames of Mind* (1983) by Howard Gardner; and *Brain and Intelligence in Vertebrates* (1982) by Euan Macphail.

CHAPTER 6. On growth and allometry, see: *Ontogeny and Phylogeny* (1977) by Stephen J. Gould; *Heterochrony: The Evolution of Ontogeny* (1991) by Michael McKinney and K. J. MacNamara; *The Human Primate* (1982) by Richard Passingham; *Scaling: Why Is Animal Size So Important?* (1984) by Knut Schmidt-Nielsen; and *On Growth and Form* (1917) by D'Arcy Thompson. On recent developmental genetics and evolution, see *The Shape of Life: Genes, Development, and the Evolution of Animal Form* (1996) by Rudolph Raff.

CHAPTER 7. On the Darwinian nature of brain development, see: *Neural Darwinism* (1987) by Gerald Edelman; *Body and Brain* (1988) by Dale Purves; and *Neuronal Man* (1985) by J. Changeaux.

CHAPTER 8. For a review of research on the control of bird and mammal vocalizations, see *The Evolution of Communication* (1996) by Marc Hauser; for a more detailed treatment of comparative neuroanatomy, including vocal systems, see *Comparative Vertebrate Neuroanatomy* (1996) by A. B. Butler and W. Hodos.

CHAPTER 9. On prefrontal cortical functions, see: *Descartes' Error* (1996) by Antonio Damasio; *The Prefrontal Cortex* (1980) by Joachin Fuster; *The Frontal Lobes Revisited* (1987), edited by Ellen Perecman; and *The Frontal Lobes* (1986) by D. Stuss and D. Benson.

CHAPTER 10. For general background, see *Fundamentals of Human Neuropsychology* (1996) by Brian Kolb and Ian Whishaw; *Higher Cortical Functions in Man* (1980) by A. R. Luria; *Images of Mind* (1994) by M. I. Posner and M. E. Raichle. One of the most complete historical reviews of aphasia literature up to the last decade can be found in *Aphasiology* (1983) by André-Roch Lecours, F. Lhermitte, and B. Bryans.

CHAPTER 11. For summaries of topics in hominid evolution, see *The Cambridge Encyclopedia of Human Evolution* (1992) edited by Stephen Jones, D. Pilbeam, and R. Martin. For summaries of dual inheritance theories of social and genetic evolution, see *Darwin, Sex, and Status* (1989) by Jerome Barkow, and *Coevolution* (1991) by William Durham. For historical review of co-evolutionary theories (especially for James Mark Baldwin), see *Darwin and the Emergence of Evolutionary Theories of Mind* (1987) by Robert Richards.

CHAPTER 12. For scenarios of human socioecological evolution, see: *The Biology of Moral Systems* (1987) by Richard Alexander; *Darwin, Sex, and Status* (1989) by Jerome Barkow; *Primate Social Systems* (1988) by Robin Dunbar; *The Sex Contract* (1982) by Helen Fisher; and *Demonic Males* (1996) by Richard Wrangham.

CHAPTER 13. On autism and savant syndromes, see: *An Anthropologist on Mars* (1995) by Oliver Sachs, and *Extraordinary People: Understanding Savant Syndrome* (1989) by Darold Treffert.

CHAPTER 14. To sample a few recent contributions to the consciousness debate, try: *The Conscious Mind* (1996) by David Chalmers; *Neurophilosophy: Toward a Unified Science of the Mind-Brain* (1986) by Patricia Churchland; *The Astonishing Hypothesis* (1994) by Francis Crick; *Consciousness Explained* (1991) by Daniel Dennett; *The Emperor's New Mind* (1989) by Roger Penrose; and *The Rediscovery of the Mind* (1992) by John Searle. For recent discussion of Darwinian models of consciousness, see: *The Cerebral Code* (1996) by William Calvin; *Darwin's Dangerous Idea* (1991) by Daniel Dennett; *Bright Air, Brilliant Fire* (1994) by Gerald Edelman; *Darwin Machines and the Nature of Knowledge* (1994) by Henry Plotkin; and from a somewhat different approach, *Mind, Brain and Consciousness* (1977) by Jason Brown.

Bibliography

Aiello, Leslie, and P. Wheeler (1995) The expensive tissue hypothesis. *Current Anthropology 36*, 199–221.

Alexander, Richard (1987) *The Biology of Moral Systems*. Hawthorne, NY: Aldine de Gruyter.

Alleva, Enrico, Aldo Fasolo, Hans-Peter Lipp, Lynn Nadel, and Laura Ricceri, eds. (1995) *Behavioral Brain Research in Naturalistic and Semi-Naturalistic Settings*. Dordrecht: Kluwer Academic Publishers.

Arensberg, Baruch, and A. M. Tillier (1991) Speech and the Neanderthals. *Endeavour 15*, 26–28.

Arensberg, Baruch, et al. (1989) A middle Paleolithic hyoid bone. *Nature 338*, 758–760.

Armstrong, Este (1980a) A quantitative comparison of the hominoid thalamus: II. Limbic nuclei anterior principalis and lateralis dorsalis. *American Journal of Physical Anthropology 52*, 43–54.

—— (1980b) A quantitative comparison of the hominoid thalamus: III. A motor substrate—The ventrolateral complex. *American Journal of Physical Anthropology 52*, 405–419.

—— (1990) Brains, bodies and metabolism. *Brain, Behavior and Evolution 36*, 166–176.

—— (1995) *Expansion and Stasis in Human Brain Evolution: Analyses of the Limbic System, Cortex and Brain Shape.* 65th James Arthur Lecture on the Evolution of the Human Brain, New York: American Museum of Natural History.

——, M. R. Clarke, and E. M. Hill (1987) Relative size of the anterior thalamic nuclei differentiates anthropoids by social system. *Brain and Behavioral Science 30*, 263–271.

Armstrong, Este, and Dean Falk, eds. (1982) *Primate Brain Evolution: Methods and Concepts.* New York: Plenum Press.

Armstrong, Este, A. Schleicher, and K. Zilles (1995) Cortical folding and the evolution of the human brain. *Journal of Human Evolution 25*, 387–392.

Armstrong, Este, A. Schleicher, H. Omran, M. Curtis, and K. Zilles (1995) The ontogeny of human gyrification. *Cerebral Cortex 5*, 56–63.

Balaban, Evan, M. A. Teillet, and N. Le Douarin (1988) Application of the quail-chick chimera system to the study of brain development and behavior. *Science 241*, 1339–1342.

Baldwin, James Mark (1895) Consciousness and evolution. *Science 2*, 219–223.

—— (1896a) Consciousness and evolution. *Psychological Review 3*, 300–308.

—— (1896b) Heredity and instinct (I). *Science 3*, 438–441.

—— (1896c) Heredity and instinct (II). *Science 3*, 558–561.

—— (1896d) On criticisms of organic selection. *Science 4*, 727.

—— (1902) *Development and Evolution.* New York: Macmillan.

Barbas, Helen, and M.-M. Mesulam (1981) Organization of afferent input to subdivisions of area 8 of the rhesus monkey. *Journal of Comparative Neurology 200*, 407–431.

Barinaga, Marcia (1996) The cerebellum: Movement coordinator or much more? *Science 272*, 482–483.

Barkow, Jerome (1989) *Darwin, Sex, and Status.* Toronto: University of Toronto Press.

Barkow, J. H., L. Cosmides, and J. Tooby, eds. (1992) *The Adapted Mind: Evolutionary Psychology and the Generation of Culture.* New York: Oxford University Press.

Bates, Elizabeth (1992) Language development. *Current Opinion in Neurobiology 2*, 180–185.

Bates, Elizabeth, and B. Wulfeck (1989) Comparative aphasiology: A cross-linguistic approach to language breakdown. *Aphasiology 3*, 111–142.

Bates, Elizabeth, and B. MacWhinney (1991b) Crosslinguistic research in aphasia: An overview. *Brain and Language (special issue on crosslinguistic aphasia) 41*, 123–148.

Bates, Elizabeth, D. Thal, and V. Marchman (1991c) Symbols and syntax: A Darwinian approach to language development. In N. Krasnegor, D. Rumbaugh, et al., eds., *Biological and Behavioral Determinants of Language Development*. Hillsdale, NJ: Lawrence Erlbaum.

Bates, Elizabeth, D. Thal, D. Aram, J. Eisele, R. Nass, and D. Trauner (1994) From first words to grammar in children with focal brain injury. In D. Thal and J. Reilly, eds., *Special Issues on Origins of Communication Disorders, Developmental Neuropsychology* (in press). [data summarized in Elman, Bates, et al., 1996, pp. 306–307].

Bateson, Gregory (1972) *Steps to an Ecology of Mind.* New York: Ballantine Books.
—— (1979) *Mind and Nature.* New York: E. P. Dutton.

Bechara, A., A. R. Damasio, H. Damasio, and S. Anderson (1994) Insensitivity to future consequences following damage to human prefrontal cortex. *Cognition 50*, 7–12.

Bellugi, Ursula, and Edward S. Klima (1982) From gesture to sign: Deixis in a visual gestural language. In R. J. Jarvella and W. Klein, eds., *Speech, Place and Action: Studies of Language in Context.* New York: John Wiley, 297–313.

Bellugi, Ursula, A. Bihrle, H. Neville, T. L. Jernigan, and S. Doherty (1991) Language, cognition and brain organization in a neurodevelopmental disorder. In M. Gunnar and C. Nelson, eds., *Developmental Behavioral Neuroscience.* Hillsdale, NJ: Lawrence Erlbaum, 201–232.

Bellugi, Ursula, P. P. Wang, and T. L. Jernigan (1994) Williams syndrome: An unusual neuropsychological profile. In S. Broman and J. Grafman, eds., *Atypical Cognitive Deficits in Developmental Disorders: Implications for Brain Function.* Hillsdale, NJ: Lawrence Erlbaum, 23–56.

Bellugi, Ursula, A. Bihrle, T. L. Jernigan, D. Trauner, and S. Doherty (1991) Neuropsychological, neurological, and neuroanatomical profile of Williams Syndrome. *American Journal of Medical Genetics Supplement 6*, 115–125.

Berlin, Brent, and Paul Kay (1969) *Basic Color Terms: Their Universality and Evolution.* Berkeley: University of California Press.

Bickerton, Derek (1981) *The Roots of Language.* Ann Arbor, MI: Karoma.
—— (1983) Pidgin and creole languages, *Scientific American 249*, 116–122.
—— (1984) The language bioprogram hypothesis. *Behavioral and Brain Sciences 7*, 173–221.
—— (1990) *Language and Species.* Chicago: University of Chicago Press.
—— (1995) *Language and Human Behavior.* Seattle: University of Washington Press.

Bitterman, M. E. (1975) The comparative analysis of learning. *Science 188*, 699–709.

—— (1988) Vertebrate-invertebrate comparisons. In H. Jerison and I. Jerison, ed., *Intelligence and Evolutionary Biology*. Berlin: Springer-Verlag, 251–276.

Blinkov, S., and I. Glezer (1968) *The Human Brain in Figures and Tables*. New York: Plenum Press.

Bogen, J., and G. M. Bogen (1976). Wernicke's region: Where is it? *Annals of the New York Academy of Science 280*, 834–843.

Boncinelli, E., M. Gulisano, and V. Broccoli (1993a) Emx and Otx homeobox genes in the developing mouse brain. *Journal of Neurobiology 24*, 1356–1366.

Boncinelli, E., M. Gulisano, and M. Pannese (1993b) Conserved homeobox genes in the developing brain. *Comptes Rendus de l'Académie des Sciences—Série III, Sciences de la Vie. 316*, 972–984.

Boncinelli, E., and A. Mallamaci (1995) Homeobox genes in vertebrate gastrulation. *Current Opinion in Genetics and Development 5*, 619–627.

Bonin, Gerhard von (1937) Brain-weight and body-weight in mammals. *Journal of General Psychology 16*, 379–389.

Boysen, Sally, G. Bernston, T. Shreyer, and K. Quigley (1993) Processing of ordinality and transitivity by chimpanzees (Pan troglodytes). *Journal of Comparative Psychology 107*, 208–215.

Boysen, Sally, and G. Bernston (1995) Responses to quantity: perceptual versus cognitive mechanisms in chimpanzees (Pan troglodytes). *Journal of Experimental Psychology and Animal Behavior Processes 21*, 82–86.

Boysen, Sally, G. Bernston, M. Hannan, and J. Cacioppo (1996) Quantity-based inference and symbolic representation in chimpanzees (Pan troglodytes). *Journal of Experimental Psychology and Animal Behavior Processes 22*, 76–86.

Brandt, A. (1867) Sur le rapport du poids du cerveau à celui du corps chez différents animaux. *Bull. Soc. impè r. Naturalistes, Moscou 40*, 525–543.

Broca, Paul (1865) Sur la faculté du langage articule. *Bulletin de la Société d'Anthropologie, Paris 6*, 337–393.

Brodie, Richard (1996) *Virus of the Mind*. Seattle. Integral Press.

Brodmann, K. (1912) Neue Ergebnisse uber die Vergleichende histologische Localisation der Grosshirnrinde mit besonderer Berucksichtigung des Stirnhirns. *Anatomischer Anzeiger, Suppl. 41*, 157–216.

Brown, Jason (1977) *Mind, Brain and Consciousness*. New York: Academic Press.

Buchler, J., ed. (1955) *The Philosophical Writings of Peirce*. New York: Dover Books.

Burke, James (1978) *Connections*. Boston: Little, Brown.

Burling, R. (1986) The selective advantage of complex language. *Ethology and Sociobiology 7*, 1–16.

Butler, A. B., and W. Hodos (1996) *Comparative Vertebrate Neuroanatomy: Evolution and Adaptation*. New York: Wiley-Liss.

Caird, Rod (1994) *Ape Man: The Story of Human Evolution*. Edited by Robert Foley. New York: Simon & Schuster.

Calvin, William (1996) *The Cerebral Code*. Cambridge, MA: MIT Press.

Chagnon, Napolean (1983) *Yanomamö: The Fierce People*. New York: Holt, Rhinehart & Winston.

Chalmers, David (1996) *The Conscious Mind: In Search of a Fundamental Theory*. Oxford: Oxford University Press.

Changeaux, J. (1985) *Neuronal Man*. New York: Pantheon Books.

Cheney, Dorothy, and Robert Seyfarth (1990) *How Monkeys See the World*. Chicago: University of Chicago Press.

—— (1992) Meaning, reference, and intentionality in the natural vocalizations of monkeys. In T. Nishida, W. C., McGrew, P. Marler, M. Pickford, and F. de Waal, eds., *Topics in Primatology. Vol. 1. Human Origins*. Tokyo: Tokyo University Press.

Chomsky, Noam (1972) *Language and Mind*. New York: Harcourt Brace Jovanovich.

—— (1975) *Reflections on Language*. New York: Pantheon.

—— (1980) *Rules and Representations*. New York: Columbia University Press.

—— (1988) *Language and Problems of Knowledge: The Managua Lectures*. Cambridge, MA: MIT Press.

—— (1984) *Modular Approaches to the Study of the Mind*. San Diego: San Diego State University Press.

Chomsky, Noam, and G. Miller (1963) Introduction to the formal analysis of natural language. In *Handbook of Mathematical Psychology, Vol. 2*, edited by R. D. Luce, R. Bush, and E. Galanter. New York: John Wiley.

Christiansen, Morton (in press). Language as an organism—Implications for the evolution and acquisition of language. *Cognition*.

Churchland, Patricia (1986) *Neurophilosophy: Toward a Unified Science of the Mind-Brain*. Cambridge, MA: MIT Press.

Clark, A., and C. Thornton (in press) Trading spaces: Computation, representation and the limits of uninformed learning. *Brain and Behavioral Sciences*.

Corballis, Michael C. (1992) On the evolution of language and generativity. *Cognition 44*, 197–126.

—— (1991) *The Lopsided Ape*. New York: Oxford University Press.

Coren, Stanley (1995) *The Intelligence of Dogs*. New York: Bantam Books.

Count, Earl W. (1947) Brain and body weight in man: Their antecedents in growth and evolution. *Annals of the New York Academy of Sciences 46*, 993–1122.

Cowan, W. M., J. W. Fawcett, D. D. M. O'Leary, and B. B. Stanfield (1984) Regressive events in neurogenesis. *Science 255*, 1258–1265.

Crick, Francis (1994) *The Astonishing Hypothesis: The Scientific Search for the Soul*. New York: Charles Scribner's Sons.

Crick, Francis, and C. Koch (1990) Towards a neurobiological theory of consciousness. *Seminars in the Neurosciences 2*, 263–275.

Crystal, David, ed., (1989) *Cambridge Encyclopedia of Language.* Cambridge, UK: Cambridge University Press, 334–339.

Daly, Martin, and Margo Wilson (1978) *Sex, Evolution and Behavior.* North Scituate, MA: Duxbury.

Damasio, Antonio R. (1989) The brain binds entities and events by multiregional activation from convergence zones. *Neural Computation 1*, 123–132.

—— (1989) Time-locked multiregional retroactivation: A systems level proposal for the neural substrates of recall and recognition. *Cognition 33*, 25–62.

—— (1990) Category related recognition defects as a clue to the neural substrates of language. *Trends in Neuroscience 13*, 95–98.

—— (1994) Cortical systems for retrieval of concrete knowledge: The convergence zone framework. In C. Koch, ed., *Large-Scale Neuronal Theories of the Brain.* Cambridge, MA: MIT Press.

—— (1994) *Descartes' Error.* New York: Grosset/Putnam.

Damasio, H., T. J. Grabowski, D. Tranel, R. D. Hichwa, and A. R. Damasio (1996) A neural basis for lexical retrieval. *Nature 380*, 499–505.

Darwin, Charles (1871) *The Descent of Man, and Selection in Relation to Sex.* 2d ed., revised and augmented. New York: Appleton.

Deacon, Terrence W. (1984) *Connections of the Inferior Periarcuate Area in the Brain of* Macaca Fascicularis. *An Experimental and Comparative Investigation of Language Circuitry and Its Evolution.* Unpublished Ph.D. thesis, Harvard University.

—— (1988) Human brain evolution: I. Evolution of human language circuits. In H. Jerison and I. Jerison, eds., *Intelligence and Evolutionary Biology.* New York: Springer-Verlag.

—— (1988) Human brain evolution: II. Embryology and brain allometry. In H. Jerison, and I. Jerison eds., *Intelligence and Evolutionary Biology.* Berlin: Springer-Verlag, 383–415.

—— (1990a) Fallacies of progression in theories of brain size evolution. *International Journal of Primatology.*

—— (1990b) Rethinking mammalian brain evolution. *American Zoologist 30*, 629–705.

—— (1992a) Brain-language co-evolution. In J. Hawkins and M. Gel-Man, eds., *The Evolution of Human Languages.* Redwood City, CA: Addison-Wesley, 49–83.

—— (1992b) Cortical connections of the inferior arcuate sulcus cortex in the macaque brain. *Brain Research 573*, 8–26.

—— (1992c) Impressions of ancestral brains. In S. Jones, R. Martin, and D. Pilbeam, eds., *Cambridge Encyclopedia of Human Evolution.* Cambridge, UK: Cambridge University Press, 117–118.

—— (1992d) The neural circuitry underlying primate calls and human language.

In J. Wind, B. Chiarelli, B. Bichakjian, and A Nocentini, eds., *Language Origin: A Multidisciplinary Approach.* Proceedings of NATO Advanced Institute, Cortona, Italy, 1988; Amsterdam: Kluwer, 1992, 121–162.

Deacon, Terrence, P. Pakzaban, L. Burns, J. Dinsmore, and O. Isacson (1994) Cytoarchitectonic development, axon-glia relationships and long distance axon growth of porcine striatal xenografts in rats. *Experimental Neurology 130,* 151–167.

De Valois, R., and K. De Valois (1975) Neural coding of color. In E. Charterette and M. Friedman, eds., *Handbook of Perception,* Vol. V: *Seeing.* New York: Academic Press.

Dehaene, S. (1992) Varieties of numerical abilities. *Cognition 44,* 1–42.

Demb, J., J. Desmond, A. Wagner, C. Vaidya, G. Glover and J. Gabrieli (1995) Semantic encoding and retrieval in the left inferior prefrontal cortex: A functional MRI study of task difficulty and process specificity. *Journal of Neuroscience 15(9),* 5870–5878.

Dennett, Daniel (1991) *Consciousness Explained.* Boston: Little, Brown.

—— (1995) *Darwin's Dangerous Idea.* New York: Simon & Schuster.

Descartes, René (1637) *The Philosophical Works of Descartes,* rendered into English by Elizabeth S. Haldane and G. R. T. Ross (1970). New York: Cambridge University Press.

D'Esposito, M., J. Detre, D. Alsop, R. Shin, S. Atlas, and M. Grossman (1995) The neural basis of the central executive system of working memory. *Nature 378,* 279–281.

Donald, Merlin (1991) *Origins of the Modern Mind.* Cambridge, MA; Harvard University Press.

Doron, N., and Z. Wollberg (1994) Cross-modal neuroplasticity in the blind mole rat *Spalax Ehrenbergi:* A WGA-HRP tracing study. *NeuroReport 5,* 2697–2701.

Dubois, E. (1913) On the relation between the quantity of brain and the size of the body in vertebrates. *Verhandlungen des Koninklijke Academie voor Wetenschappen Amsterdam 16,* 647.

Dunbar, Robin (1988) *Primate Social Systems.* London: Goom Helen.

—— (1992a) Co-evolution of neocortex size, group size and language in humans. *Behavioral and Brain Sciences.*

—— (1992b) Neocortex size as a constraint on group size in primates. *Journal of Human Evolution 20,* 469–493.

—— (1997) *Grooming, Gossip and the Evolution of Language.* Cambridge, MA: Harvard University Press.

Durham, William (1994) *Coevolution: Genes, Culture and Human Diversity.* Stanford, CA: Stanford University Press.

Edelman, Gerald (1987) *Neural Darwinism: The Theory of Neuronal Group Selection.* New York: Basic Books.

—— (1994) *Bright Air, Brilliant Fire.* New York: Basic Books.

Elman, Jeffrey (1991) Incremental learning, or the importance of starting small. In *13th Annual Conference of the Cognitive Science Society*. Hillsdale, NJ: Lawrence Erlbaum, 443–448.

—— (1993) Learning and development in neural networks: The importance of starting small. *Cognition 48*, 71–99.

Elman, Jefferey, E. Bates, M. Johnson, A. Karmiloff-Smith, D. Parisi, and K. Plunkett (1996) *Rethinking Innateness: A Connectionist Perspective on Development*. Cambridge, MA: MIT Press.

Fabbro, Franco (1992) Cerebral lateralization of human languages: clinical and experimental data. In J. Wind, B. Chiarelli, B. Bichakjian, and A. Nocentini, eds. (1988), *Language Origin: A multidisciplinary approach*. Proceedings of NATO Advanced Institute, Cortona, Italy. Amsterdam: Kluwer, 195–224.

Falk, Dean (1983) Cerebral cortices of East African early hominids. *Science 221*, 1072–1074.

—— (1989) Ape-like endocast of "ape-man" Taung. *American Journal of Physical Anthropology 80*, 335–339.

—— (1990) Brain evolution in *Homo:* The "radiator" theory. *Behavioral and Brain Sciences 13*, 333–381.

Felleman, D. J. and D. C. Van Essen (1991) Distributed hierarchical processing in the primate cerebral cortex. *Cerebral Cortex 1*, 1–47.

Fessler, Daniel M. T. (1996) *Towards an Understanding of the Universality of Second Order Emotions* (in press).

Finkelstein, R., and E. Boncinelli (1994) From fly head to mammalian forebrain: The story of otd and Otx. *Trends in Genetics 10*, 310–315.

Finlay, B. L., and R. B. Darlington (1995) Linked regularities in the development and evolution of mammalian brains. *Science 268*, 1578–1584.

Fodor, J. A. (1983) *The Modularity of Mind*. Cambridge, MA: MIT Press/Bradford Books.

Frangiskakis, J. M., et al. (1996) LIM-kinase1 hemizygosity implicated in impaired visuospatial constructive cognition. *Cell 86*, 59–69.

Franzen, E. A., and R. E. Myers (1973) Neural control of social behavior: Prefrontal and anterior temporal cortex. *Neuropsychologia 11*, 141–157.

Frege, Gottlob (1879) *Begriffsschrift, a Formula Language Modeled on That of Arithemetic, for Pure Thought*. English translation in J. van Heijenoort, ed. (1970), *Frege and Gödel: Two Fundamantal Texts in Mathematical Logic*. Cambridge, MA: Harvard University Press, 1–82.

Friedman, G., and D. O'Leary (1996) Retroviral misexpression of engrailed genes in the chick optic tectum perturbs the topographic targeting of retinal axons. *Journal of Neuroscience 16*, 5490é-5509.

Frisch, Karl von (1967) *The Dance Language and Orientation of Bees*. Cambridge, MA: Harvard University Press.

Frith, Uta (1989) *Autism: Explaining the Enigma.* New York: Cambridge University Press.

——, ed. (1991) *Autism and Ansperger Syndrome.* New York: Cambridge University Press.

Frost, D. O., and C. Metin (1985) Induction of functional retinal projections to the somatosensory system. *Nature 317,* 162.

Fuster, J. (1980) *The Prefrontal Cortex: Anatomy, Physiology and Neuropsychology of the Frontal Lobe.* New York: Raven Press.

Gainotti, G. (1972) Emotional behavior and hemispheric side of the lesion. *Cortex 8,* 41–55.

Galaburda, A. M., P. P. Wang, U. Bellugi, and M. Rosen (1994) Cytoarchitectonic anomalies in a genetically based disorder: Williams syndrome. *NeuroReport 5,* 753–757.

Gardner, Howard (1983) *Frames of Mind: The Theory of Multiple Intelligences.* New York: Basic Books.

Gardner, Howard, H. Brownell, W. Wapner, and D. Michelow (1983) Missing the point: The role of the right hemisphere in processing of complex linguistic materials. In E. Perecman, ed., *Cognitive Processes and the Right Hemisphere.* New York: Academic Press.

Gallup, G. G. (1982) Self-awareness and the emergence of mind in primates. *American Journal of Primatology 2,* 237–248.

Georgopoulos, A. P., A. Ashe, N. Smyrnis, and M. Taira (1992) The motor cortex and the coding of force. *Science 256,* 1692–1695.

Gibson, K. R. and T. Ingold, eds. (1993) *Tools, Language and Cognition in Human Evolution.* Cambridge, UK: Cambridge University Press.

Gold, E. (1967) Language identification in the limit. *Information and Control 16,* 447–474.

Goldman-Rakic, Patricia R. (1987) Circuitry of the primate prefrontal cortex and regulation of behavior by representational memory. *Handbook of Physiology,* 373–418.

—— (1992) Working memory and the mind. *Scientific American 267,* 110–117.

Goldschmidt, Richard (1952) Evolution, as viewed by one geneticist. *American Scientist 40,* 84–135.

Goodall, Jane (1986) *The Chimpanzees of Gombe: Patterns of Behavior.* Cambridge, MA: Harvard University Press.

Goodglass, H. (1968) Studies on the grammar of aphasics. In S. Rosenberg and J. Kaplan, eds., *Developments in Applied Psycholinguistics Research.* New York: Macmillan.

—— (1973) Studies on the grammar of aphasics. In H. Goodglass and S.E. Blumstein, eds., *Psycholinguistics and Aphasia.* Baltimore: Johns Hopkins University Press.

Goodglass, H., J. B. Gleason, N. A. Bernholtz, and M. R. Hyde (1972) Some linguistic structures in the speech of a Broca's aphasic. *Cortex* 8, 191–212.

Goodman, Nelson (1955) *Fact, Fiction, and Forecast.* London: University of London.

Gould, J. L., and C. G. Gould (1994) *The Animal Mind.* New York: Scientific American Library.

Gould, Stephen Jay (1981) *The Mismeasure of Man.* New York: W. W. Norton.

Gould, Stephen Jay, and R. C. Lewontin (1979) The spandrels of of San Marco and the Panglossian program: A critique of the adaptationist program. *Proceedings of the Royal Society of London* 205, 281–288.

Gould, Stephen Jay, and Vrba, E. (1982) Exaptation: A missing term in evolutionary theory. *Paleobiology* 8, 4–15.

Grandin, Temple (1995) *Thinking in Pictures and Other Reports from My Life with Autism.* New York: Doubleday.

Grandin, Temple, and Margaret M. Scariano (1986) *Emergence: Labeled Autistic.* Novato, CA: Arena Press.

Greenfield, Patricia, and E. Sue Savage-Rumbaugh (1990) Grammatical combination in *Pan paniscus:* Processes of learning and invention in the evolution and development of language. In S. Parker and K. Gibson, eds., *'Language' and Intelligence in Monkeys and Apes: Comparative Developmental Perspectives.* Cambridge, UK: Cambridge University Press.

——— (1991) Imitation, grammatical development, and the invention of protogrammar by an ape. In N. Krasnegor, D. M. Rumbaugh, M. Studdert-Kennedy, and D. Scheifelbusch, eds., *Biobehavioral Foundations of Language Development.* Hillsdale, NJ: Lawrence Erlbaum, 235–258.

Grice, H. P. (1969) Utterers' meaning and intentions. *Philosophical Review* 78, 147–177.

Grossman, M. (1980) A central processor for hierarchically structured material: Evidence from Broca's aphasia. *Neuropsychologia* 18, 299–308.

——— (1982) Reversal operations after brain damage. *Brain and Cognition* 1, 258–265.

Guilford, J. (1967) *The Nature of Human Intelligence.* New York: McGraw-Hill.

Hartshorne, C., and P. Weiss, eds. (1978) *Collected Papers: Charles Sanders Peirce.* Vols. I–VIII. Cambridge, MA: Belknap.

Hauser, Marc D. (1996) *The Evolution of Communication.* Cambridge, MA: MIT Press.

Hauser, Marc D., P. MacNeilage, and M. Ware (1996) Numerical represenatations in primates. *Proceedings of the National Academy of Sciences USA* 93, 1514–1517.

Hebb, Donald (1949) *The Organization of Behavior: A Neuropsychological Theory.* New York: John Wiley.

Heil, P., G. Bronchti, and H. Scheik (1991) Invasion of visual cortex by the auditory system in the naturally blind mole rat. *NeuroReport* 2, 735–738.

Heilman, K., R. T. Watson, and D. Bowers (1983) Affective disorders associated with hemispheric disease. In K. Heilman and P. Satz, eds., *Neuropsychology of Human Emotion*. New York: The Guilford Press.

Herrnstein, Richard (1980) Symbolic communication between two pigeons *(Columba domestica). Science 210*.

Hockett, C. F., and R. Ascher (1964) The human revolution. *Current Anthropology 5*, 135–168.

Holland, P., P. Ingham, and S. Krauss (1992) Development and evolution. Mice and flies head to head. *Nature 358, 627–628*.

Holloway, Ralph (1968). The evolution of the primate brain: Some aspects of quantitative relations. *Brain Research 7, 121–172*.

—— (1979) Brain size, allometry, and reorganization: toward a synthesis. In M. Hahn, C. Jensen, and B. Dudek, eds., *Development and Evolution of Brain Size*. New York: Academic Press.

—— (1980) Within-species brain-body weight variability: A reexamination of the Danish data and other primate species. *American Journal of Physical Anthropology 53*, 109–121.

—— (1981) Volumetric and asymmetry determinations on recent hominid endocasts: Spy I and II, Djebel Ihroud I, and the Sale *Homo erectus* specimens, with some notes on Neanderthal brain size. *American Journal of Physical Anthropology 55*, 385–393.

—— (1983) Human paleontological evidence relevant to language behavior. *Human Neurobiology 2*, 105–114.

—— (1985) The past, present, and future significance of the lunate sulcus in early hominid evolution. In P. Tobias, ed., *Hominid Evolution: Past, Present and Future*. New York: Alan R. Liss.

—— (1995) [Commentary to Aiello and Wheeler (1995)] *Current Anthropology 36*, 213–214.

Holt, A. B., D. B. Cheek, E. D. Mellits, and D. E. Hill (1975) Brain size and the relation of the primate to the nonprimate. In Cheek, D. B., ed., *Fetal and Postnatal Cellular Growth: Hormones and Nutrition*, New York: John Wiley.

Hublin, J.-J., F. Spoor, M. Braun, F. Zonneveld, and S. Condemi (1996) A late Neanderthal associated with upper Paleolithic artefacts. *Nature 381, 224–226*.

Huttenlocher, P. R. (1990) Morphometric study of human cerebral cortex development. *Neuropsychologia 28*, 517–527.

Isacson, O., T. Deacon, P. Pakzaban, W. Galpern, J. Dinsmore, and L. Burns, (1995) Transplanted xenogeneic neural cells in neurodegenerative disease models exhibit remarkable axonal target specificity and distinct growth patterns of glial and axonal fibres. *Nature Medicine 1*, 1189–1194.

Isacson, O., and T. Deacon, (1996) Specific axon guidance factors persist in the adult brain as demonstrated by pig neuroblasts transplanted to the rat. *Neuroscience 75*, 827–837.

Jackendoff, Ray (1992) *Languages of the Mind.* Cambridge, MA: MIT Press.

—— (1994) *Patterns in the Mind: Language and Human Nature.* New York: Basic Books.

Jacobsen, C. (1936) Studies of cerebral function in primates. *Comparative Psychology Monographs 13,* 1–68.

Jakobson R. (1956) Two aspects of language and two types of aphasic disturbances. In R. Jakobson and M. Halle, eds., *Fundamentals of Language.* The Hague: Mouton.

Jerison, Harry (1973) *The Evolution of the Brain and Intelligence.* New York: Academic Press.

Jones, Stephen, D. Pilbeam, and R. Martin, eds. (1993) *Cambridge Encyclopedia of Human Evolution.* Cambridge, UK: Cambridge University Press.

Jürgens, Uwe (1976a) Projections from cortical larynx area in the squirrel monkey. *Experimental Brain Research 25,* 401–411.

—— (1976b) Reinforcing concomitants of electrically elicited vocalization. *Experimental Brain Research 26,* 203–214.

—— (1979a) Neural control of vocalization in non-human primates. In H. D. Steklis and M. J. Raleigh, eds., *Neurobiology of Social Communication in Primates.* New York: Academic Press.

—— (1979b) Vocalization as an emotional indicator. A neuroethological study in the squirrel monkey. *Behaviour 69,* 88–117.

Jürgens, Uwe, and R. Pratt (1979a) Cingular vocalization pathway: Squirrel monkey. *Experimental Brain Research 34,* 499–510.

—— (1979b) The role of the peri-aqueductal grey in vocal expression of emotion. *Brain Research 167,* 367–378.

Jürgens, Uwe, A. Kirzinger, and D. von Cramon (1982) The effects of deep-reaching lesions in the cortical face area on phonation. A combined case report and experimental monkey study. *Cortex 18,* 125–139.

Karmiloff-Smith, Annette (1992) *Beyond Modularity.* Cambridge, MA: MIT Press.

Kato, N., D. Price, J. M. R. Ferrer, and C. Blakemore (1993) Plasticity of an aberrant geniculocortical pathway in neonatally lesioned cats. *NeuroReport 4,* 915–918.

Katz, J. M., and R. J. Lasek (1983) Evolution of the nervous system: Role of ontogenetic mechanisms in the evolution of matching populations. *Proceedings of the National Acadamy of Science USA 75,* 1349–1352.

Kean, Mary-Louise (1977) The linguistic interpretation of aphasic syndromes: Agrammatism in Broca's aphasia, an example. *Cognition 5,* 9–46.

Keller, Helen (1903) *The Story of My Life.* New York: Doubleday.

Killackey, Herbert P., N. L. Chiaia, C. A. Bennett-Clarke, M. Eck, and R. W. Rhoades (1994) Peripheral influences on the size and organization of somatotopic representations in the fetal rat cortex. *Journal of Neuroscience 14,* 1496–1506.

Kim, S.-G., K. Ugurbil, and P. L. Strick (1994) Activation of cerebellar output nucleus during cognitive processing. *Science 265*, 949–951.

Kinsbourne, Marcel (1978) Evolution of language in relation to lateral action. In M. Kinsbourne, ed., *Asymmetrical Function of the Brain*. New York: Cambridge University Press.

Klima, Edward, and Ursula Bellugi (1979) *The Signs of Language*. Cambridge, MA: Harvard University Press.

Köhler, Wolfgang (1927) *The Mentality of Apes*. New York: Harcourt, Brace.

Kolb, Brian, and Ian Whishaw (1990) *Fundamentals of Human Neuropsychology*. 3d ed. New York: W. H. Freeman & Co.

Krassnegor, N. A., D. Rumbaugh, R. Schiefelbusch, and M. Studdert-Kennedy, eds. (1991) *Biological and Behavioral Determinants of Language Development*. Hillsdale, NJ: Lawrence Erlbaum.

Kripke, Saul (1972) Naming and necessity. In G. Harmon and D. Davidson, eds., *The Semantics of Natural Language*. Dordrecht: Riedl, 254–355, 763–769.

Kruska, Dieter (1988) Mammalian domestication and its effect on brain structure and behavior. In H. Jerison and I. Jerison, eds., *Intelligence and Evolutionary Biology*. Berlin: Springer-Verlag, 211–250.

Kuhl, P., and J. D. Miller (1975) Speech perception by the chinchilla. *Science 190*, 69–72.

Landau, Misia (1991) *Narratives of Human Evolution*. New Haven, CT: Yale University Press.

Lapicque, L. (1907a) Tableau du poids somatique et encé phalique dans les espèces animales. *Bulletin de la Société d'Anthropologie Paris 8*, 248–262.

—— (1907b) Le poids encé phalique en fonction du poids corporal entre individus d'une même espèce. *Bulletin de la Société d'Anthropologie Paris 8*, 313.

Larsen, B., E. Skinhoj, and N. A. Lassen (1978) Variations in regional cortical blood flow in the right and left hemispheres during automatic speech. *Brain 101*, 193–209.

Lassen, N. A., D. H. Ingvar, and E. Skinhöj (1978) Brain function and blood flow. *Scientific American 239*, 62–71.

Lassen, N. A., and B. Larsen (1980) Cortical activity in the left and right hemispheres during language-related brain functions. *Phonetica 37*, 27–37.

Law, M., and M. Constantine-Paton (1981) Anatomy and physiology of experimentally induced striped tecta. *Journal of Neuroscience 1*, 741–759.

Le Douarin, N. (1993) Embryonic neural chimeras in the study of brain development. *Trends in Neuroscience 16*, 64–72.

Lecours, André-Roch, F. Lhermitte, and B. Bryans (1983) *Aphasiology*. London: Baillière Tindall.Leiner, H. C., A. L. Leiner, and R. S. Dow (1989) Reappraising the cerebellum: What does the hindbrain contribute to the forebrain? *Behavioral Neuroscience 103*, 998–1008.

—— (1993) Cognitive and language functions of the human cerebellum. *Trends in Neurosciences 16,* 444–447.

Lennenberg, Eric H. (1967) *Biological Foundations of Language.* New York: John Wiley.

Liberman, A., F. Cooper, D. Shankweiler, and M. Studdert-Kennedy (1967) Perception of the speech code. *Psychological Review 74,* 431–461.

Liberman, A., and I. Mattingly (1989) A specialization for speech perception. *Science 243,* 489–494.

Lieberman, Philip (1984) *The Biology and Evolution of Language.* Cambridge, MA: Harvard University Press.

—— (1991) *Uniquely Human: The Evolution of Speech, Thought and Selfless Behavior.* Cambridge, MA: Harvard University Press.

Lieberman, Philip, and E. S. Crelin (1971). On the speech of Neanderthal man. *Linguistic Inquiry 2,* 203–222.

Linden, R. (1990) Control of neuronal survival by anomalous targets in the developing brain. *Journal of Comparative Neurology 294,* 594–606.

Lisker, L., and A. S. Abramson (1971) Distinctive features and laryngeal control. *Language 47,* 767–785.

Lipp, Hans-Peter, and D. P. Wolfer (1995) New paths towards old dreams: Microphrenology. In E. Alleva, et al., eds. *Behavioral Brain Research in Naturalistic and Semi-Naturalistic Settings.* Dordrecht: Kluwer Academic Publishers, 3–46.

Luria, A. R. (1980) *Higher Cortical Functions in Man* (2nd edn., English translation). New York: Basic Books (original Russian text published by Moscow University Press, 1962).

MacDonald, M. C. (1989) Priming effects from gaps to antecendents. *Language and Cognitive Processes 4,* 1–72.

MacNeilage, Peter (1991) The "postural" origins of primate neurobiological asymmetries. In Krassnegor, N. A., D. Rumbaugh, R. Schiefelbusch, and M. Studdert-Kennedy, eds. *Biological and Behavioral Determinants of Language Development.* Hillsdale, NJ: Lawrence Erlbaum.

Macphail, Euan M. (1982) *Brain and Intelligence in Vertebrates.* Oxford: Clarendon Press.

Marcus, G. F. (1993) Negative evidence in language acquisition. *Cognition 46,* 53–85.

Marler, Peter (1959) Developments in the study of animal communication. In P. R. Bell, ed., *Darwin's Biological Work.* Cambridge, UK: Cambridge University Press, 150–206.

Mattingly, I., and M. Studdert-Kennedy, eds. (1991) *Modularity and the Motor Theory of Speech Perception.* Hillsdale, NJ: Lawrence Erlbaum.

McCrone, John (1991) *The Ape That Spoke.* New York: William Morrow.

McGinnis, W., and M. Kuziora (1994) The molecular architects of body design. *Scientific American 270*, 58–66.

McKinney, Michael, and K. J. MacNamara (1991) *Heterochrony: The Evolution of Ontogeny.* New York: Plenum Press.

Middleton, Frank A., and Peter L. Strick (1994) Anatomical evidence for cerebellar and basal ganglia involvement in higher cognitive function. *Science 266*, 458–461.

Miller, George (1991) *The Science of Words.* New York: Scientific American Library.

Mishkin, M., and F. Manning (1978) Nonspatial memory after selective prefrontal lesions in monkeys. *Brain Research 143*, 313–323.

Molière (1953) *The Imaginary Invalid (Le Malade Imaginaire).* In *Molière: The Misanthrope and Other Plays* (translated by John Wood). Baltimore: Penguin Books.

Molnár, Z., and C. Blakemore (1991) Lack of regional specificity for connections formed between thalamus and cortex in coculture. *Nature 351*, 475–477.

Morgan, J. L., and L. L. Travis (1989) Limits on negative information in language learning. *Journal of Child Language 16*, 531–552.

Müeller, R.-A.(1996) Innateness, autonomy, universality? Neurobiological approaches to language. *Behavioral and Brain Research* (in press).

Myers, R. E. (1976) Comparative neurology of vocalization and speech: Proof of a dichotomy. In S. R. Harnad, H. D. Steklis, and J. Lancaster, eds., *Origins and Evolution of Language and Speech, Annals of the New York Acadamy of Science 280*, 745–757.

Newman, John D., and Paul D. MacLean (1982) Effects of tegmental lesions on the isolation call of squirrel monkeys. *Brain Research 232*, 317–329.

Newmeyer, F. J. (1991) Functional explanation in linguistics and the origin of language. *Language and Communication 11*, 1–28.

Newport, Elissa L. (1990) Maturational constraints on language learning. *Cognitive Science 14*, 11–28.

—— (1991) Contrasting conceptions of the critical period for language. In S. Carey and R. Gelman, eds., *Epigenesis of Mind: Essays on Biology and Cognition.* Hillsdale, NJ: Lawrence Erlbaum.

Nichelli, P., J. Grafman, et al. (1995) Where the brain appreciates the moral of a story. *NeuroReport 6*, 2309–2313.

Nobre, A., T. Allison, and G. McCarthy (1994) Word recognition in the human inferior temporal lobe. *Nature 372*, 260–263.

Nottebohm, F., D. B. Kelly, and J. A. Paton (1982) Connections of vocal control nuclei in the canary telencephalon. *Journal of Comparative Neurology 207*, 344–357.

Nottebohm, F., and M. E. Nottebohm (1976) Left hypoglossal dominance in the control of canary and white-crowned sparrow song. *Journal of Comparative Physiology 108*, 171–192.

Nowicki, S., M. Westneat, and W. Hoese (1992) Birdsong: Motor function and the evolution of communication. *Seminars in the Neurosciences 4,* 385–390.

Nudo, R. J., D. P. Sutherland, and R. B. Masterton (1995) Variation and evolution of mammalian corticospinal somata with special reference to primates. *Journal of Comparative Neurology 358,* 181–205.

Ojemann, George A. (1979) Individual variability in cortical localization of language. *Journal of Neurosurgery 50,* 164–169.

—— (1983) Brain organization for language from the perspective of electrical stimulation mapping. *Behavioral and Brain Sciences 2,* 189–230.

—— (1991) Cortical organization of language. *Journal of Neuroscience 11,* 2281–2287.

Ojemann, George A., and C. Catherine Mateer (1979) Human language cortex: Localization of memory, syntax, and sequential motor-phoneme indentification systems. *Science 205,* 1401–1403.

O'Leary, Dennis D. M. (1989) Do cortical areas emerge from a protocortex? *Trends in Neuroscience 12,* 400–406.

—— (1992) Development of connectional diversity and specificity in the mammalian brain by the pruning of collateral projections. *Current Opinions in Neurobiology 2,* 70–77.

O'Leary, Dennis D. M., and S. E. Koester (1993) Development of projection neuron types, axon pathways, and patterned connections of the mammalian cerebral cortex. *Neuron 10,* 991–1006.

O'Leary, Dennis D. M., and B. Stanfield (1989) Selective elimination of axons extended by developing cortical neurons is dependent on regional locale experiments utilizing fetal cortical transplants. *Journal of Neuroscience 9,* 2230–2246.

Ornstein, Robert (1973) *The Nature of Human Consciousness.* San Francisco: W. H. Freeman.

Pandya, Deepak, and C. Barnes (1987) Architecture and connections of the frontal lobe. In E. Perecman, ed., *The Frontal Lobes Revisited.* New York: IRBN Press, 41–72.

Pandya, Deepak, and E. H. Yeterian (1990) Prefrontal cortex in relation to other cortical areas in rhesus monkey: architecture and connections. In H. B. M. Uylings, ed., *The Prefrontal Cortex: Its Structure, Function and Pathology.* Amsterdam: Elsevier, 63–94.

Passingham, Richard E. (1973) Anatomical differences between the neocortex of man and other primates. *Brain Behavior and Evolution 7,* 337–359.

—— (1975) Changes in the size and organisation of the brain in man and his ancestors. *Brain Behavior and Evolution 11,* 73–90.

—— (1979) Brain size and intelligence in man. *Brain Behavior and Evolution 16,* 253–270.

—— (1981) Broca's area and the origins of human vocal skill. *Philosophical Transactions of the Royal Society of London, B 292*, 167–175.

—— (1982) *The Human Primate*. San Francisco: W. H. Freeman.

—— (1985) Memory of monkeys (*Macaca mulatta*) with lesions in prefrontal cortex. *Behavioral Neuroscience 99*, 3–21.

—— (1985) Rates of brain development in mammals including man. *Brain, Behavior and Evolution 26*, 167–175.

Passingham, Richard E., and G. Ettlinger (1974). A comparison of cortical function in man and other primates. *International Review of Neurobiology 16*, 233–299.

Peirce, Charles Sanders (1897, 1903) Logic as semiotic: The theory of signs. In J. Buchler, ed., *The Philosophical Writings of Peirce* (1955). New York: Dover Books, 98–119.

—— (1978) *Collected Papers. Vol. II. Elements of Logic*. C. Hartshorne and P. Weiss, eds., Cambridge, MA: Belknap.

Penfield, W., and L. Roberts (1959) *Speech and Brain Mechanisms*. London: Oxford University Press.

Penrose, Roger (1989) *The Emperor's New Mind*. Oxford: Oxford University Press.

Pepperberg, Irene M. (1987) Acquisition of the same/different concept by an African grey parrot *Psittacus erithacus*. *Animal Learning and Behavior 15*, 423–432.

Perecman, Ellen, ed. (1987) *The Frontal Lobes Revisited*. New York: IRBN Press.

Petersen, S. E., P. T. Fox, et al. (1988) Positron emission tomographic studies of the cortical anatomy of single-word processing. *Nature 331*, 585–589.

Petrides, Michael (1982) Motor conditional associative learning after selective prefrontal lesions in the monkey. *Behavioral and Brain Research 5*, 407–413.

—— (1985) Deficits in nonspatial conditional associative learning after periarcuate lesions in monkey. *Behavioral and Brain Research 16*, 95–101.

—— (1986) The effect of periarcuate lesions in the monkey on the performance of symmetrically and asymmetrically reinforced visual and auditory go, no-go tasks. *Journal of Neuroscience 6*, 2054–2063.

Petrides, Michael, and B. Milner (1982) Deficits on subject-ordered tasks after frontal and temporal lobe lesions in man. *Neuropsychologia 20*, 249–262.

Piaget, Jean (1952) *The Origins of Intelligence in Children*. New York: International Universities Press.

Piattelli-Palmarini, M. (1989) Evolution, selection, and cognition: From "learning" to parameter setting in biology and the study of language. *Cognition 31*, 1–44.

Pinker, Steven (1991) Rules of language. *Science 253*, 530–535.

—— (1994) *The Language Instinct: How the Mind Creates Language*. New York: William Morrow.

Pinker, Steven, and P. Bloom (1990) Natural language and natural selection. *Behavioral and Brain Sciences 13*, 707–784.

Posner, M. I., S. E. Petersen, P. T. Fox, and M. E. Raichle (1988) Localization of cognitive operations in the human brain. *Science 240*, 1627–1631.

Posner and M. E. Raichle (1994) *Images of Mind*. New York: Scientific American Library.

Premack, David (1986) *Gavagai! Or the Future History of the Animal Language Controversy*. Cambridge, MA: MIT Press

Premack, David, and A. Premack (1983) *The Mind of an Ape*. New York: W. W. Norton.

Provine, Robert (1996) Laughter. *American Scientist 84*, 38–47.

Purves, Dale (1988) *Body and Brain. A Trophic Theory of Neural Connections*. Cambridge, MA: Harvard University Press.

Purves, Dale, and J. Lichtman (1980) Elimination of synapses in the developing nervous system. *Science 210*, 153–157.

—— (1985) *Principles of Neural Development*. Sunderland, MA: Sinauer Associates Inc.

Putnam, Hilary (1975) *Mind, Language and Reality*. Cambridge, UK: Cambridge University Press.

Quine, W. V. O. (1960) *Word and Object*. Cambridge, MA: MIT Press.

Raff, Rudolph (1996) *The Shape of Life: Genes, Development, and the Evolution of Animal Form*. Chicago: University of Chicago Press.

Rakic, P. (1988) Specification of cerebral cortical areas. *Science 241*, 170–176.

Ramsey, W., and S. Stich (1991) Connectionism and three levels of nativism. In W. Ramsey, S. Stich, and D. Rummelhart, eds., *Philosophy and Connectionist Theory*. Hillsdale, NJ: Lawrence Erlbaum.

Richards, Robert (1987) *Darwin and the Emergence of Evolutionary Theories of Mind and Behavior*. Chicago: University of Chicago Press.

Ridley, Mark (1993) *The Red Queen: Sex and the Evolution of Human Nature*. New York: Macmillan.

Ringo, J. L. (1991) Neuronal interconnection as a function of brain size. *Brain, Behavior and Evolution 38*, 1–6.

Roland, P. E. (1985) Cortical organization of voluntary behavior in man. *Human Neurobiology 4*, 155–167.

Romaine, Susan (1988) *Pidgin and Creole Languages*. London: Longman Group.

Rosch, Elenor (1978) Principles of categorization. In E. Rosch and B. Lloyds, eds., *Cognition and Categorization*. Hillsdale, NJ: Lawrence Erlbaum.

Roth, G., J. Blanke, and M. Ohle (1995) Brain size and morphology in miniaturized plethodontid salamanders. *Brain Behavior and Evolution 45*, 84–95.

Roth, G., K. C. Nishikawa, C. Naujoks-Manteuffel, A. Schmidt, and D. B. Wake (1993) Paedomorphosis and simplification in the nervous system of salamanders. *Brain, Behavior and Evolution 42*, 137–170.

Rubenstein, John, S. Martinez, K. Shinmamura, and L. Puelles (1994) The embryonic vertebrate forebrain: The prosomeric model. *Science 266*, 578–580.

Rumbaugh, Duane, ed. (1977) *Language Learning by a Chimpanzee: The Lana Project*. New York: Academic Press.

Rumbaugh, Duane, E. Sue Savage-Rumbaugh, and David A. Wahburn (1996) Toward a new outlook on primate learning and behavior: Complex learning and emergent processes in comparative perspective. *Japanese Psychological Research 38*, 113–125.

Rumbaugh, D. M., and Pate, J. L. (1984) The evolution of cognition in primates. A comparative perspective. In H. L. Roitblat, T. G. Bever, and H. S. Terrace, eds., *Animal Cognition*. Hillsdale, N.J.: Lawrence Erlbaum Associates, 403–20.

Salvatore, Aglioti, and Franco Fabro (1993) Paradoxical selective recovery in a bilingual aphasic following subcortical lesions. *NeuroReport 4*, 1359–1362.

Saussure, Ferdinand de (1916) *Cours de linguistique générale*. Paris: Payot. See also the 1969 translation by Wade Baskin: *Course in General Linguistics*. New York: McGraw-Hill.

Schmidt-Nielsen, Knut (1984) *Scaling: Why Is Animal Size So Important?* Cambridge, UK: Cambridge University Press.

Seyfarth, Robert, Dorothy Cheney, and Peter Marler (1980) Monkey responses to three different alarm calls: Evidence of predator classification and semantic communication. *Science 210*, 801–803.

Savage-Rumbaugh, E. Sue (1986) *Ape Language: From Conditioned Response to Symbol*. New York: Columbia University Press.

Savage-Rumbaugh, E. Sue, and Roger Lewin (1994) *Kanzi: The Ape at the Brink of the Human Mind*. New York: John Wiley.

Savage-Rumbaugh, E. Sue, D. M. Rumbaugh, and S. Boysen (1978) Symbolization, language and chimpanzees: A theoretical reevaluation based on initial language acquisition processes in four young Pan troglodytes. *Brain and Language 6*, 265.

Savage-Rumbaugh, E. Sue, D. M. Rumbaugh, S. T. Smith, and J. Lawson (1980) Reference: The linguistic essential. *Science 210*, 922–925.

Searle, John (1980) Minds, brains, and programs. *Behavioral and Brain Sciences 3*, 417–458.

—— (1992) *The Rediscovery of the Mind*. Cambridge, MA: MIT Press.

Semendeferi, K., H. Damasio, G. Van Hoesen, and R. Frank (1997) The evolution of the frontal lobes: A volumetric analysis based on three-dimensional reconstructions of magnetic resonance scans of human and ape brains. *Journal of Human Evolution* (in press).

Smith, John Maynard (1978) *Evolutionarily Stable Strategies (ESSs)*. Cambridge, UK: Cambridge University Press.

Snell, Otto (1891) Das Gewicht des Gehirns und des Hirnmantels der Saugetiere in Beziehung zu deren geistigen Fahigkeiten. *Sitz. Ges. Morph. Physiol. (Munchen) 7*, 90–94.

———— (1892) Die Obhängigkeit des Hirngewichtes von dem Körpergewicht und den geistigen Fähigkeiten. *Arch. Psychiat. Nervenkrank 23,* 436–446.

Sokoloff, Alan, and Terrence W. Deacon (1989) Direct projections from the face area of primary motor cortex to the facial nucleus in the cynomolgus monkey *(Macaca fascicularis). American Zoologist 29,* abstract.

———— (1990) Direct projections from the face area of primary motor cortex to the facial nucleus in the cynomolgus monkey but not in the cat or rat. *American Journal or Physical Anthropology 81,* 298.

———— (1992) Musculotopic organization of intrinsic tongue musculature in the Cynomolgus monkey, *Macaca fascicularis. Journal of Comparative Neurology 324,* 81–93.

Stanfield, B., and D. D. M. O'Leary (1985) Fetal occipital cortical neurons transplanted to the rostral cortex can extend and maintain a pyramidal tract axon. *Nature 313,* 135–137.

Stephan, Heinz (1969) Quantitative investigations on visual structures in primate brains. In *Proceedings of the Second International Congress on Primates 3,* Basel: Karger, 34–42.

Stephan, Heinz, H. Frahm, and G. Baron (1981) New and revised data on volumes of brain structures in insectivores and primates. *Folia Primatologica 35,* 1–29.

Stuss, D., and D. Benson (1986) *The Frontal Lobes.* New York: Raven Press.

Sur, M., P. Garraghty, and A. Roe (1988) Experimentally induced visual projections into auditory thalamus and cortex. *Science 242,* 1437–1441.

Taylor, G. M., E. Nol, and D. Boire (1995) Brain regions and encephalization in anurans: Adaptation or stability? Brain, Behavior and Evolution 45, 96–109.

Terrace, Herbert (1979) *Nim: A Chimpanzee Who Learned Sign Language.* New York: Knopf.

————, L. Petitto, R. Sanders, and T. Bever (1979) Can an ape create a sentence? *Science 206,* 891–902.

Thompson, W. D'Arcy (1917) *On Growth and Form.* Cambridge, UK: Cambridge University Press.

Tobias, Philip V. (1981) The emergence of man in Africa and beyond. *Philosophical Transactions of the Royal Society of London, B, Biological Sciences 292,* 43–56.

———— (1987) The brain of *Homo habilis:* A new level of organisation in cerebral evolution. *Journal of Human Evolution 16,* 741–761.

Todd, L. (1974) *Pidgins and Creoles.* London: Routledge & Kegan Paul.

Tramo, M. J., K. Baynes, and B. T. Volpe (1988) Impaired syntactic comprehension and production in Broca's aphasia: CT lesion localization and recovery patterns. *Neurology 38,* 95–98.

Treffert, Darold A. (1989) *Extraordinary People: Understanding Savant Syndrome.* New York: Ballantine Books.

Uylings, H. B. M., ed. (1990) *The Prefrontal Cortex: Its Structure, Function and Pathology*. Amsterdam: Elsevier, 63–94.

Vandenberghe, R., C. Price, R. Wise, O. Josephs, and R. S. J. Frackowiak (1996) Functional anatomy of a common semantic system for words and pictures. *Nature 383*, 254–256.

Vygotsky, L. S. (1978) *Mind in Society*. Initially translated by A. R. Luria and edited by M. Cole, V. John-Steiner, S. Scribner, and E. Souberman. Cambridge, MA: Harvard University Press.

Waddington, Conrad H. (1957) *The Strategy of the Genes*. London: Allen & Unwin.

Walker, Alan (1996) *The Wisdom of Bones: In Search of Human Origins*. New York: Alfred Knopf.

Wallesch, C. W., H. H. Kornhuber, et al. (1983) Language and cognitive deficits resulting from medial and dorsolateral frontal lobe lesions. *Arch Psychiatr Nervenkr. 233*, 279–296.

Warrington, E. K., and R. McCarthy (1987) Categories of knowledge: Further fractionation and an attempted integration. *Brain 106*, 1273–1296.

Watanabe, Masataka (1996) Reward expectancy in primate prefrontal neurons. *Nature 382*, 629–632.

Welker, E., and Van der Loos, H. (1986) Is areal extent in sensory cerebral cortex determined by peripheral innervation density? *Experimental Brain Research 63*, 650–654.

Werner, Heinz, and Bernard Kaplan (1963) *Symbol Formation*. New York: John Wiley & Sons.

Wernicke, Carl (1874) *Der aphasische Symptomencomplex*. Breslau: Cohn & Weigert.

Widdowson, E. M. (1981) Growth of creatures great and small. *Symposium of the Zoological Society of London 46*, 5–17.

Wilczynski, W. (1984) Central neural systems subserving a homoplasous periphery. *American Zoologist 24*, 755–763.

Willis, C. (1993) *The Runaway Brain: The Evolution of Human Uniqueness*. New York: Basic Books.

Wind, Jan (1970) *On the Phylogeny and Ontogeny of the Human Larynx*. Groningen: Wolters-Noordhoff Publishing.

Wrangham, Richard, and Dale Peterson (1996) *Demonic Males: Apes and the Origins of Human Violence*. Boston: Houghton Mifflin.

Wulfeck, B., E. Bates, and R. Capasso (1991) A cross-linguistic study of grammaticality judgements in Broca's aphasia. *Brain and Language 41*, 311–336.

Yamamoto, N., K. Yamada, T. Kurotani, and K. Toyama (1992) Laminar specificity of extrinsic cortical connections studied in coculture preparations. *Neuron 9*, 217–288.

Yamamoto, K. Toyama (1995) Repulsive and attractive mechanisms for the formation of corticofugal projections. *NeuroReport 6,* 1517–1520.

Yuasa, J., S. Hirano, M. Yamagata, and M. Noda (1996) Visual projection map specified by topographic expression of transcription factors in the retina. *Nature 382,* 632–635.

Zilbovicius, M., B. Garreau, Y. Samson, P. Rey, C. Barthelemy, A. Syrota, and G. Lelord (1995) Delayed maturation of the frontal cortex in childhood autism. *American Journal of Psychiatry 152,* 248–252.

Zilles, Karl, E. Armstrong, K. H. Moser, A. Schleicher, and H. Stephan (1989) Gyrification in the cerebral cortex of primates. *Brain, Behavior and Evolution 34,* 143–150.

Index

Page numbers in *italics* refer to figures.

men:
marriage and, 384–85
scavenging abilities of, 386
see also males
"mentalese," 26, 27, 454
mental images, word meaning
correspondences with, 26, 27
mesencephalon, 176
Mestrel, George de, 351
metabolism, brain size and, 155
metaphors, 305, 306
metonymy, 306
MGB (medial geniculate body), *210*, 211
mice, 154–55, 159
embryonic development of, 176–79, *178*
microcognition, 287
microevolution (microgenesis), 456–58
microorganisms, 111–13
midbrain, 176, 177, *178, 181*, 184, *198, 256*,
313, 422
communication pathway through, 230–32,
231
emotional and arousal states and, 232, 234,
235–36, 419
middle temporal multimodal cortex, *see*
Wernicke's area
midline, *281, 295*
mimickry:
human capacity for, 277, 278, 337
rarity of, 228–29
mind, theory of, 416–17, 423–51
animals and, 426, 428
Cartesian, 439, 441, 442
Chinese Room thought experiment and,
444–47
computer analogue of, 442–48
general problems of, 438–39
hierarchy of information and, 449
human, 23, 424–25
levels of consciousness and, 438–41
physical processes and, 460
religious traditions and, 441
misdirection, learning and, 49
Mishkin, M., *260*
misplaced correctness, 286
"mitochondrial Eve," 36
mitosis, 187–88
mockingbirds, 228–29
mole rats, 210, *210*, 248
moles, 159
Molière (Jean-Baptiste Poquelin), 37
monoamine transmitters, 422
monogamy, 385, 388
morphemes, 250
morphogenesis, 194

"motherese," 363
moths, camouflage and, 75–76
motor and premotor cortex, 216–18, *231*,
237–38, 247–48, 294, *304*
motor neurons, *196*, 216, 232–33, 235, 239
mouth, 358
muscles:
brain compared with, 149–51
facial, *231*, 241
flight, 241
geniohyoid, 239, *240*
hyoglossal, 239
jaw, *231*
laryngeal, *231, 240*, 242, 250
neurons and, *197*, 216–18
syringeal, 239–40, *240*
tongue, *231*, 238, 249
muteness, 249
myelin, myelination, 162, 251

narrative interpretation, 430
nasopharynx, *356*
natural selection, *see* Baldwinian selection;
evolution; genetic assimilation; selection
pressure; sexual selection
Neanderthals, 36, 355
demise of, 370–73
stone tool use and, 370–72
vocal abilities of, 370, 372
negative information, 263–64
neocortex, *198*, 244
nervous system, 194–96, 449
autonomic, 230–32
see also axons; brain; brain, evolution of;
brain, human; spinal cord
net strength, gross strength vs., 150
Neural Darwinism, 457
neural nets (neural networks), 129–34, *130*, 460
neural stem cells, 185
neural tube, 176, 177, *178*, 184, 189
neuroanatomy, 45, 230–36
*see also specific parts of brain and nervous
system*
neurons:
in brain stem, 234
color recognition and, *118*
connectivity and, 197–98, *198*
cortical, 203–7, *203* ·
density of, 154, 162–63
motor, *196*, 216, *231*, 232–33, 235, 239
nonspecific connectivity of, 199–200, 203–6
premotor, 232, 247
programmed death of, 195–97, *196*
structure of, 194–95
see also axons; dendrites

neurotransmission, 196, 202, 422
newborns, 236
Newport, Elissa, 128–29, 135, 137
nodes, 130–31, *130*
nöo-species designation, 341
noradrenaline, 422
"nose brain," *231*
notochord, 177
nouns, 329, 334
nucleus ambiguus, *231, 232,* 247, 248

object-object relationships, 88
object-oriented computer interfaces, 106–7
obsessive-compulsive disorders, 422
occipital cortex, *217*
Ojemann, George, 289, *290*
O'Leary, Dennis, 204, 209
olfactory bulbs, *231*
opaque meaning relationships, *61*
opponent processing, 118
oral cavity, volume of, *357*
oral tract, 232–33, 246
orangutans, 30
organisms:
    as analogue for language, 111–15
    design of, 159–60
Orwell, George, 105
"Otx" genes, 177, *178,* 179–80, *181,* 185, *186,*
    188
output nodes, *130,* 131
Owen, Richard, 175

pair bonding, 383–84, 397–401, *409*
    of carnivores, 389–93
    in human groups, 388
    symbolic reference and, 399–401
    in wild dogs and wolves, 391
    see also marriage
Paleolithic period, 374
pallidum, 184
panic syndromes, 422
paradigmatic operations of word association,
    304–6
paralinguistics, 364–65
parallel distributed processing (PDP), 132
parallelism, 292, 293, 299–300
paralysis, 237
parasites, 111–13
parcellation, 183, 189, 203, *208,* 221–22
parietal lobes, 289, 290, 291, 302
Parkinsonism, 422
parrots, 60–61, 254
Passingham, Richard, 260, *261*
PDP (parallel distributed processing), 132
peace, mediation of, 403–5

Peirce, Charles Sanders, 63, 70–73
Penfield, Wilder, 288–89, *290*
perception, difficulties in, 40
periaqueductal (central) gray area of
    midbrain, 230–32, *231,* 234–35
peripheral nervous system, 207–9, 216–18
personal experience, indirect knowledge of,
    424–25
PET (positron emission tomography), 275,
    293, 294, *296,* 303, *304, 305*
Petrides, Michael, *260*
pharynx, enlargement of, *356,* 357–58, *357*
pheromones, 391, 392
philosophy, 23, 71, 438
    of other minds, 423–32
    of reference, 70–83
    see also mind, theory of
phlogiston, 37
phonation, 246, 247
phonemes, 124, 127, 242, 250, 291, 297, 300,
    306, 314, 360
phrenology, 220–21
phylogeny, 28–29
Piaget, Jean, 95
pidgin languages, 138–39
pigeons, 48, 67
pigs, *171,* 189, *191,* 199, *200*
"pilgrim soul," 452–55
Pinker, Steven, 27, 38, 141, 327, 328
Planet of the Apes (film), 28
Pleistocene period, 394
Ploog, Detlev, 249
plosives, 242
polygyny, 381, 385, 387, 388, 391
pons, 176, *203*
pontine reticular, *231*
positron emission tomography (PET), 275,
    293, 294, *296,* 303, *304, 305*
predation, alarm calls and, 54–56, 58–59
prefrontal cerebral cortex, *217,* 218–20, *219,*
    255–78, 302
    adaptation of, 351–52, *353,* 413–17
    cingulate cortex in, 267
    cognitive demands and, 255–56, *257*
    connections overtaken by, 247–53, 255,
        417, 418–19
    cortical outputs from, 256–57, *256*
    damage to, 259–64, *260,* 266, 267–69, 415,
        427
    disorders of, 421–23
    functioning of, 259
    hominid fossil endocasts and, 342–43
    humor and, 421
    inhibiting role of, 264–65
    language and, 264–78